Lecture Notes in Computer Science 3158

Commenced Publication in 1973
Founding and Former Series Editors:
Gerhard Goos, Juris Hartmanis, and Jan van Leeuwen

Ioanis Nikolaidis   Michel Barbeau
Evangelos Kranakis (Eds.)

Ad-Hoc, Mobile, and Wireless Networks

Third International Conference, ADHOC-NOW 2004
Vancouver, Canada, July 22-24, 2004
Proceedings

 Springer

Volume Editors

Ioanis Nikolaidis
University of Alberta
Computing Science Department
322 Athabasca, Edmonton, T6G 2E8, AB, Canada
E-mail: yannis@cs.ualberta.ca

Michel Barbeau
Evangelos Kranakis
Carleton University
School of Computer Science
5360 Herzberg Laboratories
1125 Colonel by Drive, Ottawa, K1S 5B6, ON, Canada
E-mail: {barbeau,kranakis}@scs.carleton.ca

Library of Congress Control Number: 2004109141

CR Subject Classification (1998): C.2, D.2, H.4, H.3, I.2.11, K.4.4, K.6.5

ISSN 0302-9743
ISBN 3-540-22543-9 Springer-Verlag Berlin Heidelberg New York

Springer-Verlag is a part of Springer Science+Business Media

springeronline.com

© Springer-Verlag Berlin Heidelberg 2004
Printed in Germany

Typesetting: Camera-ready by author, data conversion by PTP-Berlin, Protago-TeX-Production GmbH
Printed on acid-free paper   SPIN: 11304425   06/3142   5 4 3 2 1 0

Preface

The third international workshop on AD-HOC NetwOrks and Wireless was held in the downtown Vancouver facilities of Simon Fraser University. The first ADHOC-NOW was held in 2002 at the Fields Institute in Toronto and the second in 2003 in Montreal. Its purpose is to create a collaborative forum between Mathematicians, Computer Scientists and Engineers for research in the emerging field of ad-hoc networks.

The number of submissions exceeded all expectations this year. Over 150 papers were submitted of which 22 regular and 8 short papers were accepted for presentation and inclusion in the conference proceedings. The program committee consisted of Michel Barbeau, Stefano Basagni, Azzedine Boukerche, Soumaya Cherkaoui, Leszek Gasieniec, Janelle Harms, Jeannette Janssen, Christos Kaklamanis, Evangelos Kranakis, Danny Krizanc, Thomas Kunz, Ramiro Liscano, Lata Narayanan, Ioanis Nikolaidis, Stephan Olariu, Jaroslav Opatrny, Pino Persiano, Samuel Pierre, S.S. Ravi, Mazda Salmanian, Sunil Shende, Ladislav Stacho, Martha Steenstrup, Ivan Stojmenovic, Violet Syrotiuk, Ljiljana Trajkovic, Jorge Urrutia, Peter Widmayer, and Kui Wu.

We would like to thank the invited speaker Martha Steenstrup for her research presentation and the program committee for refereeing the submissions. Many thanks to Paul Boone, Jen Hall, Jo-Ann Rockwood, Zheyin Li, and Tao Wan for helping with the workshop logistics. Special thanks go to MITACS (Mathematics of Information Technology and Complex Systems) and PIMS (Pacific Institute for the Mathematical Sciences) for supporting the workshop financially, Carleton University and the University of Alberta for providing computing facilities, and Simon Fraser University for its hospitality.

July 2004

Michel Barbeau
Evangelos Kranakis
Ioanis Nikolaidis

Table of Contents

Contributed Papers

Short Contributed Papers

Approximating the Minimum Number of Maximum Power Users in Ad Hoc Networks

Errol L. Lloyd[1], Rui Liu[1], and S.S. Ravi[2]

[1] University of Delaware, Newark, DE 19716.
{elloyd, ruliu}@cis.udel.edu.
[2] University at Albany - SUNY, Albany, NY 12222.
ravi@cs.albany.edu.

Abstract. Topology control is the problem of assigning transmission power values to the nodes of an ad hoc network so that the induced graph satisfies some specified property. The most fundamental such property is that the network/graph be connected. For connectivity, prior work on topology control gave a polynomial time algorithm for minimizing the maximum power assigned to any node (such that the induced graph is connected). In this paper we study the problem of minimizing the number of maximum power nodes. After establishing that this minimization problem is **NP**-Complete, we focus on approximation algorithms for graphs with symmetric power thresholds. We first show that the problem is reducible in an approximation preserving manner to the problem of assigning power values so that the sum of the powers is minimized. Using known results for that problem, this provides a family of approximation algorithms for the problem of minimizing the number of maximum power nodes with approximation ratios of $5/3 + \epsilon$ for every $\epsilon > 0$. Unfortunately, these algorithms, based on solving large linear programming problems are not practical. The main result of this paper is a practical algorithm having a $5/3$ (exactly) approximation ratio. In addition, we present experimental results on randomly generated networks. Finally, based on the reduction to minimizing the total power problem, we outline some additional results for minimizing the number of maximum power users, both for graph properties other than connectivity and for graphs with asymmetric power thresholds.

1 Introduction

Considerable attention has been given to problems of **topology control** in ad hoc networks. Recall that an **ad hoc network** consists of a collection of transceivers for which all communication is based on radio propagation. For each ordered pair (u, v) of transceivers, there is a **transmission power threshold**, denoted by $p(u, v)$, where a signal transmitted by the transceiver u can be received by v only when the transmission power of u is at least $p(u, v)$. The transmission power threshold for a pair of transceivers depends on a number of factors including the distance between the transceivers, the direction of the antenna at the sender, interference, noise, etc. [RR00]. Further, due to those same

I. Nikolaidis et al. (Eds.): ADHOC-NOW 2004, LNCS 3158, pp. 1–13, 2004.

factors, $p(u, v)$ and $p(v, u)$ need not be identical. When $p(u, v)$ and $p(v, u)$ are equal for all u and v, the power thresholds are **symmetric**. If they are unequal for some u and v, then the power thresholds are **asymmetric**. In this paper, unless otherwise specified, all of the problems considered utilize symmetric power thresholds.

Given the transmission powers of the transceivers, an ad hoc network can be represented by an undirected graph [KK+97]. In this graph, the nodes are in a one-to-one correspondence with the transceivers. There is an edge (u, v) if and only if the transmission powers of both u and v are at least the transmission power threshold $p(u, v)$. Note that every edge in the undirected graph corresponds to a two-way communication.

The main goal of **topology control** is to assign transmission powers to transceivers so that the resulting undirected graph satisfies a specified property, while minimizing some function of the transmission powers assigned to the transceivers. Limiting the maximum power used at any node, and using a minimum amount of power at each node to achieve a given task is likely to decrease the MAC layer interference between adjacent radios. The reader is referred to [LHB+01,RMM01,WL+01,RR00,RM99] for a thorough discussion of power control issues in ad hoc networks.

The most fundamental topology control problem [RR00] is to minimize the maximum power utilized by any node such that the resulting graph is connected. For this problem polynomial time algorithms are known [RR00,LLM02]. These algorithms are based on using binary search over all of the relevant power values (i.e. those power values corresponding to the transmission power thresholds). Since the goal is to minimize the maximum power assigned to any node, these algorithms may assign the computed minimum maximum power to *every* node of the network[1]. In practice however it is desirable not only to minimize the maximum power, but also to limit the number of nodes utilizing that maximum power.

In this paper we study the problem of assigning powers to nodes such that the induced graph is connected while minimizing the maximum power used by any node and minimizing the number of nodes that utilize that maximum power. In the next section we give formal definitions and some additional background. In Section 3 we show that the problem is **NP**-hard, and describe an approximation preserving reduction to the problem of minimizing the total power assigned to all of the nodes such that the resulting graph is connected. Using known results [ACMP04] for that problem, this provides a family of approximation algorithms for the problem of minimizing the number of maximum power nodes with approximation ratios of $5/3 + \epsilon$ for every $\epsilon > 0$. Unfortunately, these algorithms, based on solving large linear programming problems are not practical. In Section 4 we give our main result, an algorithm with an approximation ratio of $5/3$ (exactly). Experimental results are given in Section 5. Finally, some further consequences of the reduction to the problem of minimizing the total power, along with open problems are discussed in Section 6.

---

[1] A method to minimalize the power at every node, once the maximum is found is discussed in [RR00], but was not implemented.

2   Problem Formulation and Background

In this section we first give a formal definition of the problem studied in this paper, along with related terminology. The section also includes a brief review of some related prior work on topology control.

2.1   The Max-Power Users Problem

A formal statement of the decision version of the problem studied in this paper is as follows.

Max-Power Users

Instance: A positive integer M, a positive number P (maximum allowable power value), a node set V, and a power threshold value $p(u, v)$ for each pair (u, v) of transceivers.

Question: Is there a power assignment where the power assigned to each node is at most P and the number of the nodes that are assigned power P is at most M, such that the resulting undirected graph G is connected?

Note that the above definition differs slightly from that described in the introduction, in that the formal statement assumes that the maximum allowable power value is given as an input. This is a reasonable assumption since for a given network, the problem of minimizing the maximum power such that the induced graph is connected can be solved in polynomial time [RR00,LLM02]. The problem considered here takes that maximum power value as an input and aims to minimize the number of nodes utilizing that power.

2.2   Some Prior Work

A notation was given in [LLM02] whereby a topology control problem is specified by a triple of the form $\langle \mathbb{M}, \mathbb{P}, \mathbb{O} \rangle$. In this notation, $\mathbb{M} \in \{\text{UNDIR}, \text{DIR}\}$ represents the graph model, $\mathbb{P}$ represents the desired graph property and $\mathbb{O}$ represents the minimization objective.

The form of topology control considered in this paper was proposed by Ramanathan and Rosales-Hain [RR00]. They presented efficient algorithms for two topology control problems, namely $\langle \text{UNDIR}, \text{CONNECTED}, \text{MAXP} \rangle$ and $\langle \text{UNDIR}, \text{2-NODE CONNECTED}, \text{MAXP} \rangle$. In addition, they presented efficient distributed heuristics for those problems.

Considerable work has been done over the past three years on a variety of topology control problems. For instance, several groups of researchers have studied the problems $\langle \text{UNDIR}, \text{CONNECTED}, \text{TOTALP} \rangle$, $\langle \text{UNDIR}, \text{2-NODE CONNECTED}, \text{TOTALP} \rangle$ and $\langle \text{UNDIR}, \text{DIAMETER K}, \text{MAXP} \rangle$ (see for example [LLM02,KL+03]). Likewise, work on $\langle \text{DIR}, \text{STRONGLY CONNECTED}, \text{TOTALP} \rangle$ may be found in [CH89,KK+97,CPS99,CPS00]. In most instances, the problems are shown to be **NP**-Hard and the focus is on the development of approximation algorithms having either $O(\log n)$ or constant approximation ratios.

The problem of minimizing the number of maximum power users was briefly addressed in [LLM02] where it was shown that minimizing the number of maximum power users is **NP**-Hard when the goal is to produce a connected graph with diameter at most 6.

3   Two Complexity Results

In this section we discuss two complexity results for the **Max-power Users** problem. The first result shows that the problem is **NP**-Complete. The second result gives an approximation preserving reduction to the ⟨UNDIR, CONNECTED, TOTALP⟩ problem.

3.1   Minimizing the Number of Max Power Nodes Is NP-Complete

As noted above, it was shown in [LLM02] that minimizing the number of maximum power users is **NP**-Hard when the goal is produce a connected graph with diameter at most 6. In this section, we show that the problem is **NP**-Complete even when all that is desired is for the graph to be connected. The following result can be proven by a reduction from the minimum set cover problem [GJ79]. The proof is omitted due to space constraints.

Theorem 1. *The problem* **Max-power Users** *is* **NP***-Complete.*   □

3.2   A Reduction to Minimizing Total Power

In this section we give an approximation preserving reduction from the **Max-power Users** problem to ⟨UNDIR, CONNECTED, TOTALP⟩. Since several approximation results are known for ⟨UNDIR, CONNECTED, TOTALP⟩this will immediately provide identical results for **Max-power Users**.

Theorem 2. *There exists a polynomial-time reduction from* **Max-power Users** *to* ⟨UNDIR, CONNECTED, TOTALP⟩ *such that any α-approximation algorithm for* ⟨UNDIR, CONNECTED, TOTALP⟩ *is also an α-approximation algorithm for* **Max-power Users**.

Proof: Consider an instance I of the **Max-power Users** problem. Let $p(x, y)$ denote the (symmetric) power threshold values for any pair of nodes x and y. Let P denote the smallest maximum power value for I (this can be computed efficiently using the algorithms of [RR00,LLM02]). Further, with that P, let P_0 = $\max\{p(x, y): p(x, y) < P\}$. Since the goal is to minimize the number of nodes assigned the power value P, it is sufficient to consider solutions where all other nodes are assigned the power value P_0.

We now construct an instance I' of ⟨UNDIR, CONNECTED, TOTALP⟩ as follows. For the instance I', the power threshold value for each pair of nodes x and y, denoted by $p'(x, y)$, is chosen as follows:

$$p'(x, y) = 1 \quad \text{if } P_0 < p(x, y) \leq P$$
$$= 0 \quad \text{if } p(x, y) \leq P_0$$
$$= \infty \quad \text{otherwise}$$

Now, for the instance I', the power value to be assigned to each node is either 0 or 1. If the power assigned to node x in I' is 1 (0), that corresponds to assigning the power value P (P_0) to the node x in I. Thus, the total assigned power value in I' is the number of nodes assigned the maximum power value in I. Further, it is clear that any α-approximation algorithm for $\langle$UNDIR, CONNECTED, TOTALP$\rangle$is also an α-approximation algorithm for **Max-power Users**. □

Combining the above theorem with the results of [ACMP04], the following are immediate:

- For any $\epsilon > 0$, there is a $(5/3 + \epsilon)$-approximation algorithm for **Max-power Users**. Unfortunately, it is noted in [ACMP04] that "this algorithm is impractical", due to its reliance on solving large linear programming instances and other related issues.
- The minimum spanning tree based algorithm of [RR00] for $\langle$UNDIR, CONNECTED, TOTALP$\rangle$is a 2-approximation algorithm for **Max-power Users**. In contrast to the prior observation, the minimum spanning tree based algorithm of [RR00] has often been the method of choice for solving both $\langle$UNDIR, CONNECTED, MAXP$\rangle$and $\langle$UNDIR, CONNECTED, TOTALP$\rangle$. In the experimental studies described in this paper we include this algorithm as a baseline method for solving **Max-power Users**.

The NP-hardness results presented in this section apply to general instances of the **Max-power Users** problem rather than geometric instances (where the nodes are points in Euclidean space and the power threshold value for each pair of nodes is a function of the distance between the nodes). However, the approximation results presented in the subsequent sections of this paper are applicable to geometric instances as well.

4   A 5/3-Approximation Algorithm

In this section we describe a 5/3-approximation algorithm for the **Max-power Users** problem, and show that the approximation ratio is tight. We begin by defining the following concept:

Definition 1. *Consider an undirected graph $G(V, E)$ and two subsets E' and H of E. For a node $u \in V$, a* **Maximal Connected Component Tree (MCCT)** *with root u for graph $G'(V, E')$ and edge subset H, is a subset T_u of H such that all of the following conditions hold:*

    (a) The restriction of G to T_u is a tree. By abuse of notation, we let T_u denote that restriction of G.
    (b) u is a node in T_u.
    (c) Each node in T_u is in a different connected component of $G'(V, E')$.

(d) T_u is maximal with respect to properties (a) and (c) above (i.e. adding any edge $(x, y) \in H$ to T_u will either destroy property (a) or property (c)).

*The number of edges in T_u is called the **size** of T_u.*

Figure 1 illustrates the definition of MCCT.

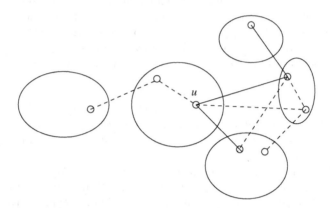

Fig. 1. MCCT of u: The ellipses illustrate the different connected components in G', H consists of all of the edges shown (solid and dashed), and T_u, a Maximal Connected Component Tree (MCCT) with root u for graph $G'(V, E')$ and edge subset H, is the set of all solid edges. The size of T_u is 3.

An MCCT based approximation algorithm for **Max-power Users** is given in Algorithm 1. For an understanding of the ideas behind this algorithm, we first consider the following method for finding a solution to **Max-power Users**: Using the power threshold graph, assign every node a power level equal to the greatest adjacent threshold that is less than p_{max} (the minimum maximum power). The graph induced by this power assignment will consist of several connected components. Then, one way to make this graph connected is to add maximum power edges (i.e. those having a power threshold of p_{max}) to this graph until the graph is connected. This is the essential idea underlying the minimum spanning tree based algorithm of [RR00] that was mentioned in the prior section. Clearly, in this approach if there are k connected components prior to the addition of any maximum power edges, then $k - 1$ maximum power edges must be added. The key issue is how many distinct nodes are adjacent to these edges. Clearly, k is a lower bound on that number of nodes. And, in [RR00] the number of distinct nodes might be as large as $2 \cdot (k - 1)$, hence the approximation ratio of 2.

How can we improve upon the ratio of 2? One approach would be to select maximum power edges that share adjacent nodes. Alternatively, we could tighten the lower bound of k on the number of maximum power nodes. Obviously, one or both approaches might apply for a particular instance of the **Max-power Users** problem. Algorithm 1 attempts to exploit the first approach, by focusing on MCCTs of particular sizes as it selects maximum power edges that result in a connected graph. In particular, the algorithm first finds MCCTs of size at least 3, then finds MCCTs of size 2, and finally includes single edges. Further, our proof on the approximation ratio associated with Algorithm 1 will tighten the lower bound as well.

Algorithm 1 A 5/3-Approximation Algorithm for **Max-power Users**

Input: A complete power threshold graph $G(V, E)$ and the minimum maximum power p_{max}.

Output: Power assignment A for each node u in V.

1: $E' \leftarrow \{(u, v) \in E| \ p(u, v) < p_{max}\}$
2: $H \leftarrow \{(u, v) \in E| \ p(u, v) = p_{max}\}$
3: **while** $\exists u \in V$, for which a MCCT T_u for $G'(V, E')$ and H has size greater than 2 **do**
4:     $E' \leftarrow E' \cup T_u$
5: **end while**
6: **while** $\exists u \in V$, for which a MCCT T_u for $G'(V, E')$ and H has size 2 **do**
7:     $E' \leftarrow E' \cup T_u$
8: **end while**
9: Compute the connected components $C_1, ..., C_L$ of $G'(V, E')$.
10: Add $L - 1$ edges in H to E' such that $G'(V, E')$ is connected.
11: **for** each u in V **do**
12:     $A(u) \leftarrow$ the weight of the largest edge in E' that is incident on u.
13: **end for**
14: Return A

It is easy to see from the algorithm's specification that it produces a power assignment for the nodes in V such that the induced graph is connected. In regard to the quality of that power assignment, we have:

Theorem 3. *Algorithm 1 is a 5/3-approximation algorithm.*

Proof: Suppose before the execution of the first *while* loop, that $G'(V, E')$ has K connected components. In iteration i of the *while* loop (lines 3–5), let s_i be the size of T_u. Note that $s_i \geq 3$, and that by adding T_u to E' the number of connected components in $G'(V, E')$ is reduced by s_i because T_u connects $s_i + 1$ different connected components.

We claim that T_u is node disjoint from any edges in H that were added to E' in the previous iterations. The reason is as follows. Suppose there exist edges (t, v) in T_u and (v', t) in T', a MCCT chosen in some previous iteration (i.e. these two edges have node t in common). Since (t, v) is in T_u, it follows that in iteration i, nodes t and v must be in different connected components of $G'(V, E')$. Also in iteration i all nodes in T' are in the same connected component. This component then includes node t, but not node v. But this means that in the prior iteration, $T' \cup (t, v)$ is a larger $MCCT$ than T'. This contradicts T' being a $MCCT$ (it was not maximal). Thus, the claim is proved.

As a result of the above, in both *while* loops, adding T_u to E' causes $s + 1$ additional nodes to be assigned power p_{max} in A, where s is the size of T_u.

Now, suppose before the execution of the first *while* loop, that $G'(V, E')$ has K connected components, that before the execution of the second *while* loop, $G'(V, E')$ has M connected components, and that after the execution of the second *while* loop, $G'(V, E')$ has P connected components. Let s_i be the size of T_u in iteration i of the first *while* loop and let m_i be the size of T_u in iteration i of the second *while* loop. Let $N(A)$ be the number of nodes that are assigned power p_{max} in A. Then, since each $s_i \geq 3$ and each $m_i = 2$, we have:

$$N(A) = \sum_i (s_i + 1) + \sum_i (m_i + 1) + 2(P - 1)$$

$$= \sum_i (\frac{s_i + 1}{s_i} \cdot s_i) + \sum_i (\frac{m_i + 1}{m_i} \cdot m_i) + 2(P - 1)$$

$$\leq \frac{4}{3} \sum_i s_i + \frac{3}{2} \sum_i m_i + 2(P - 1)$$

$$= \frac{4(K - M)}{3} + \frac{3(M - P)}{2} + 2(P - 1)$$

Consider an optimum solution for **Max-power Users**. It is obvious that:

Claim. 1 In any optimum solution, at least K nodes are assigned power p_{max}.

Let $G_{opt}(V, E_{opt})$ be the graph induced by an optimum solution. Consider the graph $G'(V, E')$ after the first *while* loop of Algorithm 1. Recalling that there are M connected components in G', we have:

Claim. 2 There exist node disjoint MCCTs for graph $G'(V, E')$ and edge set E_{opt} such that:

– Each MCCT[2] is of size 1 or 2.
– The sum of the sizes of these MCCTs is at least $M - 1$.

Proof of Claim 2: Let D_0 be the nodes in some connected component of $G'(V, E')$. Since $G_{opt}(V, E_{opt})$ is connected, for each cut of $G_{opt}(V, E_{opt})$ there must exist an edge in E_{opt} that crosses the cut. Let e_1 be an edge in E_{opt} that crosses the cut $(D_0, V - D_0)$. Note that e_1 must be a maximum power edge. Further, there must be a MCCT T_1 for $G'(V, E')$ and edge set E_{opt} that includes e_1. The size of that MCCT may be either 1 or 2 (it cannot be larger, since then it would have been discovered in the first *while* loop of Algorithm 1). Consider $G_1(V, E_1)$ where $E_1 = E' \cup T_1$. Now, let D_1 be the nodes in some connected component of G_1. As above, let e_2 be an edge (necessarily of maximum power) that crosses the cut $(D_1, V - D_1)$, and let T_2 be a MCCT for G_1 and edge set E_{opt} that includes e_2. Again, the size of that MCCT must be 1 or 2. Continuing in

---

[2] Trivially, an MCCT of size 1 is a singleton edge. In the remainder of this section we will sometimes find it convenient to refer to edges that are not part of a larger MCCT in this fashion (i.e. as MCCTs of size 1).

this fashion, we enumerate MCCTs $T_1, T_2, ..., T_g$, ending when the corresponding graph G_g is connected. Note that each of these MCCTs is of size 1 or 2, and that it follows from the definition of a MCCT that these MCCTs are node disjoint. It also follows that each of these MCCTs is present in graph G'. Finally, since G' had M connected components, the sum of the sizes of these MCCTs is $M - 1$. □

It follows from the claim that if L of the MCCTs referenced there are of size 1, then there are at least $3/2(M - L - 1) + 2L = 3M/2 + L/2 - 3/2$ nodes using power p_{max} in an optimum solution.

To complete the proof we need to relate the number of MCCTs of size 2 selected in Algorithm 1 to the values L and M. Specifically we have:

Claim. 3 In the second *while* loop of Algorithm 1 (lines 6–8), at least $(M - L)/4$ MCCTs are selected. Note that those MCCTs are all of size 2.

The proof of Claim 3 is omitted.

Let $N(OPT)$ be the number of nodes that are assigned power p_{max} in an optimum solution. From Claims 1 and 2, we have that $N(OPT) \geq \text{Max}(K, 3M/2 + L/2 - 3/2)$. And, using Claim 3, $(M - P)/2 \geq (M - L)/4$, hence $P \leq M/2 + L/2$. We then have:

$$\frac{N(A)}{N(OPT)} \leq \frac{4/3(K - M) + 3/2(M - P) + 2(P - 1)}{\text{Max}(K, 3M/2 + L/2 - 3/2)}.$$

Using straightforward algebraic manipulations, it can be shown that the above ratio is bounded by $5/3$. Thus, the approximation ratio of Algorithm 1 is $5/3$. □

Due to space constraints, we state the following corollary without proof.

Corollary 1. *The $5/3$ approximation ratio of Algorithm 1 is tight.*

An interesting question is what happens in Algorithm 1 if we do not give priority to MCCTs of size 3 or more, but rather consider equally acceptable any MCCT of size 2 or more. That is, in Algorithm 1, in line 3, replace "H has size greater than 2" with "H has size greater than 1", and we omit lines 6 to 8. We will call this *Algorithm 2*. Then we have the following corollary:

Corollary 2. *Algorithm 2 is a $7/4$-approximation algorithm for* **Max Power Users**, *and the bound is tight.*

Finally, we state without proof the running times of our algorithms:

Proposition 1. *The running times of Algorithms 1 and 2 are $O(n^3)$ and $O(n^2)$ respectively.*

5 Experimental Results

In this section we consider the experimental performance of Algorithms 1 and 2 along with the algorithm of [RR00].

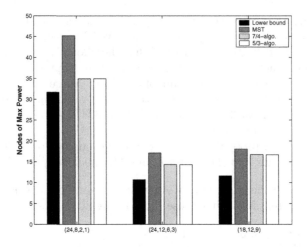

Fig. 2. Max Power Nodes

The experimental environment used here is derived from the one described in [RR00]. The radio wave propagation model is the *Log-distance Path Loss Model*:

$$PL(d) = -10 \log_{10} \left[\frac{G_t G_r \lambda^2}{(4\pi)^2 d_0^2} \right] + 10\eta \log_{10} \left[\frac{d}{d_0} \right]$$

where η is the path loss exponent, d_0 is the close-in reference distance, λ is the radio wavelength, G_t is the transmitter antenna gain, G_r is the receiver antenna gain, and d is the separation distance between transmitter and receiver (see [Ra96] for detailed descriptions of these parameters). All of the parameters are chosen to emulate a 2.4 GHz wireless radio, and if d is less than a certain threshold, the transmission power threshold is set to the minimum threshold of 1 dBm.

Experiments were conducted on networks with 200 nodes by varying the geographical distribution of the nodes and the power levels available to the nodes. Nodes were placed using a uniform random distribution, and with 200 nodes in each network, the node density was 12.5 nodes/sq mile in a 4 mile by 4 mile area. Three sets of power levels were used, with a varying ratio between the top two power levels. Those three sets of power levels were $(24, 8, 2, 1)$, $(24, 12, 6, 3)$ and $(18, 12, 9)$, corresponding to ratios between the top two power levels of 3, 2 and 1.5. Each data point represents the average taken over 19 trials.

In each experiment, after generating a placement of the nodes, three algorithms were run on the network consisting of those nodes. The three algorithms were our Algorithms 1 and 2, along with the minimum spanning tree algorithm from [RR00] that was briefly discussed in Section 3 (for brevity, we will refer to this as MST [RR00]). Each algorithm assigns powers to nodes such that the resulting network is connected. For each algorithm we measure the number of nodes that are assigned maximum power. In addition, we record the aver-

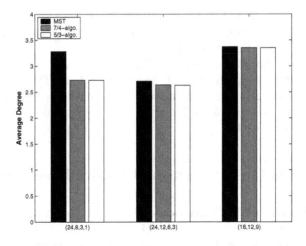

Fig. 3. Node Degree

age power assigned, and the maximum and average degrees of the nodes in the resulting network.

The experimental results on nodes of maximum power are shown in Figure 2. There, in addition to the number of maximum power nodes produced by each algorithm, we provide a lower bound on the optimal number of maximum power nodes. This lower bound is calculated as in the proofs of the performance bounds in the prior section, and is based on the number of connected components, and the numbers of MCCTs of size 2 and size 3. Relative to the results in Figure 2 we make the following observations:

- The two MCCT based algorithms (Algorithms 1 and 2) outperform the MST based algorithm (MST [RR00]) in regard to the number of maximum power nodes. The reductions obtained by using the MCCT based algorithms range from 7% to 23%, with larger reductions occurring when there is a larger spread in the top two power levels.

- There is virtually no difference in the performance of Algorithm 1 (the 5/3 approximation ratio) as compared with the performance of Algorithm 2 (the 7/4 approximation ratio). While they often obtain their MCCTs in different orders (in particular, the MCCTs of size 2 come last in Algorithm 1), it seems to be rare for the selection of one MCCT to eliminate another MCCT. Further, in most cases, the number of MCCTs found by either algorithm is fairly small. Usually there were fewer than 5 MCCTs, and in a surprisingly large number of cases there was only one, very large, MCCT. That single MCCT involved a node from the majority of the connected components that existed prior to MCCT selection in the *while* loops of Algorithms 1 and 2.

- On the average, Algorithms 1 and 2 use about 22% more maximum power nodes than the lower bound.

The results on average power are omitted due to space considerations, however we note that the improvements in average power due to Algorithms 1 and 2 are modest, ranging from under 1% to just over 10%.

Figure 3 shows the results on average degree. Here, some reductions in average degree occur when using the MCCT based algorithms. The results are most significant when the ratio between the top two power levels is high. Specifically, when the power levels are $(24, 8, 2, 1)$, the MCCT based algorithms reduce the average node degree by 17%.

6   Additional Observations and Future Research

Theorem 2 showed that the **Max-power Users** problem can be reduced in an approximation preserving manner to to ⟨UNDIR, CONNECTED, TOTALP⟩. Actually, that result is not restricted to power assignments producing connected graphs, but can be generalized to apply for any monotone graph property as well as for asymmetric power thresholds. As a consequence, approximation algorithms for several (generalized) **Max-power Users** problems can be obtained. These include: for symmetric power thresholds, constant factor approximations for the properties 2-connected and 2-edge-connected, as well as an $O(\log n)$ approximation for connectivity with asymmetric power thresholds. In addition, we can show using a reduction from minimum set cover that there is a matching $\Omega(\log n)$ lower bound for the connectivity problem under asymmetric thresholds.

The primary open question is whether there are algorithms with approximation ratios lower than 5/3 for the **Max-power Users** problem for connectivity. We conjecture that such algorithms exist. Specifically, if Algorithm 1 is extended in the obvious way by finding larger MCCTs before smaller ones (i.e. much as Algorithm 1 is an extension of Algorithm 2), it is likely that a smaller approximation ratio can be found. The difficulties in making this extension are: 1) Determining the relationship between the MCCTs of a given size found by the algorithm and the existence of MCCTs in optimal solutions; 2) Evaluating the resulting equations; and 3) The running time would seem to be dependent on the size of the MCCTs that are sought, hence it is likely that some kind of approximation scheme would result (though we suspect that even so, the achievable bound will not fall below 1.5).

References

[ACMP04]   E. Althaus and G. Calinescu and I. Mandoiu and S. Prasad and N. Tchervenski and A. Zelikovksy. "Power Efficient Range Assignment for Symmetric Connectivity in Static Ad Hoc Wireless Networks", Private communication

[CH89]   W. Chen and N. Huang. "The Strongly Connecting Problem on Multihop Packet Radio Networks", *IEEE Trans. Communication*, Vol. 37, No. 3, Mar. 1989, pp. 293-295.

[CPS99]   A. E. F. Clementi, P. Penna and R. Silvestri. "Hardness Results for the Power Range Assignment Problem in Packet Radio Networks", *Proc. Third International Workshop on Randomization and Approximation in Computer Science* (APPROX 1999), Lecture Notes in Computer Science Vol. 1671, Springer-Verlag, July 1999, pp. 195–208.

[CPS00]   A. E. F. Clementi, P. Penna and R. Silvestri. "The Power Range Assignment Problem in Packet Radio Networks in the Plane", *Proc. 17th Annual Symposium on Theoretical Aspects of Computer Science* (STACS 2000), Feb. 2000, pp. 651–660.

[GJ79]   M. R. Garey and D. S. Johnson. *Computers and Intractability: A Guide to the Theory of NP-completeness*, W. H. Freeman and Co., San Francisco, CA, 1979.

[GK99]   S. Guha and S. Khuller, "Improved Methods for Approximating Node Weighted Steiner Trees and Connected Dominating Sets", *Information and Computation*, Vol. 150, 1999, pp. 57–74.

[KK+97]   L. M. Kirousis, E. Kranakis, D. Krizanc and A. Pelc. "Power Consumption in Packet Radio Networks", *Proc. 14th Annual Symposium on Theoretical Aspects of Computer Science* (STACS 97), Lecture Notes in Computer Science Vol. 1200, Springer-Verlag, Feb. 1997, pp. 363–374.

[KL+03]   S. O. Krumke, R. Liu, E. L. Lloyd, M. V. Marathe, R. Ramanathan and S. S. Ravi, "Topology Control Problems Under Symmetric and Asymmetric Power Thresholds", *Proc. International Conference on Ad hoc and Wireless Networks* (ADHOC-NOW'03), Lecture Notes in CS, Vol. 2865, (Edited by S. Pierre, M. Barbeau and E. Kranakis), Montreal, Canada, Oct. 2003, pp. 187–198.

[LLM02]   E. Lloyd, R. Liu, M. Marathe, R. Ramanathan, and S. S. Ravi, "Algorithmic Aspects of Topology Control Problems for Ad hoc Networks", *Proc. Third ACM International Symposium on Mobile Ad Hoc Networking and Computing (MobiHoc)* (2002), 123–134. (Complete version to appear in *Mobile Networks and Applications*.)

[LHB+01]   L. Li, J. Y. Halpern, P. Bahl, Y. Wang and R. Wattenhofer, "Analysis of Cone-Based Distributed Topology Control Algorithm for Wireless Multihop Networks", *Proc. ACM Principles of Distributed Computing Conference* (PODC'01), Aug. 2001, pp. 264–273.

[Ra96]   T. S. Rappaport. *Wireless Communications: Principles and Practice*, Prentice-Hall, Inc., Englewood Cliffs, NJ, 1996.

[RM99]   V. Radoplu and T. H. Meng. "Minimum Energy Mobile Wireless Networks", *IEEE J. Selected Areas in Communications*, 17(8), Aug. 1999, pp. 1333-1344.

[RMM01]   E. M. Royer, P. Melliar-Smith and L. Moser, "An Analysis of the Optimum Node Density for Ad hoc Mobile Networks", *Proc. IEEE Intl. Conf. on Communication* (ICC'01), Helsinki, Finland, June 2001, pp. 857–861.

[RR00]   R. Ramanathan and R. Rosales-Hain. "Topology Control of Multihop Wireless Networks Using Transmit Power Adjustment", *Proc. IEEE INFOCOM 2000*, Tel Aviv, Israel, March 2000, pp. 404–413.

[WL+01]   R. Wattenhofer, L. Li, P. Bahl and Y. Wang. "Distributed Topology Control for Power Efficient Operation in Multihop Wireless Ad Hoc Networks", *Proc. IEEE INFOCOM 2001*, Anchorage, Alaska, April 2001, pp. 1388–1397.

Multicast Versus Broadcast in a MANET

Thomas Kunz

Carleton University, Ottawa, Ont., Canada K1S 5B6
tkunz@sce.carleton.ca

Abstract. Multicasting is frequently proposed to efficiently support one-to-many and many-to-many communication. Consequently, many multicast routing protocols have been developed, mostly for fixed networks. This paper explores the benefits of multicast routing protocols in mobile ad-hoc networks (MANETs). Using on-demand routing protocols developed for such networks with their dynamic network topology, we compare the use of unicast, multicast, and broadcast protocols to support a variety of many-to-many communication scenarios. The results show that multicast protocols such as ADMR or ODMRP do indeed provide superior performance compared to the use of unicast protocols. However, broadcast protocols, which do not attempt to build and maintain efficient multicast distribution structures, fare equally well or better, in particular with respect to packet delivery ratio, with little to no additional network overhead.

1 Introduction

Multicasting is the transmission of datagrams (packets) to a group of zero or more hosts identified by a single destination address. Multicasting is intended for group-oriented computing. There are many applications where one-to-many or many-to-many dissemination is an essential task [11, 13]. The multicast service is critical in applications characterized by the close collaboration of teams (e.g. rescue patrol, battalion, scientists, etc) with requirements for audio and video conferencing and sharing of text and images. In the Internet (IPv4), multicasting facilities were introduced via the Multicast Backbone (MBone), a virtual overlay network on top of the Internet. Support for multicasting is an integral component of IPv6, so it can be assumed that multicasting applications will become even more popular with the increased popularity and acceptance of IPv6.

The use of multicasting within a network has many potential benefits. Multicasting can reduce the communication costs for applications that send the same data to multiple recipients [11, 15]. Instead of sending via multiple unicasts, multicasting minimizes the link bandwidth consumption, sender and router processing, and delivery delay. In addition, multicasting provides a simple yet robust communication mechanism whereby a receiver's individual address is unknown or changeable transparently to the source.

Maintaining group membership information and building an efficient multicast distribution structure (typically in the form of a routing tree) is challenging even in wired networks. A detailed survey of the work done in that area and a discussion of various design trade-offs can be found in [8]. However, nodes are increasingly

I. Nikolaidis et al. (Eds.): ADHOC-NOW 2004, LNCS 3158, pp. 14–27, 2004.

mobile. One particularly challenging environment for multicast is a mobile ad-hoc network (MANET). The network topology can change randomly and rapidly, at unpredictable times. Since wireless links generally have lower capacity, congestion is typically the norm rather than the exception. The majority of nodes will rely on batteries, thus routing protocols must limit the amount of control information that is passed between nodes. [5] surveys some best-effort IP multicast routing protocols for fixed networks and the evolution of these protocols as hosts and finally all nodes (including intermediate routers) become mobile.

This paper addresses whether multicasting in a MANET context is worthwhile. Due to the dynamic nature of the network, building and maintaining an efficient multicast distribution structure may be inefficient. To support many-to-many communication, we could alternatively use dedicated unicast communication or broadcasting. In the first instance, if a sender communicates with N receivers, it would open N point-to-point flows to the receivers. Note that this requires the sender to know the identities of all receivers. Furthermore, these N routes now have to be maintained as well, so we do not really expect the unicast solution to perform better. However, we include it for the sake of completeness and to establish a baseline performance. In the latter instance (i.e., using broadcast protocols), the identity and number of receivers can remain unknown to the sender(s). Rather, a data packet is delivered to all nodes in the network, with those nodes interested in the data simply passing it up the protocol stack. A broadcast solution may require little or no routing overhead. However, broadcasting may be inefficient: every node in the network receives every data packet.

To explore whether multicasting can really achieve the claimed advantages, we conducted a thorough simulation study, comparing and evaluating 7 different routing protocols to support the communication between N senders and M receivers. In particular, we experimented with 2 unicast routing protocols (DSR and AODV), 3 multicast routing protocols (ADMR, ODMRP, and the multicast extensions to AODV), and 2 broadcast protocols (FLOOD and BCAST). The next section briefly describes the individual protocols. Section 3 outlines the simulation environment and Section 4 discusses the simulation results. Section 5 discusses related work and Section 6 summarizes our work and provides our conclusions.

2   Protocol Descriptions

All protocols discussed in this section have been developed for MANETs. The first two protocols are well-known on-demand unicast routing protocols, DSR[4] and AODV[12]. The multicast protocols follow a design similar to the unicast routing protocol: a packet distribution structure is created and maintained on-demand, the differences are primarily in the nature of the multicast distribution structure. The multicast extensions for the AODV (Ad-hoc On-Demand Distance Vector) routing protocol [14] discover multicast routes on demand using a broadcast route-discovery mechanism. The protocol builds a shared multicast tree based on hard state, repairing broken links and explicitly dealing with network partitions. The first member of the multicast group becomes the leader for that group. The multicast group leader is responsible for maintaining the multicast group sequence number and broadcasting this number to the multicast group. This is done through a Group Hello message. The

Group Hello contains extensions that indicate the multicast group IP address and sequence numbers (incremented every Group Hello) of all multicast groups for which the node is the group leader. Nodes use the Group Hello information to update their request table.

The second multicast protocol that builds and maintains a multicast distribution structure is ODMRP: On-Demand Multicast Routing Protocol [7]. ODMRP is mesh-based, and uses a forwarding group concept (only a subset of nodes forwards the multicast packets). In ODMRP, group membership and multicast routes are established and periodically updated by the source on demand. These meshes are source-based, in an environment with many senders, a number of these meshes will have to be built and maintained. On the other hand, no work is required to update the mesh as the topology changes or nodes join/leave the multicast group: such changes get reflected the next time the mesh is rebuilt.

The Adaptive Demand-Driven Multicast Routing (ADMR) protocol [3] eliminates any periodic elements that exist in protocols such as ODMRP (the Join-Query) or MAODV (the Group Hello message). A source-based forwarding tree is created whenever there is at least one source and one receiver in the network (i.e., there is a multicast distribution tree for each active source). AMDR monitors the traffic pattern of the sources to detect link breakages and inactive sources. In addition, ADMR resorts to flooding data packets as preferred data delivery mechanism if ADMR detects a high mobility rate.

Broadcasting protocols deliver data to all nodes in a network, independent of whether they are interested in that data or not. Since even unicast and multicast routing protocols often have a broadcast component (for example, the route discovery phase in on-demand unicast routing protocols), efficient broadcast protocols have been investigated heavily. [16] surveys various categories of broadcast protocols and provides simulation results under various mobility scenarios. For the purpose of this study, we selected two broadcast protocols: a very simple protocol (FLOOD) and one of the more complex protocols (BCAST). FLOOD is the standard flooding approach: each node, upon receiving a packet for the first time, will re-broadcast it over its wireless interface (i.e., using MAC-layer broadcasting). BCAST implements a scalable broadcast algorithm similar to the algorithm described in [9]. BCAST uses 2-hop neighbor knowledge that is exchanged by periodic "Hello" messages. Each "Hello" message contains the node's identifier (IP address) and its list of known neighbors. After a node receives a "Hello" packet from all its neighbors, it has two-hop topology information. If node B receives a broadcast from node A, B knows all neighbors of A. If B has neighbors not covered by A, it schedules the broadcast packet with a random delay. If, during the delay, B receives another copy of this broadcast from C, it can check whether its own broadcast will still reach new neighbors. If this is no longer the case, it will drop the packet. Otherwise, the process continues until B's timer goes off and B itself rebroadcasts the packet.

3 Simulation Environment

To compare the performance of the various "multicast" solutions, we studied the above protocols in NS2. We set up a rather challenging simulation environment, using the following parameters (similar to other setups reported in the literature): Area size

1500 x 300 meters, 50 mobile nodes, 910 seconds simulation length, 10 different mobility patterns per scenario, IEEE 802.11 MAC at 2 Mbps, 250 meter transmission range, and random waypoint model with no pause time, maximum speed either 20 m/s (high mobility scenarios) or 1 m/s (low mobility scenarios).

All nodes are in constant movement in our experiments. The only traffic is the multicast traffic. We study a range of multicast senders (1, 2, 5, or 10), sending to a number of multicast receivers (10, 20, 30, 40, or 50). We keep the sender and receiver sets disjoint. For example, in a scenario with 10 senders and 30 receivers, nodes 0 through 9 are the senders and nodes 20 through 49 are the receivers. Only in scenarios with 50 multicast receivers will some nodes act as both sender and receiver. In these scenarios, we expect the packet delivery ratio to be slightly better, since packet delivery within a single node is not subject to network problems.

All receivers join the single multicast group at the beginning of the simulation, the sender(s) start sending data 30 seconds into the simulation (so where appropriate, the routing protocol can start building its multicast distribution structure, for example in MAODV). After 900 seconds, all senders stop transmitting data, during the remaining 10 seconds, data packets still in flight have a chance to be delivered. We allow for this additional 10 seconds in our simulations to avoid the problem of how to account correctly for packets that are in flight at simulation end.

Each sender sends 2 packets per second; each packet is 256 bytes long. We also explored other traffic loads (more and longer packets per sender), but in these cases the MANET is almost always heavily congested, resulting in very poor protocol performance.

The basic performance metrics are Packet Delivery Ratio (PDR) and Packet Latency. Packet delivery ratio is defined as the percentage of received packets, relative to the total number of packets ideally received. Counting the total number of received packets is straightforward in the simulations. The total number of packets ideally received is the number of packets sent by the sender(s) in the case of unicast routing protocols. In the case of multicast or broadcast protocols, the number of intended multicast receivers multiplies the number of packets sent. Packet latency is the elapsed time between a packet being transmitted and ultimately received. We only measure packet latency for those packets that are received at a multicast receiver. Also, both metrics are calculated as averages over all senders and receivers, we do not break them down by a specific sender or receiver.

An ideal protocol will achieve high packet delivery ratio and low packet latency. It will also do this with little overhead. Traditionally, counting the number of control messages and relating them to the number of received packets measures protocol overhead. However, FLOOD does not generate any dedicated control messages. And the overheads in any broadcast protocol are not only related to any control messages, but also to the waste of delivering packets to nodes not interested in this data. So we generalize the protocol overhead, defining metrics that capture the "network efficiency of the protocol" as the Packet Send Ratio (PSR): the number of packet transmissions (at the MAC layer) per data packet received by a multicast receiver. This metric captures the normalized total traffic in the network. The PSR, ideally, could be very small. Since for many protocols we broadcast packets at the MAC layer, multiple receivers pick up that packet (depending on the geographical distribution of senders and receivers), so that metric could be as low as $1/n$, where n is the number of multicast receivers. However, control messages and suboptimal re-broadcasting of packets will increase this metric. In flooding, for example, each node will re-broadcast a data packet. If the number of multicast receivers is smaller than the number of nodes in the network, this metric will exceed 1.

4 Simulation Results

The following sections outline the results for the various unicast, multicast, and broadcast protocols. For the most part, we focus on packet delivery ratio (PDR) and packet latency. All results are presented in table form, with the number of senders across the columns and the number of receivers across the rows.

Table 1. PDR and Latency for AODV, 1 m/s maximum speed.

	1 Sender		2 Sender		5 Sender		10 Sender	
	PDR	Lat.	PDR	Lat.	PDR	Lat.	PDR	Lat.
10 Receiver	0.995	0.022	0.996	0.024	0.756	1.430	0.277	4.543
20 Receiver	0.996	0.033	0.966	0.208	0.328	2.957	0.128	3.676
30 Receiver	0.988	0.091	0.730	0.942	0.209	2.839	0.078	3.081
40 Receiver	0.915	0.321	0.537	1.159	0.147	2.725	0.055	2.650
50 Receiver	0.855	0.420	0.452	1.118	0.135	2.217	0.061	1.734

Table 2. PDR and Latency for AODV, 20 m/s maximum speed.

	1 Sender		2 Sender		5 Sender		10 Sender	
	PDR	Lat.	PDR	Lat.	PDR	Lat.	PDR	Lat.
10 Receiver	0.974	0.038	0.972	0.040	0.499	1.430	0.199	2.092
20 Receiver	0.974	0.049	0.784	0.558	0.219	1.920	0.095	1.915
30 Receiver	0.939	0.167	0.492	1.213	0.135	1.932	0.060	1.781
40 Receiver	0.826	0.434	0.342	1.361	0.092	1.926	0.041	1.688
50 Receiver	0.688	0.734	0.279	1.298	0.089	1.516	0.050	1.071

Due to space limitations, we only discuss one unicast protocol, AODV. Its performance by and large is similar to that of DSR: as the number of sender and/or multicast receiver increases (resulting in an increase in the number of unicast connections), the PDR drops and packet latency increases. For 10 sender scenarios, with 50 receivers (i.e., all nodes receive the multicast data), PDR and latency are slightly improved, since now about 2% of all packets are delivered locally (i.e., within a node). The performance under high-speed scenarios is generally worse than under low-speed scenarios.

AODV does perform better than DSR and behaves more consistently (i.e., PDR drops slower and packet latency increases at a slower rate). The network efficiency, in terms of MAC packets transmitted per data packet delivered (PSR), ranges from 11 to 52 packets under low mobility scenarios and from 15 to 68 packets under high mobility scenarios. Exploring the trace files in more depth, roughly 70% of all packet transmissions at the MAC layer are MAC layer control packets (RTS, CTS, and ACK). If we ignore these control packets, the revised PSR (counting only network layer packets) ranges from 3.3 to 15.6 for low mobility scenarios and 4.5 to 20.8 for high mobility scenarios.

Table 3. PDR and Latency for MAODV, 1 m/s maximum speed.

	1 Sender		2 Sender		5 Sender		10 Sender	
	PDR	Lat.	PDR	Lat.	PDR	Lat.	PDR	Late.
10 Receiver	0.985	0.033	0.990	0.041	0.940	0.067	0.544	0.510
20 Receiver	0.990	0.048	0.975	0.059	0.843	0.154	0.433	0.837
30 Receiver	0.972	0.064	0.977	0.079	0.721	0.507	0.363	1.238
40 Receiver	0.981	0.080	0.952	0.101	0.608	0.800	0.309	1.611
50 Receiver	0.981	0.089	0.943	0.115	0.583	0.890	0.321	1.583

Table 4. PDR and Latency for MAODV, 20 m/s maximum speed.

	1 Sender		2 Sender		5 Sender		10 Sender	
	PDR	Lat.	PDR	Lat.	PDR	Lat.	PDR	Lat.
10 Receiver	0.888	0.039	0.916	0.049	0.845	0.110	0.461	0.643
20 Receiver	0.903	0.056	0.925	0.072	0.708	0.275	0.358	0.982
30 Receiver	0.882	0.074	0.908	0.096	0.598	0.514	0.301	1.215
40 Receiver	0.898	0.091	0.881	0.124	0.509	0.833	0.261	1.382
50 Receiver	0.902	0.101	0.863	0.148	0.491	0.852	0.275	1.295

MAODV performs significantly better than either of the unicast routing protocols. The MAODV performance is particularly high under low mobility for few (1 or 2 senders), resulting in PDRs of around 98% and latencies of a few tens of milliseconds. Exploring the trace files in more detail, we noticed that the single shared tree, with all the operations to maintain its hard state, becomes the bottleneck in our scenarios at high mobility and/or for a large number of multicast senders.

Looking at the network efficiency, scenarios with relatively few multicast receivers generate more relative overhead than scenarios with many multicast receivers, where the cost of building and maintaining the multicast tree is amortized over an increased number of packet deliveries. However, with only one, shared, multicast tree being build and maintained, the overhead grows relatively slow with an increase in the number of multicast senders (more multicast senders will incur a higher overhead since they all need to find and maintain a path to the multicast tree). In low mobility scenarios, PSR ranges from 7.77 for 1 sender/10 receiver to 11.0 for 10 sender/10 receiver. This number drops to 4.6 for all scenarios with 50 multicast receivers (all nodes are part of the multicast tree). In high mobility scenarios, the overhead is higher, ranging from 11.65 in the 1 sender/10 receiver scenarios to 13.75 for the 10 sender/10 receiver cases. For the 50 receiver scenarios under high mobility, this ratio drops to around 6. However, since almost all MAC transmissions are unicast, again a high fraction (approximately 70%) of these packets are MAC layer control packets (RTS, CTS, and ACK).

ODMRP shows better protocol performance than either of the unicast protocols: the packet delivery ratio is consistently higher and latency is significantly reduced (often only a few milliseconds). ODMRP also outperforms MAODV, resulting in higher

Table 5. PDR and Latency for ODMRP, 1 m/s maximum speed.

	1 Sender		2 Sender		5 Sender		10 Sender	
	PDR	Lat.	PDR	Lat.	PDR	Lat.	PDR	Lat.
10 Receiver	0.995	0.009	0.991	0.010	0.973	0.015	0.937	0.031
20 Receiver	0.995	0.010	0.991	0.011	0.974	0.017	0.936	0.039
30 Receiver	0.993	0.011	0.990	0.012	0.972	0.017	0.934	0.043
40 Receiver	0.992	0.011	0.990	0.012	0.972	0.018	0.932	0.049
50 Receiver	0.973	0.011	0.971	0.012	0.953	0.018	0.915	0.056

Table 6. PDR and Latency for ODMRP, 20 m/s maximum speed.

	1 Sender		2 Sender		5 Sender		10 Sender	
	PDR	Lat.	PDR	Lat.	PDR	Lat.	PDR	Lat.
10 Receiver	0.987	0.010	0.992	0.011	0.977	0.016	0.940	0.033
20 Receiver	0.991	0.010	0.992	0.011	0.979	0.017	0.937	0.039
30 Receiver	0.991	0.010	0.994	0.012	0.980	0.017	0.935	0.046
40 Receiver	0.993	0.010	0.994	0.012	0.979	0.018	0.934	0.052
50 Receiver	0.973	0.010	0.974	0.012	0.961	0.018	0.916	0.060

packet delivery ratio and lower latency. This is true even though our implementation is based on an improved version of MAODV that pro-actively maintains the single, shared multicast tree. For 1 or 2 senders, as many as 99% or more of the packets are delivered, even in high mobility scenarios. The PDR drops for 5 senders to around 97% and is significantly lower at about 93% or so for 10 senders. There is also a noticeable increase in packet latency for the 10 sender scenarios. Mobility seems to have little impact on overall performance, with PDR and latency roughly similar across comparable scenarios.

In terms of network efficiency, ODMRP has a lower packet send ratio (PSR) than MAODV or the unicast protocols. In general, the higher the number of multicast receivers, the lower the PSR. The highest PSR value is observed for 10 senders and 10 receivers, at about 5.4. This can be explained by the need to build and maintain 10 multicast meshes. As the number of receivers increases, these costs are amortized over an increased number of successfully delivered packets, reducing the PSR to 1.4 for 50 receivers. For a single multicast sender, the PSR ranges from 2.7 (10 multicast receivers) to 0.88 (50 multicast receivers).

Table 7. PDR and Latency for ADMR, 1 m/s maximum speed.

	1 Sender		2 Sender		5 Sender		10 Sender	
	PDR	Lat.	PDR	Lat.	PDR	Lat.	PDR	Lat.
10 Receiver	0.979	0.045	0.975	0.048	0.966	0.046	0.957	0.045
20 Receiver	0.977	0.052	0.981	0.048	0.969	0.052	0.963	0.047
30 Receiver	0.980	0.048	0.981	0.046	0.975	0.047	0.968	0.048
40 Receiver	0.983	0.050	0.982	0.042	0.976	0.045	0.971	0.044
50 Receiver	0.983	0.057	0.986	0.040	0.979	0.047	0.973	0.047

Table 8. PDR and Latency for ADMR, 20 m/s maximum speed.

	1 Sender		2 Sender		5 Sender		10 Sender	
	PDR	Lat.	PDR	Lat.	PDR	Lat.	PDR	Lat.
10 Receiver	0.957	0.039	0.956	0.038	0.946	0.041	0.930	0.043
20 Receiver	0.968	0.040	0.965	0.042	0.956	0.045	0.945	0.048
30 Receiver	0.975	0.043	0.970	0.044	0.965	0.046	0.954	0.050
40 Receiver	0.977	0.044	0.973	0.048	0.969	0.048	0.962	0.051
50 Receiver	0.981	0.042	0.978	0.044	0.973	0.048	0.966	0.051

AMDR performs consistently better than MAODV. For few multicast senders, it performs worse than ODMRP, and it also appears to be more sensitive to the mobility rate in the network. It does, however, perform better for the 10 multicast sender scenarios, under both low and high mobility. Overall, it is the most stable protocol, giving very consistent performance across all scenarios. Under low mobility, for example, the PDR ranges from 95.7% to 98.6%, with packet latencies varying from 42 to 57 milliseconds. Under high mobility, PDR ranges from 93.0% to 98.1%, and packet latency varies from 38 to 51 milliseconds.

This consistent performance is also shown by the network efficiency metric. Similar to all multicast protocols, the PSR is reduced as the number of multicast receivers increases. This is mostly due to two reasons: since packets are broadcast at the MAC layer, a higher number of multicast receivers increases the chances that this packet broadcast is received by multiple receivers. Also, protocol overheads are amortized over an increased number of packet deliveries with a higher number of receivers. However, unlike the previous two protocols, the PSR varies only to a small degree: from 1.17 to 0.35 under low mobility and from 1.8 to 0.46 for high mobility scenarios.

The last two protocols implement broadcasting: all data packets are delivered to all nodes. We applied these protocols to multicast scenarios (i.e., only a subset of nodes is interested in this data), the metrics reported here are based on the protocol performance with respect to the identified multicast receivers. The overhead caused by the delivery of data packets to nodes that are not multicast receivers is captured by the network efficiency metric. For FLOOD, under both low and high mobility, the packet delivery ratio stays high (at or above 99%) for one or two multicast senders, dropping to 95.5% to 96.5% for 5 multicast senders and to 81.5% to 82.7% for 10 multicast senders. The packet latencies are small (a few tens of milliseconds) for most scenarios, but increase drastically for the 10 multicast sender scenarios. The number of multicast receivers, as expected, does not impact the performance metrics.

Table 9. PDR and Latency for FLOOD, 1 m/s maximum speed.

	1 Sender		2 Sender		5 Sender		10 Sender	
	PDR	Lat.	PDR	Lat.	PDR	Lat.	PDR	Lat.
10 Receiver	0.998	0.023	0.991	0.027	0.956	0.051	0.828	2.084
20 Receiver	0.998	0.025	0.991	0.029	0.956	0.052	0.827	2.099
30 Receiver	0.996	0.025	0.989	0.029	0.954	0.052	0.826	2.091
40 Receiver	0.996	0.026	0.989	0.029	0.954	0.052	0.826	2.078
50 Receiver	0.996	0.025	0.990	0.028	0.955	0.051	0.831	2.041

Table 10. PDR and Latency for FLOOD, 20 m/s maximum speed.

	1 Sender		2 Sender		5 Sender		10 Sender	
	PDR	Lat.	PDR	Lat.	PDR	Lat.	PDR	Lat.
10 Receiver	0.999	0.023	0.993	0.029	0.965	0.053	0.815	1.684
20 Receiver	0.999	0.023	0.993	0.028	0.965	0.052	0.815	1.683
30 Receiver	0.999	0.023	0.993	0.029	0.965	0.052	0.815	1.687
40 Receiver	0.999	0.023	0.993	0.028	0.965	0.052	0.815	1.686
50 Receiver	0.999	0.022	0.993	0.028	0.965	0.051	0.818	1.652

Some packet losses can be explained by transmission collisions. However, the protocol implementation takes great care to avoid such collisions, randomly jittering re-broadcasts by 10 ms. Also, the underlying MAC protocol is based on "carrier sense" (i.e., listen to the media and apply random backoff when media is busy), reducing the chances of collision errors even further. In addition, a receiver can receive a specific data packet over multiple different "paths". Loosing the flooded data packet over all such paths due to collisions is rare indeed, as shown by the high packet delivery ratios for 1 and 2 senders. However, with 5 or 10 senders, the network starts to experience congestion. The MAC protocol starts dropping a significant number of packets due to queue overflow.

The network efficiency is largely independent of the number of multicast senders and mobility rate. In FLOOD, every packet is transmitted up to 50 times. Therefore, the PSR is 5 (50 packet transmissions for 10 packet receptions) for 10 multicast receivers and 1 (50 packet transmissions for 50 packet receptions) for 50 multicast receivers. This is approximately true for all multicast sender scenarios. Even when the packet delivery ratio falls to significantly below 100% in the 10 multicast sender cases, the total number of packet transmissions at the MAC layer falls proportionally.

Table 11. PDR and Latency for BCAST (100 ms), 1 m/s maximum speed.

	1 Sender		2 Sender		5 Sender		10 Sender	
	PDR	Lat.	PDR	Lat.	PDR	Lat.	PDR	Lat.
10 Receiver	0.997	0.116	0.996	0.108	0.987	0.118	0.970	0.126
20 Receiver	0.997	0.119	0.996	0.112	0.987	0.121	0.970	0.129
30 Receiver	0.995	0.121	0.994	0.113	0.985	0.121	0.968	0.129
40 Receiver	0.995	0.121	0.994	0.113	0.984	0.121	0.967	0.129
50 Receiver	0.995	0.118	0.994	0.110	0.985	0.118	0.968	0.126

Table 12. PDR and Latency for BCAST (100 ms), 20 m/s maximum speed.

	1 Sender		2 Sender		5 Sender		10 Sender	
	PDR	Lat.	PDR	Lat.	PDR	Lat.	PDR	Lat.
10 Receiver	0.993	0.104	0.992	0.108	0.986	0.114	0.968	0.126
20 Receiver	0.992	0.103	0.991	0.107	0.986	0.114	0.968	0.127
30 Receiver	0.993	0.103	0.992	0.107	0.986	0.113	0.968	0.127
40 Receiver	0.993	0.103	0.992	0.107	0.986	0.113	0.968	0.126
50 Receiver	0.993	0.101	0.992	0.105	0.986	0.110	0.969	0.124

When scaling the packet delay factor by 100 ms, BCAST achieves similar performance to FLOOD for one or two multicast sender, but with significantly increased PDR for 5 and 10 multicast sender. For 1 multicast sender, the PDR is slightly below the PDR in FLOOD, in particular for high mobility scenarios. This is due to the fact that BCAST has less redundancy, dynamically selecting only a subset of nodes to re-broadcast a packet. This shows in the packet send ratio as well: PSR ranges from 2.5 to 0.5 for 10 and 50 multicast receivers, respectively. The resulting lower network traffic is beneficial in the 2 and 5 sender cases, where fewer collisions occur at the MAC layer, resulting in improved PDRs, compared to FLOOD. And scenarios with 10 multicast senders benefit the most, with PDRs of about 97%. Overall, the protocol performance is very consistent across all scenarios, with PDR ranging from 96.7% to 99.7% and packet latency ranging from 104 ms to 129 ms.

Packet latency is rather high, consistently above 100 ms. Running experiments with a smaller scaling factor, such as 10 ms, shows that a large scaling factor is beneficial, achieving high packet delivery ratio at the expense of some latency. With a 10 ms scaling factor, PDR for the 1-sender cases is around 99.6% to 99.7%, dropping to 98.3% to 98.6% for the 2 sender cases. It drops even more drastically to 91% to 92% for the 5 sender cases, and to 69% for the 10 sender cases. Except for the 10 sender cases, though, packet latency is reduced to about 30 ms in all 1 and 2 sender scenarios, increasing to about 75 ms in the 5 sender cases. All in all, however, for BCAST to perform well, a relatively large delay factor is beneficial.

In summary, there are a number of alternatives when delivering data from one or a few senders to a group of receivers: setting up dedicated unicast connections from each sender to each receiver, employing a multicast protocol, and broadcasting the packet to every node. The experiments reported here show that reducing the multicast problem to an n-fold unicast case is the worst solution: except for scenarios with only one or two senders and a small number of receivers, packet delivery ratio is low and packet latencies are high. Any of the three multicast protocols we studied improve the multicast performance. Among them, MAODV had the poorest performance. Based on our analysis, this is due to the shared multicast tree, for two reasons:

1. The tree is based on hard state, requiring explicit control messages to maintain it, in particular under high mobility scenarios. ODMRP and AMDR, by contrast, maintain their multicast distribution structure in soft state.
2. With all data flowing through a shared tree, the queues along interior tree nodes are more likely to overflow, even for a small number of multicast senders.

ADMR and ODMRP show better performance. ODMRP in particular works well for relatively few multicast senders; ADMR works well for 10 multicast sender scenarios, has the most consistent performance among all multicast protocols across all scenarios, and is very network efficient.

The broadcasting protocols work very well in most scenarios, and are more robust with respect to number of multicast receivers and mobility rate. BCAST improves on FLOOD in most cases and is more network efficient, indicating that it pays off to explore more complicated broadcast protocols.

None of the broadcast/multicast protocols outperforms the other protocols in all scenarios. The following tables summarize the best protocol for each scenario, ranked by PDR (with packet latency as tie-breaker). Any ranking, by necessity, will have to weigh different performance metrics such as PDR or latency. We chose PDR as the more significant metric for the following reasons:

1. First, the two metrics are not mutually exclusive. The above results show that significant drops in packet delivery ratio are usually accompanied by large increases in packet latency, indicating that the network starts to experience congestion. So certainly protocols can exhibit simultaneously poor performance under both metrics.
2. As long as packet latency is bound by a relatively small number, we assume that further improvements in latency do not really contribute to the overall user satisfaction. For example, the ITU recommends that VoIP applications should not experience latencies beyond 400 ms (roundtrip) for voice service. All entries in the tables below have one-way latencies well below 200 ms, meeting the roundtrip upper bound.
3. Many applications are sensitive to PDR. For example, data download times directly benefit from improvements in PDR. Furthermore, the throughput of a TCP-like data stream will suffer with every packet loss, so even small improvements in PDR can yield significant improvements in application performance and user satisfaction.

Table 13. Protocol Ranking, 1 m/s maximum speed.

	1 Sender		2 Sender		5 Sender		10 Sender	
	PDR	Lat.	PDR	Lat.	PDR	Lat.	PDR	Lat.
10 Receiver	FLOOD		AODV		BCAST		BCAST	
	0.998	0.023	0.996	0.024	0.987	0.118	0.970	0.126
20 Receiver	FLOOD		BCAST		BCAST		BCAST	
	0.998	0.025	0.996	0.112	0.987	0.121	0.970	0.129
30 Receiver	FLOOD		BCAST		BCAST		BCAST	
	0.996	0.025	0.994	0.113	0.985	0.121	0.968	0.129
40 Receiver	FLOOD		BCAST		BCAST		ADMR	
	0.996	0.026	0.994	0.113	0.984	0.121	0.971	0.044
50 Receiver	FLOOD		BCAST		BCAST		ADMR	
	0.996	0.025	0.994	0.110	0.985	0.118	0.973	0.047

Table 14. Protocol Ranking, 20 m/s maximum speed.

	1 Sender		2 Sender		5 Sender		10 Sender	
	PDR	Lat.	PDR	Lat.	PDR	Lat.	PDR	Lat.
10 Receiver	FLOOD		FLOOD		BCAST		BCAST	
	0.999	0.023	0.993	0.029	0.986	0.114	0.968	0.126
20 Receiver	FLOOD		FLOOD		BCAST		BCAST	
	0.999	0.023	0.993	0.028	0.986	0.114	0.968	0.127
30 Receiver	FLOOD		ODMRP		BCAST		BCAST	
	0.999	0.023	0.994	0.012	0.986	0.113	0.968	0.127
40 Receiver	FLOOD		ODMRP		BCAST		BCAST	
	0.999	0.023	0.994	0.012	0.986	0.113	0.968	0.126
50 Receiver	FLOOD		FLOOD		BCAST		BCAST	
	0.999	0.022	0.993	0.028	0.986	0.110	0.969	0.124

Overall, the following major trends can be identified:

- Broadcast protocols work rather well in the scenarios we studied. BCAST and FLOOD work almost always as good as or better than other protocols, though sometimes impose higher packet latency.
- For a single multicast sender, FLOOD is the obvious choice, for increasing number of multicast senders BCAST has the edge over FLOOD
- ADMR performs very well in the presence of many multicast senders, and is indeed the optimal choice in two scenarios under low mobility, with BCAST being runner-up. All other protocols perform poorly in these scenarios.
- The choice of an optimal multicasting solution is largely independent of the mobility rate. This is true for the 1 and 5 multicast sender scenarios, but also largely true in the 2 multicast sender scenarios. In the latter case, it is often the tie-breaking criterion that results in the selection of a specific protocol in Tables 13 and 15. In all cases, BCAST achieves the same maximal PDR, but at higher packet latency. For the 10 sender scenarios, BCAST and ADMR are the only two protocols to provide consistently high performance, with BCAST's PDR being higher in the vast majority of cases.

5 Related Work

Many papers present performance results, based on simulations, to study and evaluate protocol behavior under a range of scenarios. Almost any paper presenting a new routing protocol will contain an evaluation section that compares the proposed protocol against a (typically small) set of related protocols. Good examples are [2], which is co-written by the designers of AODV and compares this protocol with DSR, and [3], which introduces ADMR and compares it with ODMRP.

A number of papers are dedicated to performance comparisons of various routing protocols, typically based on simulation. [1] is one of the earliest comparative studies of MANET unicast routing protocols in NS2, indicating the superior performance of AODV and DSR. [6] simulates several multicast routing protocols developed specifically for MANET and evaluate them under diverse network scenarios using the GloMoSim library. The reported results show that mesh protocols performed significantly better than the tree protocols in mobile scenarios. Finally, a number of papers have studied the performance of broadcast protocols, one of the more recent papers is [16], which categorizes various broadcast protocols into a small set of categories, implements a representative protocol from each category in a simulator and conducts an in-depth analysis of the performance across a range of scenarios. The results show that all protocols will eventually suffer from low packet delivery ratio as the mobility rate increases, and that BCAST is one of the protocol that will "break" the latest.

All these papers compare only similar protocols with each other: [1] focuses on unicast protocols, [6] on multicast protocols, and [16] on broadcast protocols. Only [10] compares multicast protocols with broadcast protocols, showing that flooding results in higher packet delivery ratios then ODMRP, which in turn outperforms AODV. However, the simulation scenarios were all based on the assumption that every node in the MANET was interested in the data packets (i.e., a broadcast scenario). And [7] provides simulation results that compare flooding with ODMRP and three other multicast routing protocols, but the discussion in the paper focuses mostly on the relative performance of the multicast protocols. This study is a systematic experimental comparison of various alternatives to support one-to-many and many-to-many communication in a MANET.

6 Summary and Conclusions

This report documents the results of our study into how to best support one-to-many and many-to-many communication in a MANET. We conducted extensive simulations to evaluate the performance of a number of distinct solutions to this problems: applying two unicast routing protocols (DSR and AODV), employing one of three multicast routing protocols (MAODV, ODMRP, or ADMR), or resorting to one of two broadcast protocols (FLOOD and BCAST). With packet delivery ratio the major performance metric, broadcast protocols provide the best performance, certainly for scenarios with relatively few multicast senders.

Arguably, broadcast may appear as overkill: every node will receive the data packets, even though they may not participate in the multicast group. As a result, the network around these nodes is busy relaying multicast packets. However, this is not that severe a limitation as it may appear at first sight for a number of reasons. First, as the results on PSR show, BCAST (the more efficient broadcast protocol) achieves its high packet delivery ratio with fewer packet transmissions per packet delivered (counting only interested multicast receivers) then any other protocol. Second, due to the nature of the wireless media, it is rather unlikely that nodes that are not participating in a multicast group can actually utilize the media. All multicast protocols will elect non-participating nodes to relay data if this becomes necessary to reach all multicast receivers. In addition, with a radio range of 250 m and an interference range of over 500 meters, nodes in the vicinity of a multicast receiver are likely to overhear multicast packets (or due to interference be prevented from transmitting/receiving concurrently). So assuming a relatively uniform distribution of multicast receivers among the nodes in the network, multicast traffic is likely to impact all nodes.

Of course, having to actually reach every node can result in higher overheads, as shown by FLOOD. Similarly, when multicast receivers tend to cluster in a certain area, network overheads can be reduced as well, multicast protocols will only re-broadcast packets in that network area. The first problem can be addressed by more efficient broadcast protocols such as BCAST (at least to some extent). The second problem (which we have not studied in-depth) may be resolved by applying some form of geocast (geographically limited broadcast).

Ultimately, our goal is to study and develop reliable multicast solutions, i.e., protocols that deliver packets to all multicast receivers with high probability. Work is already under way in this respect, using BCAST as a starting point. We will also explore variations on IEEE 802.11 to increase the overall network capacity. With congestion being the major reason for packet loss, increasing the network capacity should result in improved packet delivery ratios as well.

Acknowledgements. This work is funded by the Defense Research and Development Canada (DRDC) and benefited from various discussions with members of the CRC mobile networking group.

References

1. J. Broch et al., A Performance Comparison of Multi-Hop Wireless Ad Hoc Network Routing Protocols, Proc. of MobiCom 1998, pp. 85–97, Dallas, USA, Oct. 1998.
2. S.R. Das et al., Performance Comparison of Two On-demand Routing Protocols for Ad hoc Networks, IEEE Personal Comm. Magazine, pp. 16-28, Feb. 2001.
3. J.G. Jetcheva and D.B. Johnson, Adaptive Demand-Driven Multicast Routing in Multi-Hop Wireless Ad Hoc Networks, Proc. of MobiHoc 2001, pp. 33-44, Long Beach, USA, October, 2001.
4. D.B. Johnson et al., DSR: The Dynamic Source Routing Protocol for Multi-Hop Wireless Ad Hoc Networks, Ad Hoc Networking, pp. 139-172, Addison-Wesley 2001.
5. T. Kunz, Multicasting: From Fixed Networks to Ad Hoc Networks, in Handbook of Wireless Networks and Mobile Computing, pp. 495-507, John Wiley & Sons 2002, ISBN 0-471-41902-8.
6. S.-J. Lee et al., A Performance Comparison Study of Ad Hoc Wireless Multicast Protocols, Proc. of the 19th Conf. of the IEEE Comp. and Comm. Societies, Vol. 2, pp. 565-574, Tel Aviv, Israel, Mar. 2000.
7. S.-J. Lee et al., On-Demand Multicast Routing Protocol in Multihop Wireless Mobile Networks, Mobile Networks and Applications, 7(6):441-453, Dec. 2002.
8. V. Li and Z. Zhang, Internet Multicast Routing and Transport Control Protocols, Proc. of the IEEE, 90(3):360-391, March 2002.
9. W. Lou and J. Wu, On Reducing Broadcast Redundancy in Ad Hoc Wireless Networks, IEEE Transactions on Mobile Computing, 1(2):111-122, April 2002.
10. K. Obraczka et al., Flooding for Reliable Multicast in Multi-Hop Ad Hoc Networks, Wireless Networks, 7(6):627-634, 2001.
11. S. Paul, Multicasting on the Internet and its Applications, Kluwer Academic Publishers 1998, ISBN 0792382005.
12. C. E. Perkins and E. Royer, Ad hoc On-Demand Distance Vector Routing, Proc. of the 2nd IEEE Workshop on Mobile Computing Systems and Applications, pp. 90-100, New Orleans, LA, Feb. 1999.
13. B. Quinn and K. Almeroth, IP Multicast Applications: Challenges and Solutions, RFC 3170, Sept. 2001.
14. E. Royer, and C. E. Perkins, Multicast Operation of the Ad-Hoc On-Demand Distance Vector Routing Protocol, Proc. of MobiCom 1999, pp. 207-218, Seattle, USA, Aug. 1999.
15. U. Varshney, Multicast over Wireless Networks, Comm. of the ACM, 45(12):31-37, Dec. 2002.
16. B. Williams and T. Camp, Comparison of Broadcast Techniques for Mobile Ad Hoc Networks, Proc. of MobiHoc 2002, pp.194-205, Lausanne, Switzerland, June 2002.

Maximal Source Coverage Adaptive Gateway Discovery for Hybrid Ad Hoc Networks

Pedro M. Ruiz and Antonio F. Gomez-Skarmeta

University of Murcia, Murcia, E-30071, Spain

Abstract. One of the most important aspects affecting the overall performance of hybrid ad hoc networks is the efficient selection of Internet gateways. We have analytically modeled existing proposals (i.e. adaptive, reactive and hybrid gateway discovery) showing that each of them is only suitable for a particular range of network parameters. We propose a new adaptive gateway discovery scheme based on the dynamic adjustment of the scope of the gateway advertisement packets. We show through simulation that our proposed adaptive gateway discovery scheme based on the maximum source coverage outperforms existing mechanisms over a variety of scenarios and mobility rates.

1 Introduction and Motivation

The flexibility, self-configurability and easy deployment of mobile ad hoc networks (MANET) are making these networks an indispensable component in future mobile and wireless network architectures. In addition, with the advent of future wireless systems consisting of an integration of different heterogeneous wireless technologies, the interconnection of MANETs to fixed IP networks is one of the areas which are becoming of paramount importance. In such scenarios, commonly known as hybrid ad hoc networks, mobile nodes are witnessed as an easily deployable extension to the existing infrastructure. Some ad hoc nodes act as "gateways" which can used by mobile terminals to seamlessly communicate with other nodes in the fixed network. The challenge in interconnecting ad hoc networks to Internet, stems from the need to inform ad hoc nodes about available gateways in an extremely changing scenario while making a minimal consumption of the scarce network resources. So, an efficient gateway discovery for ad hoc networks becomes one of the key elements to enable the use of hybrid ad hoc networks in future mobile and wireless networks.

The different proposals to the issue of Internet connectivity for MANETs in the literature have used either a proactive gateway discovery or a reactive one. In the approaches based on a proactive gateway advertisements ([1], [2], [3]) the gateways periodically send advertisement messages which are flooded throughout the ad hoc network to inform all ad hoc nodes about available Internet gateways. Although these approaches achieve good connectivity, they have been usually criticized due to the high overhead they require and their limited scalability. In reactive approaches ([4], [5]) those nodes which require connectivity to the

I. Nikolaidis et al. (Eds.): ADHOC-NOW 2004, LNCS 3158, pp. 28–41, 2004.

Internet reactively find those gateways by means of broadcasting some kind of solicitation within the entire ad hoc network. Although these approaches have been considered to require less overhead, we show in the next section that this process of finding gateways is as costly as the proactive advertisement. In fact, we show that proactive gateway discovery mechanisms scale badly regarding the number of active sources willing to access the Internet.

There are also other works ([6], [7]) which propose hybrid gateway discovery approaches. In [6], the authors propose an scheme in which advertisements are only propagated up to a certain number of hops, and those nodes out of that scope will proactively find the gateways. However, as the authors show, the optimal TTL depends very much on the particular scenario and network conditions under consideration and so does the performance of this approach. In [7] the authors propose a more sophisticated approach in which advertisements are sent out only when changes in the topology are detected. However, they rely on a source based routing protocol, which limits very much the applicability of their approach.

In our opinion, existing approaches have neglected the huge overhead that the reactive gateway discovery scheme can have. The overall performance of the static approaches proposed so far, by strongly depending on the scenarios under consideration (e.g. number of sources, number of nodes, degree of the network, etc.) can vary dramatically as the network conditions change. We propose an adaptive gateway discovery approach based on the dynamic tuning of the scope of the gateway advertisements. Just by monitoring data packets, gateways will adaptively select the time to live of their advertisement that best suits the current network conditions. So, even when the network conditions change, the overall network overhead is reduced while still maintaining a good connectivity. In the author's opinion the main contributions of this paper are (i) an analytical study of the overhead of different gateway advertisement approaches showing the need for an adaptive scheme, and (ii) an adaptive gateway discovery approach for hybrid networks which is shown through simulation to outperform existing alternatives.

The remainder of the paper is organized as follows: section 2 provides an analytical evaluation of the different approaches and shows the need for adaptive gateway discovery alternatives. In section 3 we describe our proposed adaptive approach based on the maximal source coverage. The results of the simulations are shown in section 4. Finally, section 5 gives some conclusions and draws some future directions.

2   Analytical Evaluation of Existing Gateway Discovery Approaches

We consider as the baseline scenario for our analysis an hybrid network using the AODV [8] ad hoc routing protocol with an Internet connectivity approach similar to the reactive one proposed in [4]. We analyse that approach with three different gateway discovery variants: reactive, proactive and hybrid.

The reactive approach is the basic approach described in [4]. RREP and RREQ messages are extended with a new flag ("I") which is used to differentiate usual RREP and RREQ messages from those used to discover routes to the Internet. We refer to the new messages as RREP_I and RREQ_I. A source willing to communicate with a node in the fixed network, will first attempt to contact it within the ad hoc network doing an extended ring search (as described below). If no answer is received after a network-wide search, then the source tries to find a route towards the Internet. So, it broadcasts a RREQ_I to the ALL_MANET_GW_MULTICAST address. Gateways, upon reception of this message will send out a unicast RREP_I message to the source. Then the source will select one of the gateways (based on the hop count) and will send the data towards the fixed node through that gateway.

To implement the proactive approach, a new message called GWADV ("Gateway Advertisement") is introduced. Gateways will periodically broadcast within the ad hoc network these messages in order to inform all the nodes about the availability of that gateway. Upon reception of a GWADV message, mobile nodes will select their preferred gateway based on the hop count, and they will store a default route entry in their routing table. When a source wants to communicate with a destination, it tries first to find a direct route within the MANET, and if it does not manage to do it, it then uses its default route.

Finally, the hybrid approach we have implemented is basically the one described in [6]. Gateways will periodically send GWADV messages only a few hops away. The sources within that range will behave as in the proactive approach, and those beyond that range will find default routes proactively using the same RREQ_I-based reactive scheme described before.

2.1 Analytical Model

In our model, we assume that the nodes are uniformly distributed in a rectangular lattice covering a certain area. Each vertex of the lattice is a possible location for a node, but only one node can be at a concrete vertex. An example of such a rectangular lattice is shown in figure 1. Given a node n in the lattice (not in the boundary) there are $4k$ nodes at a distance of k hops from n. These nodes are placed in the kth concentric ring centered on the node n. It is easy to show that the total number of nodes including n at a distance of k hops is given by equation (1). We also give the relation between k and N, in which $\lceil x \rceil$ is the standard ceiling operation meaning completion to the next integer. It is used in the expression for obtaining k because the last concentric ring might not be complete. So, given a broadcast message with time to live (TTL) equal to x, $N_r(x)$ will be the number of nodes forwarding that message if $x \leq (\sqrt{2N-1}-1)/2$ and N otherwise.

$$N_r(k) = 1 + \sum_{j=1}^{k} 4j = 1 + 2k(k+1), k = \lceil (\sqrt{2N-1}-1)/2 \rceil \qquad (1)$$

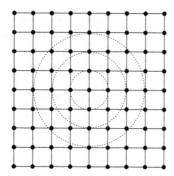

Fig. 1. Rectangular lattice

Regardless of our gateway discovery mechanism, the approach used in [4] detects that a destination is a fixed node when the source does not receive any answer after a network wide search. This network wide search is done using an expanding ring search. That is, the first route request message is only sent to the nodes at TTL_START hops. If no answer is received, a new message is sent with the previous TTL plus TTL_INCREMENT. This process is repeated up to a TTL of TTL_THRESHOLD. If no answer is received, then the last request message is sent with a TTL equal to NETWORK_DIAMETER. The typical values defined for these constants in AODV specification are TTL_START=1, TTL_INCREMENT=2, TTL_THRESHOLD=7 and NET-WORK_DIAMETER=30. Although we think that these values are not appropriate for hybrid networks, we have obeyed the original specification.

Whenever an ad hoc source tries to find a route towards a fixed node it never gets any answer within the ad hoc network. Thus, for each source S, the number of messages associated to realizing that a destination is a fixed node can be calculated according to equation (2).

$$\Omega_{FN} = \sum_{j \in \{1,3,5,7,30\}} N_r(j) \qquad (2)$$

Similarly, whenever a source wants to reactively discover a gateway there is an overhead which is the sum of the number of messages required to do a network-wide distribution of the RREQ_I packet addressed to the ALL_MANET_GATEWAYS multicast address, plus the number of messages required to send an unicast RREP_I reply from every gateway to the source. Assuming the the gateways are in the borders of the lattice, it is easy to demonstrate that the mean path length is $\sqrt{N} - 1$. Thus, if we denote the number of gateways by N_{GW}, the overhead of the reactive discovery of the gateways by one source can be computed as it is shown in equation (3).

$$\Omega_{r-gw} = N + N_{GW} + N_{GW} \cdot (\sqrt{N} - 1) = N + N_{GW} \cdot \sqrt{N} \qquad (3)$$

Let S be the number of active sources in the hybrid network communicating with fixed nodes, λ_{adv} the rate at which GWADV messages are being sent out by the gateways and t the duration of the time interval under consideration. The overhead of delivering each of this messages to the whole ad hoc network is $N+1$ messages; one forwarding by each of the N nodes (because of the duplicate messages avoidance) plus the message sent out by the gateway itself. In addition, if we take into account that initially all the sources in the network will need to realize that the destination node is a fixed node, then the total overhead in number of messages required by the proactive approach can be obtained from equation (4).

$$\Omega_p = S \cdot \Omega_{FN} + \lambda_{adv} \cdot t \cdot (N + 1) \cdot N_{GW} \qquad (4)$$

In the same way, if we denote by R_{dur} the route duration time in AODV[1]. Then, R_{dur} obeys an exponential random distribution with parameter λ_{dur}. Let N_{break} be a random variable representing the number of route expirations during an interval of t units of time. Then, N_{break} follows a Poisson distribution with an arrival rate equal to λ_{dur} so that $P[N_{break} = k] = \frac{e^{-\lambda_{dur} \cdot \lambda_{dur}^k}}{k!}$. So, the mean number of default route expirations per source will be given by $E[N_{break}] = \lambda_{dur} \cdot t$. Accordingly, the total overhead for the proactive route discovery will consist of the initial overhead so that every source gets aware that their destination is a fixed node, plus the overhead associated to the proactive discovery of the gateways whenever their default route expires or breaks. This overhead can be computed according to equation (5).

$$\Omega_r = [\Omega_{FN} + (\Omega_{r-gw} \cdot \lambda_{dur} \cdot t)] \cdot S \qquad (5)$$

The hybrid gateway discovery scheme, has an overhead which is a combination of the overheads of the other approaches. For those sources located outside the area covered by GWADV messages, the overhead will be the similar to the overhead of the reactive approach. Thus, in order to asses the overhead of the hybrid approach it is of paramount importance, being able to calculate the mean number of sources which will be within the GWADV range.

Let's assume that the gateways are located in the corners of the lattice as in our simulated scenario. In the hybrid approach it makes no sense sending GWADV at longer TTLs than $\sqrt{N} - 1$, because other gateways will be covering the area beyond that TTL. Then its is easy to derive an expression for the number of nodes which are at t hops, $t \in [0, \sqrt{N} - 1]$ from any gateway according to equation (6).

$$N_r^{GW_i}(t) = \sum_{j=1}^{t}(t + 1) = \frac{t(t + 3)}{2} \qquad (6)$$

---

[1] configured to be 10 seconds unless the route becomes invalid before (e.g. due to mobility)

Given a node n from the ad hoc network, the probability that this node will be able to receive a GWADV message from any of the gateways can be computed as shown in equation (7).

$$P_c(t) = \frac{\sum_{i=1}^{N_{GW}} N_r^{GW_i}(t)}{N - N_{GW}} \qquad (7)$$

If we denote N_c as the number of sources being covered by any gateway when using a TTL of t, then N_c is a random variable obeying a binomial distribution $B \sim (S, P_c(t))$. Thus, the mean number of sources being covered when gateways use a TTL of t can be computed as $E[N_c] = S \cdot P_c(t)$. So, the overall overhead of the hybrid approach consists of three different parts: the overhead associated to realize that the destinations are fixed nodes, the overhead associated to the propagation of GWADV messages over t hops by each gateway, and the overhead required so that those sources not covered by the GWADV messages can find the gateways and create a default route. An expression for that overhead is shown in equation (8).

$$\Omega_h = S \cdot \Omega_{FN} + \lambda_{adv} \cdot t \cdot (N_r^{GW}(t) + 1) \cdot N_{GW} + \Omega_{r-gw} \cdot \lambda_{dur} \cdot t \cdot S \cdot (1 - P_c(t)) \quad (8)$$

To compare the overhead of the different approaches, we have used the figures in table 1. As it was expected the proactive approach is less scalable regarding the number of nodes in the ad hoc network. This is because the higher the number of nodes, the higher the number of retransmissions which are required to propagate GWADV messages to the whole network. This is why usually proactive approaches has been said in the literature to have too much overhead. However, we can also notice that the process of discovering the gateways can be as costly as the process of propagating the GWADV messages. In fact, under certain network conditions the reactive approach can incur in higher overhead than the proactive one. In particular, we have found interesting to stress the poor scalability of the reactive approach as the number of sources connecting to Internet increase. This is supported by the graphs in figure 2(a) and figure 2(b).

Table 1. Values used for the analytical evaluation graphs.

Constant	N	λ_{adv}	N_{GW}	λ_{dur}	t
Value	25	1/5	2	1/10	900 sec

As is it also shown in figure 2(a), the hybrid approach is somehow a trade-off between the reactive and the proactive approaches. Different values of TTL lead to different flavors of the hybrid approach. However, as it was also corroborated in [6], the optimal value of TTL is something that strongly varies from one scenario to another. In fact, as depicted in figure 2(b), there are situations in which a proactive approach performs better than an hybrid approach and vice versa. Thus, the definition of an universal hybrid gateway discovery approach

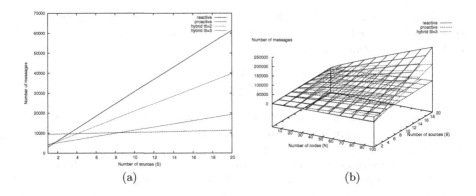

Fig. 2. Overhead vs. number of source (a) and vs. number o nodes and number of sources (b)

seems to be unrealistic without adding some degree of adaptability. We describe our proposed hybrid approach in the next section.

3 Adaptive Approach Based on Maximal Source Coverage

Given the conclusions of our analytical study, we believe that an adaptive gateway discovery mechanism being able to dynamically change its proactiveness or reactiveness can reduce the overhead of the gateway discovery without jeopardizing the overall network performance or at least reaching a good trade-off between performance and network overhead.

From the model of the hybrid approach we have learnt that the scope of the advertisements has a strong impact on the proactiveness or the reactiveness of the scheme. Thus, it seems reasonable to use the TTL of the GWADV messages as the parameter to adjust depending on the network conditions. The higher the TTL, the higher the overhead due to the periodic advertisement and the lower the overhead associated to the reactive discovery of the Internet gateways. That is, the higher the TTL the higher the proactiveness of the approach. In fact, a TTL = 0 corresponds to the totally reactive approach whereas a TTL = NETWORK_DIAMETER corresponds to a completely proactive scheme.

There are different criteria to determine when the TTL should be adjusted. For instance, the rate at which neighbors change or the mean duration of the links can be an indication of the network mobility. However, these kind of metrics are not usually easy to interpret. In addition, they do not capture one of the key parameters according to our model which is the number of sources.

For a gateway to be aware of the total number of sources communicating with nodes in the Internet it is required some kind of signaling mechanism facilitating such information to the gateway. However, that would incur in extra overhead and it is something which can require changes to the routing protocols. So, we propose to use simpler metrics being able to convey the required information without any additional overhead. In our proposal, each gateway will only know about the sources which are accessing to the Internet through them. This scheme is very convenient because that information is very easy to learn by the gateway provided that it is routing those datagrams that it would receive anyway.

In our approach, the gateways will keep track (using a structure like the one which is shown in 3) of the number of hops at which each of its active sources is located. This information is easy to extract by simply looking at the IP header. This table will be periodically purged so that stale entries do not influence the TTL of the next advertisement.

IP address	TTL	Time learnt
. . . .		

Fig. 3. Table to store source TTL entries

The second part of the problem is which heuristic to use for selecting the TTL of the next advertisement. We propose what we call the "maximum source coverage algorithm". Using this algorithm the gateway will send out the next advertisement with a TTL equal to the maximum number of hops for all its sources. So, our proposed approach uses one of the most proactive heuristics. Other feasible heuristics which behave more reactively are the selection of the TTL to cover only the first source, the TTL which covers a certain number of sources, the TTL which covers a certain percentage of its sources, etc.

The selection of the maximum source coverage is because our main concern is maintaining a high packet delivery ratio as close as possible to the proactive approach at a low overhead. The other heuristics tend to produce in some scenarios less overhead than the selected one, but would achieve a lower packet delivery ratio and would have been less scalable regarding the number of sources. As we show in the next section the proposed approach is able to obtain a similar throughput than the proactive approach while keeping the overhead in the figures of the hybrid and proactive ones.

4   Simulation Results

In order to assess the performance of the proposed scheme, we have performed a series of simulations using the NS-2 [9] network simulator. The simulated scenario consists of 25 mobile hosts randomly distributed over an area of 1200x500 m. The

radio channel capacity for each mobile node is 2Mb/s, using the IEEE 802.11b
DCF MAC layer and a communication range of 250 m. In addition, there are two
gateways; one located at the coordinates (50, 450) and (1150, 50) respectively.
In the hybrid approach both of them use a TTL = 2 for their advertisements
as it is recommended in [6] for the kind of scenarios under simulation. Each of
the gateways is connected to a router and the routers are connected one to each
other. Additionally, each router has a fixed node connected to it. All the fixed
links have a bandwidth of 10Mb/s, which is enough to accommodate all the
traffic coming from the mobile nodes. An example scenario is shown in figure 4.

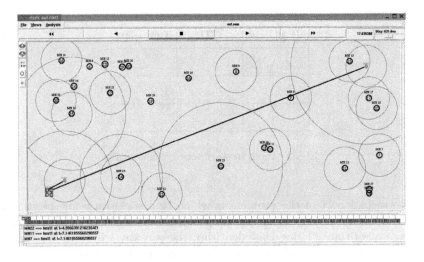

Fig. 4. Visualization of simulated scenarios

Each of the approaches has been evaluated over the same pre-generated set
of 840 scenarios with varying movement patterns and traffic loads. Mobile nodes
move using a random waypoint model with changing pause times. Nodes start
the simulation being static for *pause time* seconds. Then they pick up a random
destination inside the simulation area and start moving to the destination at a
speed uniformly distributed between 0 and 20 m/s (mean speed = 10m/s). After
reaching its destination this behavior is repeated until the end of the simulation.
Seven different pause times were used: 0, 30, 60, 120, 300, 600, and 900 seconds.
A pause time of 0 seconds corresponds to a continuous motion whereas a pause
time of 900 seconds corresponds to a static scenario. For each of these pause times
10 different scenarios where simulated. The results were obtained as the mean
values over these 10 runs to guarantee a fair comparison among the alternatives.

Four different traffic loads where tested consisting of 5, 10, 15, and 20 different
CBR sources communicating with nodes in the fixed network. Each of these CBR
sources start sending data at a time uniformly distributed between the first 10

seconds of the simulation. Each of the sources generates 512 bytes data packets at a rate of 5 packets per second (20Kb/s).

4.1   Performance Metrics

To assess the effectiveness of the different gateway discovery mechanisms, we have used the following performance metrics:

- Packet delivery ratio. Defined as the number of data packet successfully delivered over the number of data packets generated by the sources.
- Routing overhead. Defined as the total number of control packets, including gateway discovery, sent out during the simulation time.
- Normalized effectiveness. Defined as the number of packets successfully delivered minus the (weighted) number of control packets required divided by the total number of data packets generated by the sources. This metric gives a value of the overall performance by taking into account not only the packet delivery ratio but the overhead. The maximum value of 1 would only be achieved when all data packets are delivered without any overhead.

4.2   Simulation Results

The simulation results show that our proposed approach is able to offer a packet delivery ratio as higher as the proactive approach at a slightly higher overhead than the reactive and hybrid approaches. This is clearly shown in the case of 10 and 15 sources in figures 5 (a) and 5 (b) respectively. This differences in overhead are due to the fact that sometimes during the simulation it is required to use higher TTLs than the hybrid approach so that the GWADV messages can reach all the sources.

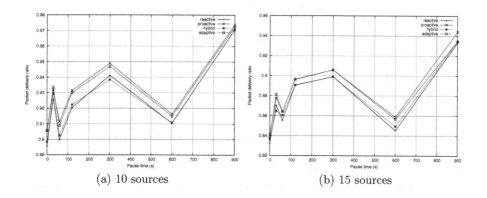

(a) 10 sources   (b) 15 sources

Fig. 5. Packet delivery ratio for different number of sources.

As shown when comparing figures 5 (a) and 5 (b), the higher the number of sources, the best performs our approach compared to the others. In addition, the higher the mobility of the nodes, the best the performance of the adaptive approach. For 10 sources the proposed approach is almost obtaining the same packet delivery ratio than the proactive scheme and much better than the hybrid and reactive ones. For 15 sources the proposed approach outperforms all the others. The reason is that with 15 sources the reactive and hybrid approaches require too much overhead due to the need for sources to reactively discover the gateways. The proactive approach also starts working worse because its high control packet load does not leave enough resources to carry all the data packets generated by the sources. However, the proposed approach is able to find a good trade-off between the signaling overhead and the proactivity of the protocol.

Regarding the routing overhead, a similar trend is observed. As it is depicted in figures 6 (a) and 6 (b), the proposed approach has a lower overhead than the proactive approach and a little bit more than the reactive and hybrid ones. The differences in overhead are also lesser as the number of sources increase. As explained in our analytical model, this is due to the cost required in the reactive appraoch in which the sources are required to perform a network-wide search of the gateways.

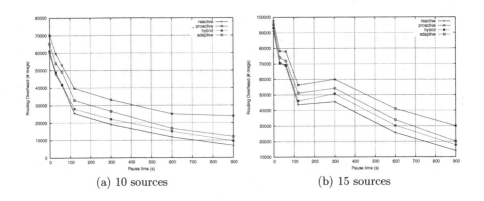

(a) 10 sources   (b) 15 sources

Fig. 6. Overhead for different number of sources sources.

The packet delivery ratio is a good metric to evaluate the performance of the protocol from the outside. That is, what is the performance obtained by the applications. Conversely, the routing overhead is a good internal metric of how much network resources does the protocol need to do its work. So, in order to evaluate the overall performance of the different solutions we need a metric taking into account both internal and external performance. As explained in the previous subsection, the metric that we will use is the normalized effectiveness. Of course, we adjust the weighting factor to give more importance to achieving a good packet delivery ratio, but there is a penalization for not doing it at a low overhead.

In figure 7 we show the normalized effectiveness for different number of sources and different mobility rates. As the mobility of the nodes decrease (higher pause times) the performance of all the approaches improve. The cause is that the number of link breaks decreases and so does the control overhead required to re-establish the routes to to the gateways. It is also worth mentioning that the low performance of all the protocols at a pause time of 600 seconds is due to the fact that the random scenario generator produced several scenarios with very bad connectivity and by no means that performance is related with the mobility of the network.

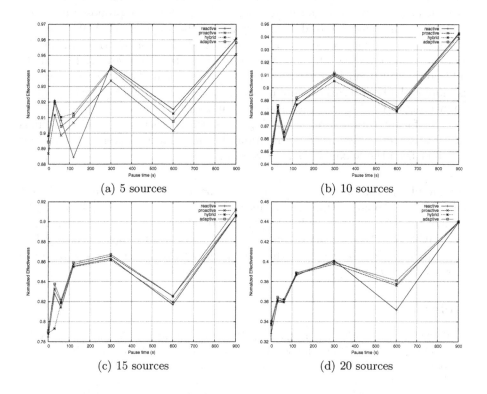

(a) 5 sources

(b) 10 sources

(c) 15 sources

(d) 20 sources

Fig. 7. Normalized effectiveness for increasing number of sources.

However, the most important result is that, as our analytical model predicted, an adaptive approach can obtain a good trade-off between the efficiency of the protocol in terms of packet delivery and the signaling overhead. As our model also anticipated the performance of the approaches is highly dependent on the number of sources. In fact, the adaptive approach has shown to be the one which is less affected by the increase of the sources compared to the others. Whereas for 5 sources most of the protocols obtain a high effectiveness, in the rest of the experiments the adaptive approach outperforms the other approaches. In addition, the proposed scheme also tends to be better than the others as the mobility of the nodes increase, which is precisely when the conditions are more demanding.

5   Conclusions and Future Work

We have analytically modeled existing alternatives for gateway discovery in hybrid ad hoc networks. The evaluation has shown that previous approaches do not behave well as the number of sources increase. Furthermore, the proposed model shows hybrid approaches based on limiting the scope of GWADV messages as a good trade-off between performance and overhead. We have proposed an adaptive approach being able to dynamically adjust the scope of GWADV messages to reach the maximal number of active sources. We have shown through simulation that the proposed scheme outperforms the approaches proposed so far. In addition, as our model anticipated, we have shown that the proposed approach is more scalable in terms of mobility of the nodes and number of active sources connecting to the Internet than the other approaches.

As a future work we are considering the experimentation with different kinds of data sources as well as the evaluation of other adaptive coverage algorithms different from the maximal source coverage.

Acknowledgments. This work has been partially funded by the Spanish Science and Technology Ministry by means of the "Ramon y Cajal" workprograme, the SAM (MCYT, TIC2002-04531-C04-03) project and the grant number TIC2002-12122-E. Part of this work was performed under a Visiting Fellowship of Pedro M. Ruiz at the International Computer Science Institute (www.icsi.berkeley.edu).

References

1. U. Jonsson, F. Alriksson, T. Larsson, P. Johansson, and G. Maguire, Jr., "MIP-MANET - Mobile IP for Mobile Ad hoc Networks," *in Proceedings of IEEE/ACM Workshop on Mobile and Ad Hoc Networking and Computing,* Boston, MA, USA, August 1999.
2. Y. Sun, E. M. Belding-Royer, and C. E. Perkins, "Internet Connectivity for Ad Hoc Mobile Networks." *International Journal of Wireless Information Networks,* vol. 9, issue 2, 2002, pp. 75–88.
3. C. Jelger, T. Noel, and A. Frey, "Gateway and address autoconfiguration for IPv6 adhoc networks," Internet-Draft, draft-jelger-manet-gateway-autoconf-v6-01.txt, Work in progress, October 2003.
4. R. Wakikawa, J. Malinen, C. Perkins, A. Nilsson, and A. Tuominen, "Internet Connectivity for Mobile Ad hoc networks," Internet-Draft, draft-wakikawa-manet-globalv6-02.txt, Work in progress, Nov. 2002.
5. J. Broch, D. Maltz, and D. Johnson, "Supporting Hierarchy and Heterogeneous Interfaces in Multi-hop Wireless Ad Hoc Networks," *in Proceedings of the IEEE International Symposium on Parallel Architectures, Algorithms, and Networks,* June 23-25, Perth, Australia, pp. 370–375.

6. P. Ratanchandani and R. Kravets, "A Hybrid Approach to Internet Connectivity for Mobile Ad hoc Networks," *in Proc. of the IEEE WCNC 2003*, Vol. 3, pp. 1522–1527. New Orleans, USA, March 2003.
7. J. Lee, D. Kim, J.J. Garcia-Luna-Aceves, Y. Choi, J. Choi, and S. Nam, "Hybrid Gateway Advertisement Scheme for Connecting Mobile Ad Hoc Networks to The Internet," *in Proc. of the 57th IEEE VTC 2003*, Vol. 1, pp. 191–195. Jeju, Korea, April 2003.
8. C. Perkins and E. M. Belding-Royer, "Ad hoc On-Demand Distance Vector (AODV) Routing," *in Proc. of the 2nd IEEE Workshop on Mobile Computing Systems and Applications*, February 1999, pp. 90–100.
9. K. Fall, K. Varadhan, "ns, Notes and Documentation", The VINT Project, UC Berkeley, LBL, USC/ISI, and Xerox PARC, November 2003.

A Coverage-Preserving Density Control Algorithm for Wireless Sensor Networks

Jie Jiang and Wenhua Dou

School of Computer Science
National University of Defense Technology
410073, Changsha, China
jiangjie@nudt.edu.cn

Abstract. In wireless sensor networks that consist of a large number of low-power, short-lived, unreliable sensors, one of the important design challenges is to reduce energy consumption and prolong network lifetime. In this paper, we propose a decentralized and localized density control algorithm that prolongs network lifetime by keeping a minimal number of sensors in active mode while not scarifying any sensing coverage. Experimental results show that our algorithm outperforms the PEAS algorithm and the sponsor area based algorithm with respect to the number of working sensors needed (about 30% more off-duty eligible nodes), thus longer system lifetime can be obtained, while preserving completely the sensing coverage of the original network.

1 Introduction

Wireless sensor networks (WSN) consist of a large number of ad-hoc networked, low-power, short-lived and unreliable micro-sensors, which are limited in computation, memory capacity and radio range. Lots of sensors are deployed in region of interest to collect related information or report that some event has taken place in that area. WSN enhances our capability of sensing and controlling the physical world. WSN have many potential applications, such as battlefield surveillance, environment monitoring and biological detection [10,2,11].

Since micro sensors are limited in resources and vulnerable in nature, they are usually deployed with high density (up to 20 $nodes/m^3$ [4]) in order to enhance the accuracy of collected data and the reliability of the network. In such a high-density network, it is highly probable that a sensor's coverage is completely covered by other neighboring sensors. The sensing redundancy will result in an excessive mount of energy to be wasted because data collected in this area would be highly correlated and redundant. Moreover, excessive packet collision would occur because many nodes try to send packets simultaneously when certain triggering event occurs, and more energy is wasted as a result of wireless communication collision and interference. On the other hand, micro-sensors are usually supported by battery or other exhaustible energy source and it is impossible or undesirable to replace or recharge the battery power of all sensors, when sensors are deployed in hostile or remote environment, such as

I. Nikolaidis et al. (Eds.): ADHOC-NOW 2004, LNCS 3158, pp. 42–55, 2004.

battlefield or remote desert. So it is an important design challenge to prolong the lifetime of wireless sensor networks by utilizing the inherent redundancy of densely deployed sensors.

One effective method to realize the above object is density control, which only keeps part of sensors in active mode while maintaining complete sensing coverage of original network (where all sensors are active). When active sensors die or fail, those non-active sensors can replace them to keep network functioning, thus network lifetime can be prolonged. Density control must satisfy two requirements when turning off redundant sensors [9]: maintaining sensing coverage of original network and maintaining connectivity of network. As pointed out in [9], if the communication range is at least twice the sensing range, a complete coverage of a convex area implies connectivity of the working nodes. So here we assume that this condition is satisfied and will only focus on the problem of preserving the original coverage.

In this paper, we address the issue of finding a minimum subset of sensor nodes while these working nodes can preserve the sensing coverage of the original network. We propose an effective, distributed and localized density control algorithm for sufficiently densely deployed wireless sensor network. Our algorithm can reduce energy consumption caused by redundantly collected data and radio collision, thus can prolong system lifetime while preserving the original sensing coverage. Our algorithm is an improvement over the sponsor area based node scheduling algorithm in [7]. In our approach, we extend the concept of neighbor and present a new approach to effectively increase the number of redundant sensors and reduce the number of working nodes needed to maintain the original sensing coverage. We also introduce the concept of "effective neighbor" to reduce the communication overhead largely. Experimental results show that our algorithm outperforms the PEAS algorithm and the sponsor area based node scheduling algorithm. When compared to the sponsor area based algorithm, our algorithm can obtain about 30% more redundant sensors (that is to say, working nodes needed by our algorithm are 30% less than that of sponsor area based node scheduling algorithm).

The rest of this paper is organized as follows: section 2 reviews the related work in the literature. In section 3, we analyze the flaws of the sponsor area based node scheduling algorithm (we also call it "sponsor area based off-duty eligibility rule"). And after that, we present our density control algorithm in detail in section 4. Section 5 presents the experimental results and section 6 concludes the paper.

2   Related Work

How to prolong system lifetime by reducing nodes' energy consumption is an important research challenge for both Mobile Ad Hoc Networks (MANET) and wireless sensor networks (WSN). GAF [13] and SPAN [1] are typical energy-efficient protocols for MANET. The energy conserving methods for MANET mainly consider the communication connectivity among mobile nodes, which belong to point coverage problem [3]. On the other hand, area coverage should be emphasized in WSN to avoid sensing holes. Therefore, those methods for MANET cannot be directly adapted into WSN.

In paper [12], Slijepcevic and Potkonjak consider the problem of just keeping a subset of sensor nodes in working state to reduce power consumption. They try to divide sensors into disjoint sets while each node set can maintain the sensing coverage. Then each node set is activated by turn, thus network lifetime can be prolonged. The problem of finding maximum number of disjoint node sets is defined as set k-cover problem and it is NP-complete. Slijepcevic et al. propose a centralized solution to this problem. But the centralized solution is not scalable and not suitable for large scale distributed sensor networks.

In paper [6,5], Ye et al. propose a distributed, probing based density control mechanism called PEAS. In PEAS, only a subset of sensors are in working state to ensure desired sensing coverage, and other nodes are put into low-powered sleep state. The active nodes keep on working through its lifetime. A sleeping node wakes up occasionally to probe its local neighborhood within a predefined probing range. If there is at least one working sensor within its probing range, it returns to sleep state. Otherwise, it should start working. PEAS cannot preserve the complete sensing coverage of the original network after turning off some nodes.

In paper [9], Zhang et al. present a decentralized and localized density control algorithm called OGDC. OGDC gives rules (R1–R4) that specify what action one node should adapt and how to change state. Simulation results show that OGDC does well in reducing number of working nodes. But OGDC assumes that a sensor can be found at any desirable point in the sensing field to ensure the coverage of crossing points. The performance of OGDC is questionable for practical deployment. And it does not take into consideration the edge effect of the sensing field. Finally, OGDC cannot preserve the original sensing coverage completely (as shown by its simulation results).

In paper [7], Tian et al. propose a distributed and localized, coverage preserving node scheduling algorithm. (In following description, we refer it as "sponsor area based node scheduling algorithm", also as "sponsor area based off-duty eligibility rule".) This algorithm divides network lifetime into rounds, where each round has a self-scheduling phase followed by a sensing phase. In self-scheduling phase, the off-duty eligibility rule determines whether a node is eligible to be put into sleep state. If a node's sensing area is completely covered by its neighbors' sensing coverage, then it is safe to turn itself off without scarifying any coverage performance of network. Otherwise, the node should be kept active to perform sensing task. We will describe its basic idea and analyze its flaws in section 3.

In paper [8], Tian et al. also propose some lightweight node scheduling schemes, which do not depend on exact location service, for those applications that can tolerate certain level of loss in sensing coverage. These schemes include neighbor-number based, nearest-neighbor based and probability based node scheduling algorithm.

3   Flaws of Sponsor Area Based Off-Duty Eligibility Rule

The idea of the coverage-preserving node scheduling algorithm in [7] is the base of our work. We also try to identify the redundant node whose sensing area is covered by neighbor nodes. Before presenting our density control algorithm, we analyze the flaws of the method of calculating coverage redundancy used by the off-duty eligibility rule in [7].

The off-duty eligibility rule is evaluated by each sensor to decide whether it is redundant in coverage. It is assumed that each sensor has the same sensing range and its detection area is modelled as a disk, whose center is the sensor and radius is the sensing range. The set of sensor s_i's neighbors is defined as $N(i) = \{s_j \in \aleph | d(s_i, s_j) \leq r_s, j \neq i\}$, where $\aleph$ is the set of all sensors deployed in network and r_s is sensor's sensing radius. Obviously, each sensor in $N(i)$ will overlap s_i's sensing coverage (as in Fig.1(a)). The overlapped crescent is called "sponsor area" of neighbors, denoted as $s_j \cap s_i$. As the area of the crescent is difficult to calculate, the sector within the crescent-shaped intersection (called sponsor sector, denoted as $s_{j \to i}$) is used to replace the crescent (as in Fig.1(b)), since the area of a sector can be represented by its central angle (denoted as $\theta_{j \to i}$) accurately and uniting two sectors is equivalent to merging two central angles (as in Fig.1(c)). If the union of central angles of all neighbors' sponsor sectors can cover the whole 2π, sensor s_i's sensing area is fully included in neighbors' sponsor sectors, thus s_i is eligible for off-duty. It is pointed out that a node must have at least three such neighbors to get a chance to be turned off (as in Fig.1(d)).

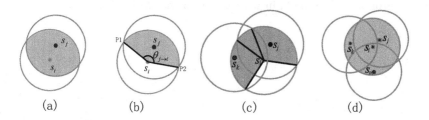

(a)   (b)   (c)   (d)

Fig. 1. Sponsor area based off-duty eligibility rule

The above off-duty eligibility rule has the following flaws:

(1) The area of sponsor sector is always smaller than area of the crescent intersection, some overlapped area is not considered in the coverage calculation;
(2) It considers only neighboring sensors within sensing range r_s. Nodes far away than r_s, but within $2r_s$(equal to the communication range r_c) are ignored. In fact, those sensors can also do help in finding redundant nodes. Fig.2 shows that two "neighbor" nodes (s_j, s_k) and one "non-neighbor" node (s_m) can cooperate to completely cover sensor s_i. And surely this scenario is ignored by the sponsor area based off-duty eligibility rule.

4   Decentralized Density Control Algorithm

4.1   System Model and Assumptions

In designing our density control algorithm, we use the following model and assumptions:

(1) In two-dimension plane, all sensors have the same sensing radius and the sensing area of a sensor is modelled as a circle, whose center is the sensor and radius is the sensing radius.

(2) To ensure connectivity, sensor's communication range r_c is at least twice the sensing range r_s, i. e. , $r_c \geq 2r_s$. Since it is possible to control and adjust sensor's radio power, this assumption can be easily satisfied. In our following description, we assume that $r_c = 2r_s$.

(3) Each sensor knows its own position. The location information can be obtained by means of GPS or other localization service. Many research efforts into localization problem in wireless sensor network [14,15] make this assumption realistic.

(4) No more than one sensor is located at the same position. If there is more than one sensor at the same location, in each time round, one sensor can be randomly chosen as active and others go into sleep state. Then the active sensor runes our algorithm to decide its state.

(5) All sensors in the network are time-synchronized. This assumption is not impractical, since many research efforts have been made to address the time synchronization problem in wireless sensor networks [17,18].

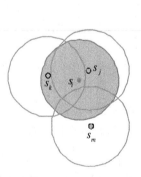

Fig. 2. Scenario ignored by off-duty eligibility rule in [7]

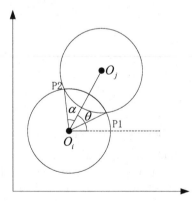

Fig. 3. Calculation of central angle

4.2   Related Definitions and Theorems

Definition 1 (Coverage). *Consider a point p located at (x_p, y_p). If the Euclid distance between p and sensor s_i is less than or equal to s_i's sensing radius, that is, $d(p, s_i) \leq r_s$, point p is covered by sensor s_i. Similarly, consider an area A and a set of sensors $S = \{s_1, s_2, \ldots, s_n\}$. If every point in A is covered by at least one sensor in S, we say that area A is covered by S.*

Definition 2 (Neighbor). *Consider two sensors $s_i, s_j \in S$, located at (x_i, y_i) and (x_j, y_j). If the Euclid distance between s_i and s_j satisfies $0 < d(s_i, s_j) < 2r_s$, s_i and s_j are neighbors. The set of s_i's neighbors is denoted as $N(i) = \{s_j | 0 < d(s_i, s_j) < 2r_s, s_j \in S\}$.*

Obviously, each neighbor will overlap with s_i. Note that here $d(s_i, s_j) < 2r_s$, not $d(s_i, s_j) \leq 2r_s$, because when $d(s_i, s_j) = 2r_s$,the sensing overlap between s_i and s_j is only a point, which is not helpful to s_i's off-duty eligibility. Compared to the neighbor definition of the sponsor area based off-duty eligibility rule in [7], our definition extends the range of neighbors and will lead to more energy efficiency as shown by experimental results.

Definition 3 (Perimeter coverage). *Sensor s_i's sensing area is modelled as a circle centered at s_i and its radius equal to the sensing radius r_s. If a point on the perimeter of the sensing circle is covered by sensor s_j, we say the point is perimeter covered by s_j [16]. If every point on the circle perimeter is covered by at least one neighbor, then the whole circle is perimeter covered by neighbors. Similarly, if every point on a segment of the circle perimeter is covered by at least one neighbor, then the segment is perimeter covered.*

Consider sensors s_i and s_j , located at (x_i, y_i) and (x_j, y_j) individually as shown in Fig.3. O_i and O_j are centers of their sensing circles and the angle between line $\overline{O_iO_j}$ and X axis is $\theta = \arctan \dfrac{y_j - y_i}{x_j - x_i}$. And it is easily proved that $\angle P_1O_iO_j = \angle O_jO_iP_2 = \alpha = \arccos \dfrac{d(s_i, s_j)}{2r_s}$. Since $0 < d(i, j) < 2r_s$, we have $0 < \alpha < \pi/2$. The range of the central angle of segment $\overparen{P_1P_2}$ (along the counterclockwise direction) covered by s_j is denoted as $[\theta - \alpha, \theta + \alpha] \triangleq \theta_{j \to i}$.

Theorem 1. *If $\bigcup\limits_{j \in N(i)} \theta_{j \to i} \neq [0, 2\pi]$, s_i's sensing area is not completely covered by neighbors.*

Proof. Suppose that s_i's sensing area is fully covered by neighbors. Then according to the "area coverage" definition in Definition 1, every point on s_i's sensing circle perimeter must be covered by at least one neighbor.

When $\bigcup\limits_{j \in N(i)} \theta_{j \to i} \neq [0, 2\pi]$, suppose $\Im = [0, 2\pi] - \bigcup\limits_{j \in N(i)} \theta_{j \to i}$. Consider a central angle $\beta \in \Im$, then the segment corresponding to β is not covered by any neighbor. Therefore, points on that segment are not covered by any neighbor. This is contrary to the previous conclusion that every point on s_i's sensing circle perimeter must be covered by at least one neighbor. $\qquad\square$

Theorem 1 gives a sufficient condition for s_i to decide whether it is eligible for off-duty. If $\bigcup_{j \in N(i)} \theta_{j \to i} \neq [0, 2\pi]$, its sensing area is not completely covered by neighbor sensors, then it is not allowed to be turned off. Otherwise sensing "blind points" will occur. This local decision making, which is carried by each sensor based on neighbor information (obtained at the beginning of system setup), can do help in energy saving because no more interactions among neighbors are needed.

We need another criteria to judge whether a sensor node is safe to be turned off, if Theorem 1 can not hold. In the following description, we will present this criteria. As it requires interactions among its one-hop neighbors, we should try to make the communication cost as low as possible. Note that in high-density environment, one sensor may have lots of neighbors. And the overlap between s_i and one neighbor may be completely included by the overlap between s_i and another neighbor. Then the first neighbor can be ignored during the decision making process. Thus we introduce the concept of "effective neighbor". We show that it is sufficient to query only the effective neighbors to decide whether it is redundant in sensing coverage.

Definition 4 (Effective neighbor). *Suppose s_{j_1} and s_{j_2} are s_i's neighbors, and their sensing circles intersect s_i at P_{j_11}, P_{j_12} , and P_{j_21}, P_{j_22} individually (as shown in Fig.4). If segment $\overparen{P_{j_11}P_{j_22}}$ (along counterclockwise direction) is completely included in $\overparen{P_{j_21}P_{j_22}}$, we say s_{j_1} is not an effective neighbor and s_{j_2} is an effective neighbor.*

From the above definition, since $\overparen{P_{j_11}P_{j_12}}$ is included in $\overparen{P_{j_21}P_{j_22}}$, we have $\theta_{j_1 \to i} \subseteq \theta_{j_2 \to i}$. Furthermore, $\bigcup_{j \in N(i)} \theta_{j \to i} \subseteq \bigcup_{j \in N'(i)} \theta_{j \to i}$, where $N'(i)$ is the set of s_i's effective neighbors. So $\bigcup_{j \in N(i)} \theta_{j \to i}$ in Theorem 1 can be replaced by $\bigcup_{j \in N'(i)} \theta_{j \to i}$. In our following discussion, we only consider effective neighbors.

Theorem 2. *Sensor s_i's sensing area is completely covered by its effective neighbor if and only if the segment of each effective neighbors, lying in sensor s_i's sensing circle, is also perimeter covered by s_i's other effective neighbors.*

Proof. Consider $s_j \in N'(i)$ (From now on, we only refer to effective neighbors.), and it intersects s_i at P_1 and P_2(as shown in Fig.5). According to the definition of effective neighbor, if the segment $\overparen{P_2P_1}$ is perimeter covered by other neighbors, then segment $\overparen{P_2P_1}$ must intersect other neighbors' sensing perimeters. Other neighbors' intersecting points within s_i's sensing area are marked as P_3, P_4, P_5, P_6 and M_1, M_2.

Consider any sub-segment in $\overparen{P_2P_1}$, such as $\overparen{P_5P_4}$. Here, $\overparen{P_5P_4}$ is only perimeter covered by s_m. So its coverage degree is 1 (not considering sensor s_i). Then the sub-area inside s_j, composed by vertex P_4, P_5 and M_1, M_2, is covered by

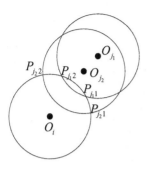

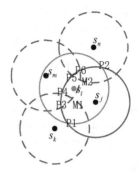

Fig. 4. Effective neighbor

Fig. 5. Coverage redundancy by effective neighbors

both s_m and s_j. So its coverage degree is 2. However, the area outside s_j is only covered by s_m and its coverage degree is only 1. We can say that if P_5, P_4 is perimeter covered by other neighbors, the coverage degree of the sub-region inside s_j is one bigger than that of the sub-region outside s_j. It is clear that sensor s_i's sensing area is divided into many sub-regions, with each sub-region is composed by at least one sub-segment of effective neighbors. Since each effective neighbor's sub-segment is completely perimeter covered, the coverage degree of each sub-segment is bigger than or at least equal to 1. So the coverage degree of each sub-region is at least 1, without considering sensor s_i. Thus we can say sensor s_i's sensing area is completely covered by its effective neighbors.   □

4.3   Description of Proposed Algorithm

From the above discussion, we present a distributed and localized density control algorithm for wireless sensor networks as in Fig. 6.

In our algorithm, a sensor is in one of the two states: "ACTIVE" and "NON-ACTIVE". At the beginning, all nodes are in ACTIVE state. Network lifetime is divided into rounds, and each round has a scheduling phase followed by a sensing phase (see Fig.7). In each time round, the ACTIVE nodes work for the sensing task and the NON-ACTIVE nodes will turn off their sensing and communication units to save energy.

To minimizing the energy overhead consumed in the scheduling phase, the sensing phase should be long enough compared to the scheduling phase. The scheduling phase is further divided into two sub-phases: neighbor discovery phase and evaluating phase.

At the beginning of the neighbor discovery phase, node must broadcast a *hello* message (containing the source node's position information and ID) to its one-hop neighbors and sets a timer to wait for neighbors' *hello* message. Upon this timer times out, node has obtained knowledge about one-hop neighbors and is able to construct its neighbor set and effective neighbor set.

1   Broadcasting *hello* message;
2   Constructing neighbor set $N(i)$ and effective neighbor set $N'(i)$;
3   If $\bigcup\limits_{j\in N'(i)} \theta_{j\to i} \neq [0, 2\pi]$, then
4   Keep ACTIVE;
5   Exit;
6   Random back-off; /*to avoid sensing blind points*/
7   If the "if part" of Theorem 2 holds, then
8   Set state to NON-ACTIVE;
9   Broadcasting *OFF* to neighbors;
10   Exit.
11   else
12   Keep ACTIVE;
13   Exit.

Fig. 6. Pseudo code for proposed density control algorithm

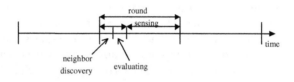

Fig. 7. Network lifetime

Once entering the evaluating phase, sensor begins to evaluate our density control algorithm to decide which state it should go to. As our algorithm is completely distributed, it is possible that two neighbor sensors evaluate the algorithm at the same time, and go to NON-ACTIVE state simultaneously, although they depend on each other to be redundant in coverage. This will cause sensing "blind point" in the sensing field. A random back-off scheme can be used to make neighboring sensors evaluate algorithm sequently. At the beginning of the evaluating phase, node should wait a random time. The node with less back-off will evaluate algorithm earlier. If it is eligible to be NON-ACTIVE, it will notify its one-hop neighbors by broadcasting an *OFF* message. Note that this broadcasting is limited in one-hop range, not much traffic cost will be caused. Then node with more back-off time, will know "early" neighbors' decision and should reconstruct its neighbor set after deleting those NON-ACTIVE neighbors from its original neighbor set. Sometimes only the random back-off scheme is not enough, we need more schemes to avoid the sensing blind points. So before really turning off sensing and communication units, the NON-ACTIVE sensor should wait another short period of time to hear from neighbors. If no *OFF* message is received during this time, it is safe to turning off all sensing and communication units and go to sleep state. If one or more message received, it can decide whether it should resume working or go to sleep according the "remaining energy level" encoded in the *OFF* message. Nodes with less energy will have more chance to be NON-ACTIVE. By this means, we can not only avoid sensing holes, but also make the energy consumption among sensors evenly, which is helpful to prolong network lifetime.

Although the third step can finally decide whether a sensor is eligible to be turned off, but considering that the energy consumed by computation is much less than that needed by communication, the second step of our algorithm can indeed help to save energy for randomly deployed sensors because the non-redundant sensors can judge their states without communicating with any neighbor.

5 Experimental Results

In this section, we present some experimental results as the performance evaluation of our algorithm. We compare our algorithm with PEAS because PEAS can maintain approximately the original sensing coverage more than 99%) when the probing rang is short enough. And since the sponsor area based off-duty eligibility rule is the base of our work, we will focus on the performance comparison with it. To show the effectiveness of our algorithm in energy efficiency, we also compare the average sensing degree before and after turning off some nodes.

5.1 Comparison with PEAS

PEAS is a probing-based off-duty eligibility protocol, whose performance is heavily affected by node's probing range. Longer probing range can turn off more nodes, however, it will result in more sensing holes (area covered by original network but not covered after turning off some nodes). On the contrary, shorter probing range can obtain higher sensing coverage but fewer off-duty eligible nodes. To compare our algorithm with PEAS, we carry some experiments in static networks. In a square field ($50m \times 50m$), we deploy some senor nodes randomly. So nodes' x and y-coordinates are random. We compare the sensing coverage and off-duty node number with PEAS, when the original deployed node number is 100 and sensing range is $10m$. To calculate the coverage, we divide the deployed area into $1m \times 1m$ unit cells. If a cell center, which is covered by original network, cannot be covered any more after turning off some nodes, we say a sensing hole occurs. The occurrence of sensing holes means that the corresponding algorithm cannot preserve the original sensing coverage. We also compute the average sensing degree before and after turning off some nodes. Table 1 shows the performance comparison of two algorithms in 100 random topologies respectively.

From Table 1, we can see that PEAS can obtain approximately equal off-duty node number to our algorithm only when the probing range is longer than $6m$. But there are 36 sensing holes in that case.

Table 1. Performance Comparison with PEAS

Algorithm	Probing range	# of off-duty nodes	Original sensing degree	Obtained sensing degree	# of topolo-gies with sensing holes	Average # of sensing holes per topology
Proposed	N/A	75	10	3	0	N/A
PEAS	3	36	10	6	14	< 1
	4	50	10	4	24	2
	5	63	10	3	63	7
	6	72	10	2	92	36
	7	80	10	1	100	103

5.2 Comparison with Sponsor Area Based Node Scheduling Algorithm

To our best knowledge, the sponsor area based node scheduling algorithm is the only scheme that can preserve the original sensing coverage after turning off some nodes. Our density control algorithm can also provide 100% sensing coverage of the original network. But our algorithm can obtain more off-duty eligible nodes than the sponsor area based scheduling algorithm, thus longer system lifetime can be expected.

Fig.8–10 shows the obtained off-duty node number with different sensing range and different deployed node number. We can see from them that our algorithm can obtain about 30% more off-duty nodes compared to the sponsor area based scheduling algortihm. The performance improvement is obtained by extending the range of neighbors (the distance between node and its neighbors is extended from r_s to $2r_s$) and the new method for coverage calculation.

Fig.11 is a 3–D surface plot of working node number in different deployment density, including our algorithm and the sponsor area based algorithm. We can see from it that the number of working nodes needed to preserve the original sensing coverage is much smaller than that of the sponsor area based algorithm. We can also see that the working node number does not keep constant when node's sensing range varies, although the deployed node number and the de-ployed area are fixed. This is due to the increasing number of edge nodes (nodes located near the boundary of the deployment area). According to our algorithm, the edge nodes have fewer neighbors, and its neighbors are located on one side of the edge nodes, thus they have less chance to be completely perimeter covered. Intuitively, the increase of edge nodes will result in increase in working node number. However, our experimental result shows that our density control algo-rithm can effectively control the working node number. When original deployed node number increases from 100 to 300, the number of working nodes increases only about 20%.

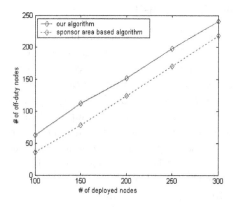

Fig. 8. # of off-duty nodes vs. node density ($r_s = 8m$)

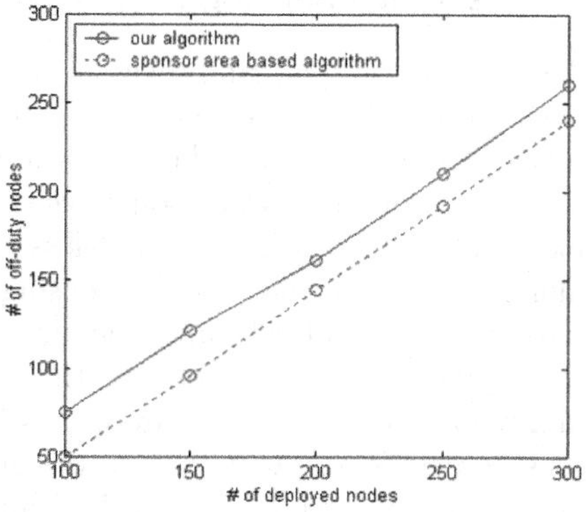

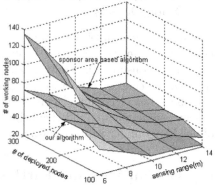

Fig. 9. # of off-duty nodes vs. node density ($r_s = 10m$)

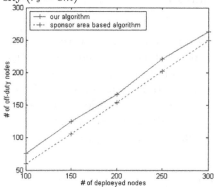

Fig. 10. # of off-duty nodes vs. node density ($r_s = 12m$)

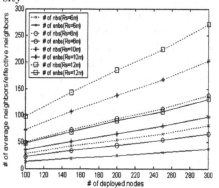

Fig. 11. # of working nodes vs. node density

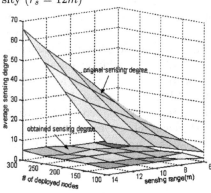

Fig. 12. Sensing degree vs. node density

Fig. 13. Average # of neighbors/effcetive neighbors vs. node density

5.3   Average Sensing Degree Versus Node Density

Another metric that can prove the effectiveness of our algorithm in energy saving is the resulted average sensing degree of the deployed area. Sensing degree can reflect the sensing redundancy of the network. Higher sensing degree will result in more report data when some event occurs, and the redundant data will cause more traffic load and more wireless communication collision, which will waste more energy. To calculate the sensing degree, we also divide the deployed area into $1m \times 1m$ unit cells and assume an event occurs in each cell, with the event source located at the center of the cell. The sensing degree is defined as how many nodes can detect the event. As illustrated in Fig.12, although the original sensing degree varies from 5 to 66, our density control algorithm can result in about 3 degree, which effectively reduces the redundant sensing.

5.4   Average Number of Neighbors and Effective Neighbors Versus Node Density

In our density control algorithm, nodes need to communicate with its active neighbors to decide whether it is eligible to be off-duty. As pointed out in [19], the energy consumed by communication is far more than sensing and computation. So the communication cost should be considered. By introducing effective neighbors, our algorithm can effectively reduce the communication cost, because the average number of effective neighbors is much less than the average number of neighbors under various sensing radius, as illustrated in Fig.13.

6   Conclusions and Further Work

In this paper, we present a coverage-preserving, distributed density control algorithm for wireless sensor networks. The algorithm can reduce overall system energy consumption, therefore increase network system lifetime, by turning off some redundant nodes. Experimental results show that our algorithm outperforms the PEAS algorithm and the sponsor area based node scheduling algorithm (about 30% more off-duty node number can be obtained while maintaining the original sensing coverage). We will further implement this algorithm as an extension to some existing routing protocol (such as LEACH [20]) in ns2 simulator to analyze its effectiveness in energy saving and system lifetime increasing. Current algorithm needs node's position information thus will rely on localization service, which is not very easy and cheap in sensor networks. Our next work is to develop new algorithm that will not depend on geographical location information.

References

1. Chen, B., Jamieson, K., Balakrishnan, H., Morris, R.: Span: An energy-efficient operation in multihop wireless ad hoc networks. In Proc. of ACM MobiCom' 01, Rome, Italy, July, 2001.

2. Estrin, D., Govindan, R., Heidemann, J.S., Kumar, S.: Next century challenges: Scalable coordination in sensor networks. In Proc. of ACM MobiCom' 99, Washington, August, 1999.
3. Gage, D.W.: Command Control for Many-Robot System. In Proc. of the Nineteenth Annual AUVS Technical Symposium, Hunstville Alabama, USA, June, 1992.
4. Shih, E., Cho, S., Ickes, N., Min, R., Sinha, A., Wang, A., Chandrakasan, A.: Physical layer driven protocol and algorithm design for energy-efficient wireless sensor networks. In Proc. of ACM MobiCom' 01, Rome, Italy, July, 2001.
5. Ye, F., Zhong, G., Lu, S., Zhang, L.: Energy Efficient Robust Sensing Coverage in Large Sensor Networks. Technical Report. UCLA. 2002.
6. Ye, F., Zhong, G., Lu, S., Zhang, L.: Peas: A robust energy conserving protocol for long-lived sensor networks. In Proc. of the 23nd International Conference on Distributed Computing Systems (ICDCS), 2003.
7. Tian, D., Georganas, N.D.: A Coverage-Preserving Node Scheduling Scheme for Large Wireless Sensor Networt. In Proc. of 1st ACM Workshop on Wireless Sensor Networks and Applications, Atlanta, Geogia, USA, 2002.
8. Tian, D., Georganas, N.D.: Location and Calculation-free Node-scheduling Schemes in Large Wireless Sensor Networks. Ad Hoc Networks Journal, Elsevier Science. 2003.
9. Zhang, H., Hou, J.C.: Maitaining Sensing Coverage and Connectivity in Large Sensor Networks. Technical Report. UIUCDCS-R-200302351. UIUC. June, 2003.
10. Akyildiz, I.F., Su, W., Sankarasubramaniam, Y., Cayirci, E.: Wireless Sensor Networks: A Survey. Computer Networks. March, 2002.
11. Kahn, J.M., Katz, R.H., Pister, K.S.J.: Next century challenges: Mobile networking for Smart dust? In Proc. of of ACM MobiCom' 99, Washington, August, 1999.
12. Slijepcevic, S., Potkonjak, M.: Power Efficient Organization of Wireless Sensor Networks. In Proc. of IEEE International Conferrence on Communications, Helsinki, Finland, June, 2001.
13. Xu, Y., Heidemann, J., Estrin, D.: Geography-informed energy conservation for ad hoc routing. In Proc. of ACM MobiCom'01, Rome, Italy, July, 2001.
14. Nasipuri, A., Li, K.: A Directionality based Location Discovery Scheme for Wireless Sensor Network. In Proc. of ACM WSNA'02, Atlanta, Georgia, USA, September 28, 2002.
15. He, T., Huang, C., Blum, B.M., Stankovic, J.A., Abdelzaher, T.F.: Range-Free Localization Schemes in Large Scale Sensor Networks. In Proc. of the Ninth Annual International Conference on Mobile Computing and Networking (MobiCom 2003), San Diego, CA, September, 2003.
16. Huang, C-F., Tseng Y-C.: The Coverage Problem in a Wireless Sensor Network. In Proc. of WSNA'03, San Diego, California, USA, September 19, 2003.
17. Elson, J., Romer, K.: Wireless Sensor Networks: A New Regime for Time Synchronization. ACM Mobile Computing and Communication Review (MC2R), Vol.6 No.4, pp.59-61. October, 2002.
18. Li, Q., Rus, D.: Global Clock Synchronization in Sensor Networks. In Proc. of IEEE Infocom 2004, Hong Kong, China, March, 2004.
19. Hill, J., Szewczyk, R., Woo, A., Holalr, S., Culler, D., Pister, K.: System Architecture Directions for Networked Sensors. In Proc. of the ninth international conference on architectural support for programming languages and operating systems, Cambridge, Massachusetts, United States, 2000.
20. Akkaya, K., Younis, M.: A Survey on Routing Protocols for Wireless Sensor Networks. Ad Hoc Networks. September, 2003.

Improved Approximation Algorithms
for Connected Sensor Cover

Stefan Funke[1], Alex Kesselman[1]*, Zvi Lotker[2], and Michael Segal[3]

[1] Max Planck Institut fur Informatik, Saarbrucken, Germany.
{funke,akessel}@mpi-sb.mpg.de
[2] Project Mascotte, INRIA, Sophia Antipolis, 06902, France.
zvilo@eng.tau.ac.il
[3] Communication Systems Engineering Dept.,
Ben Gurion University of the Negev, Beer-Sheva, Israel.
segal@cse.bgu.ac.il

Abstract. Wireless sensor networks have recently posed many new system building challenges. One of the main problems is energy conservation since most of
the sensors are devices with limited battery life and it is infeasible to replenish
energy via replacing batteries. An effective approach for energy conservation is
scheduling sleep intervals for some sensors, while the remaining sensors stay
active providing continuous service. In this paper we consider the problem of selecting a set of active sensors of minimum cardinality so that sensing coverage
and network connectivity are maintained. We show that the greedy algorithm that
provides complete coverage has an approximation factor of $\Omega(\log n)$, where n
is the number of sensor nodes. Then we present algorithms that provide approximate coverage while the number of nodes selected is a constant factor far from
the optimal solution.

1 Introduction

Recent technological advances have led to the emergence of small, low-power devices
that integrate sensors with limited on-board processing and wireless communication
capabilities [4,12]. Pervasive networks of such sensors open new perspectives for many
potential applications, such as surveillance, environment monitoring and biological detection [2,20]. A sensor network consists of multiple sensor nodes and each sensor can
sense certain physical phenomena like light, temperature or vibrations around its location. The purpose of a sensor network is to process some high-level sensing tasks and
report the data to the application.

Minimizing energy consumption to prolong the system lifetime is a major design
objective for sensor networks since sensors need to operate for a long time on battery
power. If all the sensor nodes simultaneously operate in active mode, an excessive amount
of energy is wasted and the data collected is highly correlated and redundant. In addition,

---

* The second author has done a part of this work while visiting the Dipartimento di Scienze
dell'Informazione at Universita di Roma La Sapienza, Italy. Supported in part by EYES project
and MIUR grant FIRB WebMinds.

multiple packet collisions may occur when all the sensors in a certain area try to transmit as a result of a triggering event. Thus, maximizing the network lifetime can be achieved by scheduling some nodes to sleep (a power saving mode) while the remaining active nodes can still provide continuous service.

Many existing solutions have treated the problems of sensing coverage and network connectivity separately. The former problem has been studied extensively. A protocol that uses a local geometric calculation to preserve the sensing coverage is presented in [25], where if the sensing area of a node is completely covered by its neighbors, it enters sleep mode. A distributed probing-based [29] proposes a density control algorithm for robust sensing coverage called PEAS. In PEAS a sleeping node wakes up occasionally to check if there exist working nodes in its vicinity. If so, it sleeps again, otherwise it enters active mode. Several algorithms that use linear programming techniques to select a minimal set of active nodes for maintaining coverage are designed in [8,23]. However, all these protocols do not guarantee network connectivity.

On the other hand, many protocols have been designed to maintain network connectivity. Although a wireless ad-hoc network has no physical backbone infrastructure, a virtual backbone can be formed by nodes in a connected dominating set (CDS) of the corresponding unit-disk graph. The most important benefit of virtual backbone-based routing is significant reduction in the protocol overhead, which greatly improves the network throughput. GAF [28] conserves energy by dividing a region using a rectangular grid and electing a leader in each cell while putting the other nodes into sleep. In SPAN [9] a node decides whether it should be active or sleeping based on the connectivity among its neighbors. A different approach is used in ASCENT [7], where to make such a decision each node estimates the number of active neighbors and the per-link data loss rate. However, all the protocols mentioned above do not ensure sensing coverage.

Unfortunately, satisfying only coverage or connectivity alone is not sufficient since nodes may not be able to coordinate effectively or monitor the environment with the required accuracy. As a result, the problem of reducing energy consumption by keeping a minimal number of sensor nodes in active mode while maintaining sensing coverage and connectivity has received significant attention in recent time. The work in [14] designs centralized and distributed approximation $O(\log n)$-approximation algorithms for the connected sensor cover problem, where n is the number of nodes. In [30] it is shown that if the communication range is at least twice the sensing range, a complete coverage of a convex area implies connectivity among the nodes and derive optimality conditions under which a subset of working sensor nodes can be chosen for full coverage. The work in [27] derives the Coverage Configuration Protocol (CCP) that can provide different degrees of connected coverage as well as present a geometric analysis of the relationship between coverage and connectivity.

Our results. We study the problem of providing coverage and connectivity in a unified framework. We assume that sensors have fixed locations. First we assume that the communication range is twice the sensing range. Then we show how to connect a set of sensors that already provides coverage under a more realistic assumption that the the communication range equals the sensing range. We analyze the natural greedy sector cover algorithm, which is known to have an approximation factor of $\log m$, where m is the maximal number of sectors covered by a single sensor (the formal definition of a sector is given in Section 2). We demonstrate that despite the geometric nature of the problem, greedy sector cover has an approximation factor of at least $\Omega(\log m)$. Thus, to obtain better approximation factors we focus on algorithms that always guarantee connectivity, but provide only approximate coverage. We derive the grid placement and the fine grid algorithms, which achieve approximation factors of 6π (≈ 18.84) and 12, respectively. We note that the technique used by the fine grid algorithm is of interest on its own, since it can be applied to the problem of covering a convex area with the minimum number of fat geometric objects. We also present the distributed dominating cover algorithm that has an approximation factor of 18 and provides approximate coverage.

Related Work. For energy-efficient monitoring, sensors can be partitioned into covers, which are activates iteratively. This approach takes advantage of the overlap between the sensing areas of individual sensors. The work in [24] considers the problem of maximizing the number of mutually exclusive sets of sensor nodes, where the members of each set together completely cover the monitored area. The work in [1] studies a variation of this problem, where the objective is to partition the sensors into disjoint covers such that the number of covers that include an area, summed over all areas, is maximized. In [3] the problem of finding a monitoring schedule that maximizes the network lifetime is considered.

Determining the positions of sensor nodes is impractical in most real-life scenarios. The preferred method of sensors placement is bulk dispersion from an aircraft since sensor networks are often deployed in remote or hostile areas. Unfortunately, the current state-of-the-art sensor nodes are not capable of dynamic adjustment of their positions. However, in friendly environments intelligent sensor placement algorithms can be applied prior to the deployment of the sensor network in order to optimize the underlying architecture. Sensor placement for surveillance and target location is considered in [8], where the problem of achieving the desired coverage while minimizing the cost (sensors may have different ranges and costs) and covering every grid vertex by a unique subset of the sensors are studied. The work in [11] gives algorithms for finding efficient placement of sensors that guarantee probabilistic coverage of the grid vertices. The work in [17] studies the problem of optimal node placement and provides constant approximation algorithms for covering a region or a given set of points.

Since a sensor network is usually deployed to perform surveillance and monitoring tasks, another definition of coverage is calculating a path with specific properties through a sensor network. The maximal support problem is to find a path that minimizes the maximal distance of a point on the path to the closest sensor and the maximal breach problem is to find a path that maximizes the minimal distance of a point on the path to the closest sensor. Algorithms for finding a maximal breach path and a maximal support path in a sensor network are presented in [22,19]. The work in [15] derives constant-approximation algorithms for dynamic maintenance of the best-case and the worst-case coverage distances and improves the running time of the shortest maximal support path algorithms due to [22,19].

The rest of the paper is organized as follows. Section 2 describes our model. Our algorithms and their analysis are presented in Section 3. Section 4 shows how to connect a covering set. Concluding remarks appear in Section 5.

2   Model Description

Given a set of n sensors $S = \{s_1, \ldots, s_n\}$ distributed on the plane. Each sensor s_i has a location (x_i, y_i). The locations of the sensors are given in advance. Sensor s_i can monitor objects that are within a distance of its *sensing range* R_s^i. This area is called the *sensing region* of s_i and is denoted by A_i (note that A_i is a disk with a radius R_s^i whose center is (x_i, y_i)). We define a *sector* to be a maximal region that is formed by the intersection of a number of sensing regions such that all points within the sector are covered by the same set of sensors. We denote by m the maximal number of sectors covered by a single sensor.

Sensor s_i can communicate to all its neighbors within a distance of its *communication range* R_c^i. The *communication graph* GC of the network is an undirected graph in which nodes are sensors and there is an edge between two nodes if the distance between them is at most the minimum of their communication ranges (i.e., both nodes can talk to each other). For a subset of nodes S', the *communication subgraph* is the subgraph induced by the nodes in S'.

Let P be a region of interest on the plane. A *connected cover* of P is a subset $S' = \{s_{j_1}, \ldots, s_{j_m}\}$ of sensors such that $P \subseteq A_{j_1} \cup \ldots \cup A_{j_m}$ and the communication subgraph induced by S' is connected. An example of a connected cover is presented in Figure 1.

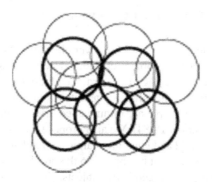

Fig. 1. A connected cover example.

Definition 1. *Given a region of interest P, the* connected coverage problem *is to find a connected cover of P that uses the minimum number of sensors. We denote by OPT an optimal connected cover.*

Definition 2. *We say that an algorithm A has an approximation factor of c, if for any instance of the problem σ the size of the solution produced by A is at most $c \cdot |OPT| + a$, where a is a constant independent of σ.*

Definition 3. *We say that A provides* complete coverage *if the set of the selected sensors always covers P provided that there exists a feasible solution. Otherwise, we say that A provides* approximate coverage.

We assume that all sensors are identical and have the same sensing range R_s and communication range R_c. We also make a few simplifying assumptions.

1. We assume that $R_c \geq 2R_s$ unless it is explicitly stated otherwise (i.e., two sensors are able to communicate if their sensing areas intersect).
2. We assume that P is convex.

Theorem 1 ([27,30]). *Under the above assumptions, complete coverage implies connectivity.*

In Section 4 we show how to replace assumption (1) by a more realistic assumption that $R_c = R_s$.

In our algorithms we will extensively use a grid covering the area of interest P.

Definition 4. *A grid is defined as a packed tiling of regular rectangles called cells. We assume that the sides for each of the cells are parallel to the x and y axes of the plane.*

For simplicity, we assume that the number of grid cells intersecting the boundary of P is negligible compared to the total number of cells and we will ignore it in the analysis.

3   Algorithms for Connected Coverage

First we consider the natural greedy sector cover algorithm. We demonstrate that even for geometric instances of the problem considered here, the approximation factor of greedy sector cover is at least $\Omega(\log m)$. In order to obtain better approximation factors we concentrate on algorithms that always guarantee connectivity, but provide approximate coverage. We derive a simple algorithm called grid placement which achieves an approximation factor of 6π (≈ 18.84). We also propose the fine grid algorithm that has an approximation factor of 12 and guarantees almost complete coverage. The technique used in the fine grid algorithm can be applied to the problem of covering a convex area with the minimum number of fat geometric objects. Finally, we present the distributed dominating cover algorithm that has an approximation factor of 18. Unfortunately, the coverage provided by dominating cover is less accurate compared to that of the centralized algorithms.

3.1   Greedy Sector Cover Algorithm

We consider the sectors produced by the sensors as elements to be covered while each sensor represents a set. We establish a tight bound of $\Theta(\log m)$ on the performance of greedy sector cover.

The greedy sector cover algorithm proceeds in two steps. Step 1: we use the algorithm of [5] to check whether the sensors cover the region of interest P and report failure if a feasible solution does not exist. Step 2: we apply greedy set cover to our problem, i.e., at each step we select a sensor that covers the maximal number of uncovered sectors.

Observation 1 *The number of sectors created by intersection of n disks is at most $n(n-1)+1$.*

The running time of greedy sector cover is $O(n^2 \log n)$: Step 1 takes $O(n^{1+\epsilon})$ time and Step 2 can be implemented in $O(n^2 \log n)$ time. The next theorem derives an upper bound on the approximation factor of the greedy sector cover algorithm.[1]

Theorem 2. *If the greedy sector cover algorithm terminates successfully, then the returned set is connected and P is completely covered. The approximation factor of the greedy sector cover algorithm is at most $\log m$.*

Proof. Step 1 of the algorithm ensures that the union of the sensing regions covers P and in Step 2 all sectors that intersect P are covered. Thus, P is completely covered. According to Theorem 1, the covering set is connected. The approximation factor is a well-known result.   □

The following theorem demonstrates that our upper bound is almost tight.

Theorem 3. *The approximation factor of the greedy sector cover algorithm is at least $\Omega(\log m)$.*

Proof. The region we wish to cover is a square. We start with some definitions. We say that the set of disks $\{D_i\}_{i=1}^n$ is a *chain* of size n if all the disks intersect and their centers lie on a ray, which defines the ordering of the disks. We denote the intersection point between disk i and disk j by $P_{i,j}$. Figure 2 shows an example of such a chain. Note that we can associate with each point $P_{i,j}$ a sector. Thus, the number of sectors in a chain is $\binom{n+1}{2}$. Denote by $\Upsilon(P_{1,2}, P_{1,n}, P_{n,n-1})$ all the sectors that are not in the first and the last disks. Observe that we can make the area of $\Upsilon(P_{1,2}, P_{1,n}, P_{n,n-1})$ as small as we like. We have that most of the sectors of the chain are in $\Upsilon(P_{1,2}, P_{1,n}, P_{n,n-1})$, only $2n-1$ sectors are outside $\Upsilon(P_{1,2}, P_{1,n}, P_{n,n-1})$. For a chain δ, we denote by $\Upsilon(\delta)$ all the sectors that are not contained in the first and the last disks of δ.

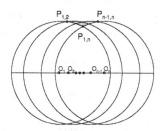

Fig. 2. t A chain of unit disk. The point $P_{1,2}$ is the intersection of the first two disks. The point $P_{n-1,n}$ is the intersection of the last two disk and the point $P_{1,n}$ is the intersection of the the first and the last two disks. The gray area is $\Upsilon(P_{1,2}, P_{1,n}, P_{n,n-1})$ contains most of the sectors of the chain.

---

[1] Observe that the analysis of greedy sector cover can be extended to the case of non-uniform sensing radii.

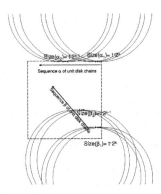

Fig. 3. The dashed square is the area we wish to cover and there are two sequences of chains α and β. The former sequence of chains is close to the upper edge of the square and the latter sequence of chains is close to the diagonal of the square.

Observation 2 *For any square S with side length ϵ and for any $n \in \mathbb{N}$, there exists a chain $\delta = \{D_i\}_{i=1}^n$ s.t. $\Upsilon(\delta) \in S$.*

We place a sequence of chains $\alpha = \{\alpha_i\}_{i=1}^k$ in such a way that the size of the i-th chain is $2^{k+1-i} \cdot t$ and disks from different chains intersect outside the relevant square (see Figure 3). We also place a sequence of chains $\beta = \{\beta_i\}_{i=1}^k$ in such a way that the size of the i-th chain is $2^{k+1-i} \cdot t$ and it is fully contained in the intersection of all the disks that are in the $(i+1)$-th chain (see Figure 3). Since both sequencers of chains are disjoint, no disk in these sequences can contain more than half of the sectors.

Now we place a chain $\gamma = \{\gamma_i\}_{i=1}^k$ so that the greedy sector cover algorithm will select this chain as a cover (see Figure 4). We add this chain in such a way that disk γ_i contains the $\Upsilon(\alpha_1), \Upsilon(\alpha_2), ..., \Upsilon(\alpha_i)$ and $\Upsilon(\beta_i)$ sectors. This means that $\Upsilon(\beta_i) \subset \gamma_j$ iff $i \neq j$ and $\Upsilon(\alpha_i) \subset \gamma_j$ iff $i \leq j$. Note that we can assure that no new sectors are added to chain α. Since disk γ_i does not intersect $\Upsilon(\beta_j)$, the number of new sectors generated by disk γ_i is linear in the number of disks in chain β_j while the number of sectors in $\Upsilon(\beta_j)$ is quadratic in the number of disks in chain β_j. Hence, if t is big enough, the number of new sectors is negligible. We get that the number of sectors contained in disk γ_i is bigger than $2\binom{2^{k+1-i}+1}{2}$.

Finally, we place two more disks δ_1, δ_2. These disks cover all the sectors that are in the square we wish to cover (see Figure 4). Since δ_1 does not contain the β sequence of disk chains and δ_2 does not contain the α sequence of disk chains, we get that the total number of sectors in those disks is less than the number of sectors in γ_1.

We will prove that the greedy sector cover algorithm chooses all γ disks. Let $t = 4^k$. The proof of the next lemma is omitted from this abstract.

Lemma 1. *The greedy sector cover algorithm selects disk γ_i at the i-th iteration.*

Now we will elaborate more on the placement of the disks. The idea is to place the γ chain first in such a way that the ray of γ is parallel to the diagonal, and $\Upsilon(\gamma)$ is close to the upper right corner of the square. By Observation 2, we can place the α sequence

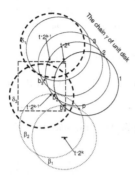

Fig. 4. The dashed disks δ_1, δ_2 form the OPT cover. The solid disks are the disks selected by the greedy sector cover algorithm. Each disk is labeled with the iteration at which it is selected. The point p is the intersection of all k initial disks that form β-chains. The point $b_i \in \gamma_i \setminus (\gamma_{i+1} \cup \gamma_{i-1})$ is the second point we use to define the i-th initial disk.

of chains as claimed. Placement of the β sequence of chains is a little bit more tricky. Note that two points on the boundary completely define a unit disk. The idea is to put k initial disks as follows. We pick one point p outside the square near the lower right corner. We have that p is on the boundary of all the initial disks. The second point of the i-th disk is a point in $\gamma_i \setminus (\gamma_{i+1} \cup \gamma_{i-1})$.

The theorem follows by Lemma 1 and the fact that the total number of disks is at most 8^k. □

3.2 Grid Placement Algorithm

We put a grid with the cell size of $R_s/\sqrt{2} \times R_s/\sqrt{2}$ over the region of interest P. A specific instance of grid is defined by its position. Then we choose exactly one sensor in each cell to be in the covering set. Finally, we add extra sensors to make the covering set connected.

Observation 3 *The selection of the cell size implies that each sensor covers its cell completely and sensors in neighboring cells are able communicate with each other.*

Observe that depending on the position of the grid, some cells may be empty. To obtain better coverage, we aim to minimize the number of such cells. The work in [6] gives an algorithm for solving the grid placement problem that minimizes the number of grid cells containing no point with running time of $O(n \log n)$. We use the algorithm of [6] to optimize the grid placement.

The grid placement algorithm and the MST connection algorithm are presented in Figure 5 and Figure 6, respectively. An example of the grid placement algorithm can be found in Figure 7. The running time of the MST connection algorithm is $O(n^3)$: Step 1 takes $O(n^3)$ time and Step 2 takes $O(n^2)$ time. The running time of grid cover is dominated by that of MST connection.

1. Define a grid with the cell size of $R_s/\sqrt{2} \times R_s/\sqrt{2}$ covering P.
2. Apply the algorithm of [6] to find the grid placement that minimizes the number of empty cells.
3. Select an arbitrary sensor in each non-empty cell intersecting P and add it to the covering set (the *basic* cover).
4. Run the MST connection algorithm and add the returned disks to the final set (the *extended* cover).

Fig. 5. The grid placement algorithm.

1. Create a weighted graph G in which nodes are the connected components in the communication subgraph induced by the nodes from the basic cover. We call a node in G a *super-node*. We say that a node s is *directly reachable* from a super-node U if there exists an edge in GC between a node $v \in U$ and s. We add an edge between two super-nodes U and V in G if there exists a path between them in GC that does not contain any node directly reachable from another super-node W but not directly reachable from either U or V. The weight of the edge equals to the number of regular nodes that are not included in the basic cover on a shortest path satisfying the above condition.
2. Compute a minimum weight spanning tree (MST) of G, T.
3. Return the set of nodes that lie on the shortest paths corresponding to the edges of T.

Fig. 6. The MST connection algorithm.

The following theorem analyzes the performance of the grid placement algorithm.

Theorem 4. *The covering set found by the grid placement algorithm is connected and the uncovered area of P is bounded by the union of the empty cells. The approximation factor of the grid placement algorithm is at most 6π.*

Proof. First we consider the MST connection algorithm. We argue that if there exists a path between two super-nodes U and V in GC, then there also exists a path between them in G. If G contains edge (U, V), we are done. Otherwise, by our construction, U can reach V in G through another super-node W. Therefore, the MST connection algorithm returns a connected set. By Observation 3, all non-empty cells are covered. Hence, the uncovered area of P is bounded by the union of the empty cells.

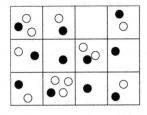

Fig. 7. An example of the grid placement algorithm.

Let k and l be the number of nodes in the basic and the extended cover, respectively. The area covered by a single sensor from the basic cover is $R_s^2/2$. On the other hand, any sensor in OPT can cover the area of at most πR_s^2. Therefore, $k \leq 2\pi|OPT|$, since P is covered by OPT. We will show that the number of sensors in the extended cover is at most $2(k-1)$. Clearly, the number of nodes in G is at most k and thus the number of edges in the MST is bounded by $k-1$. We claim that the weight of any edge in G is at most two. We say that a super-node in G *covers* a cell if it includes a node from the basic cover that is located in this cell. Suppose towards a contradiction that the weight of an edge between super-nodes U and V is greater than two. We have that at least one intermediate node on the shortest path between U to V must lie in a cell C that is not covered by the super-nodes U and V. Let W be the super-node covering C. We obtain that the node from C on the path between U and V is directly reachable from W but not directly reachable from U or V, which contradicts our construction. Therefore, we get that $l \leq 2(k-1) < 4\pi|OPT|$, which establishes the theorem.   □

3.3   Fine Grid Algorithm

We put a coarse grid with the cell size of $R_s \times R_s$ over the region of interest P and put over this grid a fine grid with the cell size of $\epsilon \times \epsilon$. Then we consider the set of disks $\{A_1, \ldots, A_n\}$. We align the center of each disk to be on the closest vertex of the fine grid and decrease its radius to $R_s - \epsilon/\sqrt{2}$. Observe that now we can cover a cell of the coarse grid using the minimum number of aligned disks since only a constant number of such disks can intersect any cell.

Observation 4 *The selection of the cell sizes implies that (i) each sensor completely covers the corresponding aligned disk (ii) each aligned disk intersects at most four cells of the coarse grid, and (iii) each cell of the coarse grid is intersected by at most $9R_s^2/\epsilon^2$ aligned disks.*

The fine grid algorithm is described in Figure 8. The running time of fine grid is dominated by Step 2 and is $O(q \cdot 2^{9R_s^2/\epsilon^2})$, where q is the number of cells of the coarse grid.

---

1. Put a grid with the cell size of $R_s \times R_s$ covering P and put over this grid a fine grid with the cell size of $\epsilon \times \epsilon$.
2. For each cell C, consider all subsets of aligned disks intersecting C and find the smallest set that covers the part of C covered by the union of all disks. Add these disks to the covering set (the *basic* cover).
3. Run the MST connection algorithm and add the returned disks to the final set (the *extended* cover).

Fig. 8. The fine grid algorithm.

Now we analyze the performance of the fine grid algorithm.

Theorem 5. *The covering set found by the fine grid algorithm is connected and the uncovered fraction of P is at most $R_s^2/(R_s - \epsilon/\sqrt{2})^2$. The approximation factor of the grid placement algorithm is at most 12, compared to OPT that uses aligned disks.*

Proof. The uncovered fraction of P is at most $R_s^2/(R_s - \epsilon/\sqrt{2})^2$. That is due to the fact that according to Observation 4 each disk completely covers the corresponding aligned disk. In Step 2 of the fine grid algorithm, we cover each cell using the minimum number of aligned disks. On the other hand, by Observation 4, OPT can use the same disk to cover at most four cells. Therefore, the size of the basic cover is at most four times the size of OPT that uses the aligned disks to cover P. The approximation factor follows since the size of the extended cover is bounded by twice the size of the basic cover. If it is not the case, the center of at least one disk is not covered, which contradicts the construction of the basic cover (see the proof of Theorem 4).   □

Note that the smaller the value of ϵ, the better the accuracy of the solution returned by the algorithm. However, the running time grows exponentially as ϵ decreases.

There is also an interesting application of the fine grid algorithm to the problem of covering a convex area with fat geometric objects. Suppose that we wish to cover a convex area using the minimum number of disks, squares or triangles. We put a coarse and a fine grid over this area so that each object intersects a constant number of coarse grid cells and each coarse grid cell is intersected by a constant number of different objects located at the vertices of the fine grid. Then we cover each cell of the coarse grid using the minimum number of objects by running exhaustive search over the sets of objects intersecting it. Finally, we return the union of the sets that cover all cells of the coarse grid.

3.4   Distributed Dominating Cover Algorithm

A connected dominating set (CDS) of GC is a subset $S' \subseteq S$ such that each node in $S \setminus S'$ is adjacent to some node in S' and the communication subgraph induced by S' is connected. It has been shown that the problem of finding a minimum CDS for unit-disk graphs is NP-hard [10]. The work in [26] gives an 8-approximation algorithm with $O(n \log n)$ message complexity.

We assume that each sensor has the communication range of $R_c = 2R_s/3$ and the communication graph GC is connected. Let GD be a unit disk graph in which each sensor corresponds to a disk of radius $R_s/3$ and two disks are connected by an edge if they intersect. The dominating cover algorithm just computes a connected dominating set in GD using the algorithm of [26]. The following observation is useful to demonstrate the coverage property.

Observation 5 *If a disk d is adjacent to another disk d' in GD, the sensor corresponding to d completely covers d'.*

The next theorem analyzes the performance of the dominating cover algorithm.

Theorem 6. *The covering set found by the dominating cover algorithm is connected and the uncovered area of P is bounded by the part of P that is not covered by the set of disks in GD. The approximation factor of the dominating cover algorithm algorithm is at most 18.*

Proof. Obviously, dominating cover returns a connected set since GD is connected. The coverage property follows by Observation 5. We have that the area covered by a single sensor from a maximal *independent set* IS computed by the CDS algorithm [26] is $\pi R_s^2/9$. On the other hand, any sensor in OPT can cover the area of at most πR_s^2. The approximation factor of 18 is due to the fact that in [26] the size of the final connected dominating set is at most twice the size of IS.   □

4   Connectivity

In this section we show how to convert a *complete* covering set B into a connected covering set under a realistic assumption that $R_c = R_s$.[1] We analyze two algorithms based on MST and Steiner Tree techniques. If the basic covering set is small, it is worth to use the former algorithm. Otherwise, it is preferable to use the latter algorithm.

Observation 6 *For each sensor from the basic cover, there is a sensor in OPT at distance of at most R_s.*

The observation follows from the fact that the sensing range is R_s and if there is no such an OPT sensor, the location of at least one sensor is not covered by OPT. Observation 6 implies the following lemma.

Lemma 2. *Adding nodes in OPT makes the basic cover connected.*

We consider the MST connection algorithm (see Figure 6). The proof of the next theorem is similar to that of Theorem 4.

Theorem 7. *The size of the connecting set returned by the MST connection algorithm is at most $2(|B| - 1)$.*

In the steiner tree connection algorithm, we apply the algorithm of [13] for the node-weighted steiner tree problem with unit weights on the communication graph GC while the basic cover B represents the set of terminal nodes. Note that Lemma 2 implies that the cost of an optimal steiner tree is bounded by $|OPT|$.

Theorem 8. *The size of the connecting set returned by the steiner tree connection algorithm is at most $O(\ln |B|) \cdot |OPT|$.*

5   Concluding Remarks

In this paper we investigate an important problem of maintaining coverage and connectivity in wireless sensor networks. The goal is to keep the minimum number of sensor nodes in active mode, thus maximizing the network lifetime. We present approximation algorithms with provable worst-case guarantees. There is a trade-off between the ease of implementation and the accuracy of the proposed algorithms, which allows one to select the proper algorithm for the specific needs. An open problem is whether it is possible to obtain a constant approximation algorithm that provides full coverage. Some interesting future research directions are to design distributed versions of our algorithms and to study the average-case rather than the worst-case performance.

---

[1] We note that even if B does not provide complete coverage, our algorithms would still return a connected set, but the approximation factor could be worse.

Acknowledgment. The third author wishes to thank Fabian Kuhn for many useful discussions.

References

1. Z. Abrams, A. Goel, and S. Plotkin, "Set K-Cover Algorithms for Energy Efficient Monitoring in Wireless Sensor Networks." *Proc. of Third International Symposium on Information Processing in Sensor Networks (IPSN)*, 2004.
2. I. F. Akyildiz, W. Su, Y. Sankarasubramaniam, and E. Cayirci, "Wireless Sensor Networks: A Survey," *Computer Networks*, March 2002.
3. P. Berman, G. Calinescu, C. Shah and A. Zelikovsky, "Power Efficient Monitoring Management in Sensor Networks," *IEEE Wireless Communication and Networking Conference (WCNC'04)*, Atlanta, March 2004.
4. B. Badrinath, M.Srivastava, K.Mills, J.S Holtz and K. Sollins, editors, "Special Issue on Smart Spaces and Environments," *IEEE Personal Communications*, 2000.
5. B. Ben-Moshe, M. J. Katz and M. Segal, "Obnoxious Facility Location: Complete Service with Minimal Harm," *International Journal of Computational Geometry and Applications*, Vol. 10, pp. 581-592, 2000.
6. P. Bose, A. Maheshwari, P Morin, and J. Morrison, "The grid placement problem," *Proceedings of the 7th Annual Workshop on Algorithms and Data Structures*, Vol. 2125 of LNCS, pp. 180-191, 2001.
7. A. Cerpa and D. Estrin, "ASCENT: Adaptive Self-Configuring Sensor Networks Topologies," *Proc. of INFOCOM '02*, New York, NY, USA, June 2002.
8. K. Chakrabarty, S. S. Iyengar, H. Qi and E. Cho, "Grid coverage for surveillance and target location in distributed sensor networks," *IEEE Transactions on Computers*, Vol. 51(12), pp. 1448-1453, December 2002.
9. B. Chen, K. Jamieson, H. Balakrishnan and R. Morris, "Span: An Energy-Efficient Coordination Algorithm for Topology Maintenance in Ad Hoc Wireless Networks," *ACM/IEEE International Conference on Mobile Computing and Networking (MobiCom 2001)*, Rome, Italy, July 2001.
10. B. Clark, C. Colbourn and D. Johnson, "Unit Disk Graphs", *Discrete Mathematics*, Vol. 86, pp. 165-177, 1990.
11. S. S. Dhillon and K. Chakrabarty, "Sensor placement for effective coverage and surveillance in distributed sensor networks," *Proc. of IEEE Wireless Communications and Networking Conference*, pp. 1609-1614, 2003.
12. D. Estrin, R. Govindan and J. Heidemann, editors, "Special Issue on Embedding the Internet," *Communications of the ACM*, Vol. 43, 2000.
13. S. Guha and S. Khuller, "Improved Methods for Approximating Node Weighted Steiner Trees and Connected Dominating Sets," *Inf. Comput.*, Vol. 150(1), pp. 57-74, 1999.
14. H. Gupta, S. R. Das and Quinyi Gu, "Connected Sensor Cover: Self-Organization of Sensor Networks for Efficient Query Execution," *Proc. of MobiHoc 03*, Annapolis, Maryland, June 2003.
15. H. Huang, A. Richa and M. Segal, "Dynamic Coverage and Related Problems in Ad-hoc Sensor Networks," *ACM Mobile Networks and Applications: Special Issue on Algorithmic Solutions for Wireless, Mobile, Ad Hoc and Sensor Networks*, 2003.
16. D. S. Johnson, "Approximation algorithms for combinatorial problems," *J. Comput. System Sci.* Vol. 9, pp. 256-278, 1974.
17. K. Kar and S. Banerjee, "Node Placement for Connected Coverage in Sensor Networks," *Proceedings of WiOpt 2003*, Sophia-Antipolis, France, March 2003.

18. D. T. Lee, M. Sarrafzadeh and Y. F. Wu, "Minimum cut for circular-arc graphs", *SIAM J. Computing* Vol. 19(6), pp. 1041-1050, 1990.
19. X. Li, P. Wan, and O. Frieder, "Coverage in wireless ad-hoc sensor networks," *ICC 2002*, New York City, 2002.
20. A. Mainwaring, J. Polastre, R. Szewczyk, and D. Culler, "Wireless sensor networks for habitat monitoring," *First ACM International Workshop on Wireless Workshop in Wireless Sensor Networks and Applications*, August 2002.
21. S. Masuda and K. Nakajima, "An optimal algorithm for finding a maximum independent set of a circular-arc graph," *SIAM Journal on Computing*, Vol. 17(1), pp. 41-52, 1988.
22. S. Meguerdichian, F. Koushanfar, M. Potkonjak and M. B. Srivastava, "Coverage Problems in Wireless Ad-hoc Sensor Networks," *Proc. of IEEE Infocom 2001*.
23. S. Meguerdichian and M. Potkonjak, "Low Power 0/1 Coverage and Scheduling Techniques in Sensor Networks," *UCLA Technical Report 030001*, January 2003.
24. S. Slijepcevic and M. Potkonjak, "Power efficient organization of wireless sensor networks," *Proc. of IEEE International Conference on Communications (ICC)*, pp. 472-476, 2001.
25. D. Tian and N.D. Georganas, "A Coverage-preserved Node Scheduling scheme for Large Wireless Sensor Networks," *Proc. of First International Workshop on Wireless Sensor Networks and Applications*, Atlanta, USA, September 2002.
26. P. J. Wan, K. Alzoubi and O. Frieder, "Distributed Construction of Connected Dominating Set in Wireless ad hoc networks," *Proc. of INFOCOM 2002*.
27. X. Wang, G. Xing, Y. Zhang, C. Lu, R. Pless and C. D. Gill, "Integrated Coverage and Connectivity Configuration in Wireless Sensor Networks," *Proc. of SenSys '03*.
28. Y. Xu, J. Heidemann, and D. Estrin, "Geography-informed Energy Conservation for Ad Hoc Routing," *ACM/IEEE International Conference on Mobile Computing and Networking (Proc. of MobiCom 2001)*, Rome, Italy, July 2001.
29. F. Ye, G. Zhong, S. Lu, and L. Zhang, "Peas: A robust energy conserving protocol for long-lived sensor networks," *Proc. of The 23nd International Conference on Distributed Computing Systems (ICDCS)*, 2003.
30. H. Zhang and J. C. Hou, "Maintaining Sensing Coverage and Connectivity in Large Sensor Networks," *Technical Report UIUC*, UIUCDCS-R-2003-2351, 2003.

Energy-Memory-Security Tradeoffs in Distributed Sensor Networks

David D. Hwang[1,2], Bo-Cheng Charles Lai[1], and Ingrid Verbauwhede[1,2]

[1] University of California—Los Angeles, Electrical Engineering Dept.,
Los Angeles CA 90095, USA
dhwang,bclai,ingrid@ee.ucla.edu
[2] Katholieke Universiteit Leuven, Dept. ESAT/SCD-COSIC,
Kasteelpark Arenberg 10, B-3001 Leuven-Heverlee, Belgium

Abstract. Security for sensor networks is challenging due to the resource-constrained nature of individual nodes, particularly their energy limitations. However, designing merely for energy savings may not result in a suitable security architecture. This paper investigates the inherent tradeoffs involved between energy, memory, and security robustness in distributed sensor networks. As a driver for the investigation, we introduce an energy-scalable key establishment protocol called cluster key grouping, which takes into account resource limitations in sensor nodes. We then define a metric (the security leakage factor) to quantify security robustness in a system. Finally, a framework called the security-memory-energy (SME) curve is presented that is used to evaluate and quantify the multi-metric tradeoffs involved in security design.

1 Introduction

Distributed sensor networks (DSNs) are a particular class of ad hoc networks that consist of (potentially) thousands of individual sensor nodes, each communicating via low-power wireless channels. The individual nodes are resource-constrained devices comprised of a microcontroller, memory, environmental sensors, and a radio transceiver. Potential application areas include military battlefield scenarios, habitat monitoring, building environmental control, and factory reliability sensing. Security plays an important role in sensor network architecture, as sensors may be deployed in enemy territory, contain private monitoring information, relay trade secrets, or possess other forms of sensitive data. Due their unique nature, DSN security is challenging in many ways: security must be scalable, be maintained without a powerful server, adapt to changes in network topology, and take into account the physical exposure of the nodes to an adversary.

However, perhaps the most challenging aspect of security in sensor nodes is coping with their severe resource constraints. Each individual node is limited in processing ability, memory, and energy. Energy in particular is the most valuable resource a node possesses; in most cases, unless energy-scavenging techniques are used [1], once energy is depleted in a node, the node is permanently offline. This is

I. Nikolaidis et al. (Eds.): ADHOC-NOW 2004, LNCS 3158, pp. 70–81, 2004.

why energy-based denial of service attacks [10] such as sleep deprivation torture [11] are particularly effective against DSNs. Hence, unlike security protocols for workstations (or even PDAs and notebook computers), energy expenditures for the security architecture of sensor networks must be kept to a minimum. By examining energy costs of security in a sensor network, it is clear the largest energy consumer is radio transmission. In the symmetric-key protocol SPINS [2], the radio transmission aspect of the protocol uses over 97% of the total security energy, while the actual encryption requires less than 3%. Studies in [12] show that for sensor networks, the energy/bit expended in radio transmission is three orders of magnitude greater than the energy/bit expended in AES symmetric-key encryption. Hence, minimizing transmission cost in a security architecture is a key design goal. However, merely reducing security transmission energy may require a tradeoff with other resources (such as memory) or even affect the robustness of the security mechanism itself. The purpose of this paper is to investigate this relationship between energy, memory, and security in sensor networks. We propose a framework called the SME (security-memory-energy) curve to quantify such tradeoffs.

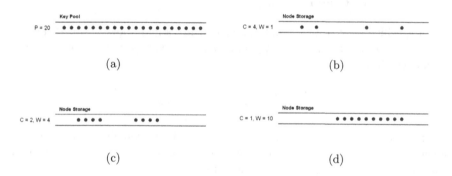

Fig. 1. Cluster Key Grouping Pre-Deployment Scenarios.

2   Cluster Key Grouping

As a driver for our investigation into energy-memory-security tradeoffs, we first present an energy-scalable protocol called cluster key grouping. In cluster key grouping, a key pool of P keys is generated off-line as shown in Fig. 1a. Prior to deployment, each node is randomly programmed with C clusters of keys, each cluster having a width of W keys per cluster; the total number of keys stored in each node is $K = C \cdot W$. All nodes are programmed with the same number of clusters C, with each cluster having the same width W. The key ring model presented in [4] is the specific case where $W = 1$ and thus $K = C$, as shown in

Fig. 1b. In the scenario in Fig. 1c, each node is programmed with $C = 2$ clusters, each of $W = 4$ keys; in the scenario in Fig. 1d, each node is programmed with only one large cluster of width $W = 10$.

Upon deployment, each node broadcasts the starting address of each of its C clusters. The remaining $W - 1$ addresses of each cluster are not broadcast since they are implicitly known from the starting address. If two nodes share at least one key between them, then a connection can be established based on the shared key and a secure link is said to be formed between them. The probability that the entire network is securely connected is related to P_c, the graph connectivity. The graph connectivity is a function of p, the probability two nodes share at least one key between them, also called the overlap probability. Given N nodes in a network, the desired graph connectivity can be calculated using the equation [4]:

$$P_c = \lim_{N \to \infty} Pr[G(N, p) \text{ is connected}] = e^{e^{-c}} \tag{1}$$

where c is a real constant, and $p = \ln(N)/N + c/N$. For an N-node network, a p can be specified to meet the required graph connectivity.

2.1   Effects of C and W on p

For a specified overlap probability p, we now investigate how C and W are used to generate p. As shown in the Appendix, p is derived as a function of the number of clusters C as:

$$p = 1 - \left(\frac{P - 2CW + 1}{P - CW + 1}\right)^{2C} \cdot \frac{(P - CW + 1)^2}{P \cdot (P - 2CW + 1)}. \tag{2}$$

This equation can be used as follows: a desired probability of overlap p is given as a specification. For a certain cluster size C, this equation determines the cluster width W required to obtain the desired p.

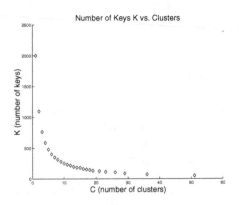

Fig. 2. Total Number of Keys in Cluster Key Grouping (to achieve $p = 0.4$).

Fig. 2 shows the total number of keys required ($K = C \cdot W$) for different C values, given $P = 10,000$ and the specified $p = 0.40$. At one extreme is the case presented in [4], where $W = 1$ and hence $K = C \cdot W = C$, as in Fig. 1b. In this case, by spreading the keys evenly throughout the entire key space, the lowest $K = 51$ total number of keys is required to obtain the specified p. At the other extreme is the case where only one wide cluster is stored, hence $C = 1$ and $K = C \cdot W = W$, as in Fig. 1d. Since all the keys are forced into one cluster and are not spread over the key pool, this case requires the highest $K = 2001$ total number of keys to obtain the specified p.

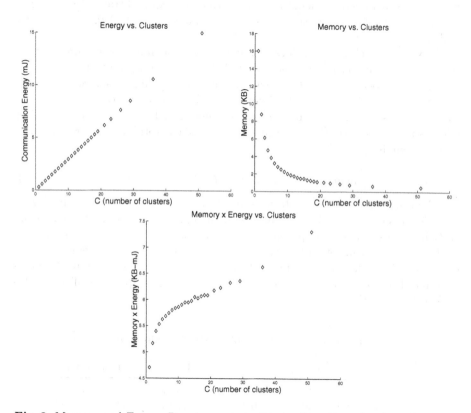

Fig. 3. Memory and Energy Requirements in Cluster Key Grouping (to achieve $p = 0.4$).

3 Energy-Memory Tradeoffs

With a driver protocol in place, we now examine the tradeoffs in security design. As stated earlier, the cluster-key grouping protocol is energy-scalable. By this we mean that the transmission energy expended to complete key agreement varies based upon the value of C chosen. Recall that upon deployment each node must broadcast C starting addresses to its neighbors at a cost of E_b Joules per bit, requiring a total transmission energy of:

$$\text{energy} = C \cdot E_b \cdot \lceil \log_2 P \rceil \quad [\text{Joules}]. \quad (3)$$

In the following simulations we assume the transmission energy of a Sensoria WINS NG node broadcasting at 10 mW RF over 900 meters, with a value of $E_b = 21$ mJ/bit [12]. (It is important to note that the energy factor does not take into account the reception energy or the energy required for re-broadcasting due to collisions in a CSMA scheme.) It is clear that the required transmission energy increases linearly as C increases, as seen in Fig. 3 for $P = 10,000$, $keysize = 64$, and the desired $p = 4$.

However, by choosing different values of C, the memory requirements of the security architecture are also affected, but with an opposite trend. In this sense, the protocol can also be considered memory-scalable. Each node requires memory to store $K = C \cdot W$ total keys, each of $keysize$ bits (i.e. 64, 128, etc.). Each node also must store the starting address of each of C clusters, with each address requiring $\lceil \log_2 P \rceil$ bits of memory. Hence the total memory requirement of each node is:

$$\text{memory} = K \cdot keysize + C \cdot \lceil \log_2 P \rceil \quad [\text{bits}]. \quad (4)$$

These memory requirements fall quickly as C increases, as seen in Fig. 3. Thus for a specified overlap probability p, though energy requirements increase as C increases, memory requirements decrease as C increases. This leads to a tradeoff between the two physical metrics called the weighted memory-energy curve, which is the memory multiplied by the energy. (A weighing factor is introduced to compensate for the effects of varying overlap probabilities.) This is also shown in Fig. 3.

Table 1. Corner Cases of Cluster Key Grouping.

W	C	K	p_{actual}	Memory	Energy	Memory·Energy
1	51	51	0.40787	497 bytes	15 mJ	7.31 Kbytes-mJ
2001	1	2001	0.40010	16,000 bytes	0.294 mJ	4.71 Kbytes-mJ

To consider these tradeoffs in more detail, the two corner cases are examined in Table 1. The corner cases are defined as the extreme scenarios of C: the $W = 1$ ($C = K$) case, where each cluster is only one key, and the $C = 1$ case, where only one wide cluster is stored. From a pure communication energy standpoint, the $C = 1$ case requires 51 times less energy than the $W = 1$ case, since each node broadcasts only one starting address rather than 51 starting addresses to its neighbors. If energy were the only design factor, it is obvious this is the optimal solution. However, from a memory standpoint the $C = 1$ case also requires 32 times more memory than the $W = 1$ case, since a greater number of keys are required to obtain the desired overlap probability p. Thus, a tradeoff between memory and energy—the memory-energy curve—is quite apparent. The values of the memory-energy curve for the $W = 1$ and $C = 1$ cases are 7.31 and 4.71 Kbytes-mJ, respectively, which are on the same order of magnitude. Thus designing for the minimum of the memory-energy function would lead to using the $C = 1$ scenario.

Between the two corner cases lies the wide range of energy and memory utilization seen in Fig. 3. By varying the factor C, the cluster key grouping protocol can be energy-scaled (or conversely, memory-scaled) to meet the resource constraints of the network, with all permutations meeting the overlap requirement p.

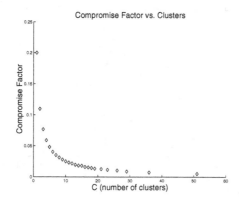

Fig. 4. Compromise Factor in Cluster Key Grouping (to achieve $p = 0.4$).

4   Security Tradeoffs

The above tradeoffs involve only the physical metrics of energy and memory. The question that remains is how security robustness is affected by such tradeoffs. Since security is an abstract concept, we must first formulate a "metric" for security that can be then traded off with the physical metrics. This metric will be called the security leakage factor, which quantifies the significance of a security breach to a system.

We begin by introducing the notion of the compromise factor, which is the number of keys compromised if a single node is compromised divided by the key pool, hence *compromise factor* $= K/P$. The compromise factor is plotted for different values of C in Fig. 4. Clearly, the greater the key spread over the key pool (i.e. the larger the number of clusters C), the fewer total keys K that are required, creating a lower total compromise factor. In order to quantify the effects of such a compromise on the security architecture, the security leakage factor is defined. The security leakage factor (SLF) is a function of the compromise factor:

$$security\ leakage\ factor = 1 + s_w \cdot compromise\ factor \qquad (5)$$

where s_w is the security weight, which is any natural number. Security weight can be any positive real number, but for the sake of illustration natural numbers are used in this paper. As stated earlier, the security leakage factor—and the security weight in particular—attempts to roughly quantify the importance a security compromise has to a network. For example, if a network for whatever reason is not affected by a compromise or assumes that one cannot be made, then $s_w = 0$ would be assigned. If a network is extremely sensitive to key compromise, then a higher security weight (e.g. $s_w = 5$) would be assigned.

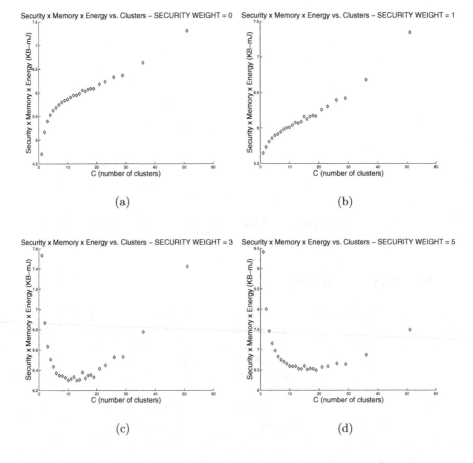

Fig. 5. Security-Memory-Energy Tradeoffs in Cluster Key Grouping (to achieve $p = 0.4$).

With a security metric in place, a comparison framework can be defined which quantifies the tradeoffs between security, memory, and energy. We call this framework the SME curve, which is defined as *security leakage factor · memory · energy*. Fig. 5 shows SME curves for different security weights. As can be seen, depending on the security weight, there is a minimum of the curve along a particular number of clusters C. The case of $s_w = 0$ (no security effects) is the memory-energy curve mentioned earlier whose minimum is at $C = 1$. When security has a low priority ($s_w = 1$), the SME follows a curve similar to the weighted memory-energy function. However, when security leakage is more important ($s_w = 3$), then the curve alters into a reverse bell with a minimum at $C = 13$ clusters. This indicates that a network can have a minimal loss of security coupled with minimal energy consumption and memory requirements (according to our definitions) by choosing this cluster size. The sudden increase in the metric at smaller values of C is due to the increased importance of the compromise factor, which is large for small cluster numbers. (This characteristic increase begins at the security weight of approximately 1.8). At a security weight of $s_w = 5$ the minimum of the curve shifts to the right (towards a lower compromise factor) and hence a minimum is achieved at $C = 19$ clusters. As s_w increases to even larger values, the "tail" of the curve occurring at larger values of C continues to decrease until eventually the curve mimics the compromise factor itself; for $s_w = 65$ and greater, the minimum SME is always at $C = 51$ clusters. Table 2 summarizes these results.

Table 2. Security Tradeoffs in Cluster Key Grouping.

s_w	C (minimum)	Security-Memory-Energy
0	1	4.71 SLF-Kbytes-mJ
1	1	5.65 SLF-Kbytes-mJ
3	13	6.40 SLF-Kbytes-mJ
5	19	6.49 SLF-Kbytes-mJ
65	51	9.73 SLF-Kbytes-mJ

Therefore, by designing for the minimum of the SME curve instead of the minimum of energy curve, tradeoffs between energy, memory, and security can be taken into account. Though the security leakage factor is specific to the cluster key grouping model, it demonstrates the notion that security can be quantified as a metric, which can then be used to perform tradeoffs with traditional metrics such as energy, memory, processing latency, etc. In security architecture design, such a metric and such a framework are necessary to measure the sum effect of dimensionally-different metrics and to allow for fair comparison between similar protocols.

5   Related Work

In this section of the paper, we evaluate prior art sensor network security models. There are two simple key management schemes that can be used in sensor network security. The first scheme is the pre-deployment programming of a single universal key into each of the N nodes of the network. In this scheme the key distribution problem is solved and nodes can communicate with one another securely by encryption via this universal key. However, security compromise is catastrophic in that if one of the N nodes is compromised, the security of the entire network is compromised. The other simple key management scheme is the pre-storage of pair-wise node keys, where each node stores a separate key for communication with each of the other $N-1$ nodes in the network, with the total number of keys in the network being $N \cdot (N-1)/2$. Clearly if $N = 10,000$, this mechanism is impractical in terms of memory constraints. Renewability is also an issue in this scheme, in that it is difficult to add new nodes to the network, unless extra renewal keys are pre-stored in each node before initial network deployment. Looking past these cases, there have been other proposals to address sensor network security.

[2] introduces the protocol suite SPINS (Security Protocols for Sensor Networks). SPINS includes two symmetric-key cryptographic protocols: μTESLA for authenticated broadcast from basestations (powerful nodes) to nodes, and SNEP for data confidentiality, authentication, integrity, and freshness. The SPINS suite uses counters and rough time stamps to ensure freshness. In SNEP, each basestation shares a secret key with each node in the network. In order for two nodes to communicate securely with one another, they must each go through a protocol with the basestation in order to obtain a node-to-node encryption key. Hence, SPINS has a potential problem with scalability, as the basestation may become a computational bottleneck if the number of nodes requesting keys overloads its bandwidth or computational capacity. SPINS also requires all nodes to be in the radio range of the basestation.

The scheme presented in [3] introduces another symmetric-key scheme in which all nodes share an initial universal key K_{GI}. This key is used as a root to generate other keys and allows for groups of nodes to elect local leaders (clusterheads) to control key management. A universal traffic encryption key (TEK) is generated from the universal key K_{GI} via hash functions and allows for all the nodes to communicate with one another in the network. This TEK is periodically updated to ensure freshness. One assumption mentioned in the paper is that tamper-proofing is assumed for each of the nodes. In some deployment scenarios (i.e. high security applications) tamper-proofing every node is possible; in other scenarios tamper-proofing may not be as feasible.

[4] presents a key distribution and establishment scheme based on the principle of a key pool and key ring. First, a large pool of P random keys is generated off-line. Prior to deployment, each node is programmed with K keys selected from the key pool using the sampling without replacement probability model. The set of K keys is called the key ring. Upon deployment, a discovery protocol begins in which each node broadcasts, in cleartext, its K key identifiers to its neighbors. If two neighboring nodes discover that a key is shared between them, then a secure connection can be established based on that shared key. This model is what was referred to as the basic key ring model.

[5] expands on the principles of [4] with three new techniques: 1) the q-composite scheme requires q keys to overlap, rather than a single key, to establish a secure connection, 2) the multipath-reinforcement scheme allows for a key update that requires correct communication from multiple paths to regenerate the key, and 3) the random-pair-wise scheme allows for node-to-node authentication. [6] presents a Blom pre-distribution scheme intended to improve resiliency of the network to node capture.

For general ad hoc network security, [8] proposes a public-key cryptography solution, in which threshold cryptography is used. Threshold cryptography is a technique in which a group of nodes form a certificate authority, which essentially verifies the link between an entity and his public key. [7] also presents an ad hoc security model using threshold cryptography and public-key encryption and decryption. The protocol mentioned in [9] uses a modified PGP (pretty good privacy) model to implement key management. However, public-key techniques, while possibly applicable for general ad hoc networks, may not be suitable for sensor networks based on energy considerations. [12] shows that public-key encryption energy requirements are orders of magnitude greater than symmetric-key encryption requirements. For example, on a MC68328 DragonBall processor, a 1024-b encryption for the symmetric-key AES algorithm requires 0.104 mJ. In contrast, a corresponding public-key encryption using RSA requires 42 mJ, which is 400 times more than AES. [13] also demonstrates that the energy cost of elliptic curve public-key cryptography is two to three orders of magnitude greater than Rijndael symmetric-key cryptography.

6 Conclusion

Energy-driven design is important in the architecture of secure sensor networks. However, energy is not the only criterion that must be factored into the development of a security architecture. This paper has demonstrated—via a protocol called cluster key grouping—that there are inherent tradeoffs involved between energy, memory, and security. By quantifying security as a metric and creating a comparison framework (the SME curve), an optimal architecture can be designed which factors in all such metrics.

Acknowledgements. The authors wish to acknowledge the support of the NSF (CCR-0098361), the Fannie and John Hertz Foundation (DH), and the Belgian National Science Foundation (FWO-G.0450.04).

References

1. A. Kansal and M. B. Srivastava. An environmental energy harvesting framework for sensor networks. *Proc. Int. Symposium on Low power Electronics and Design (ISLPED 2003)*, pp. 481-486, Aug. 2003.

2. A. Perrig, R. Szewczyk, V. Wen, D. Culler, and J. D. Tygar. SPINS: Security protocols for sensor networks. *Proc. 7th ACM Mobile Computing and Networks (MobiCom '01)*, pp. 189-199, July 2001.

3. S. Basagni, K. Herrin, E. Rosti, and D. Bruschi. Secure pebblenets. *Proc. 2nd ACM International Symposium on Mobile Ad Hoc Networking & Computing (MobiHoc '01)*, pp. 156-163, Oct. 2001.

4. L. Eschenauer and V. Gligor. A key-management scheme for distributed sensor networks. *Proc. 9th ACM Conference on Computer and Communications security (CCS '02)*, pp. 41-47, Nov. 2002.

5. H. Chan, A. Perrig, and D. Song. Random key predistribution schemes for sensor networks. *Proc. 2003 IEEE Symposium on Research in Security and Privacy*, pp. 197-213, May 2003.

6. W. Du, J. Deng, Y. S. Han, and P. K. Varshney. A pairwise key pre-distribution scheme for wireless sensor networks. *Proc. 10th ACM Conference on Computer and Communications security (CCS '03)*, pp. 42-51, Oct. 2003.

7. J. Kong, P. Zerfos, H. Luo, S. Lu, and L. Zhang. Providing robust and ubiquitous security support for mobile ad hoc networks. *Proc. 9th IEEE International Conference on Network Protocols (ICNP '01)*, pp. 251-260, Nov. 2001.

8. L. Zhou and Z. J. Haas. Securing ad hoc networks. *IEEE Network*, pp. 24-30, Nov./Dec. 1999.

9. J.-P. Hubaux, L. Buttyan, and S. Capkun. The quest for security in mobile ad hoc networks. *Proc. 2nd ACM International Symposium on Mobile Ad Hoc Networking & Computing (MobiHoc '01)*, pp. 156-163, Oct. 2001.

10. A. D. Wood and J. A. Stankovic. Denial of service in sensor networks. *IEEE Computer*, pp. 54-62, Oct. 2002.

11. F. Stajano and R. Anderson. The resurrecting duckling: security issues for ad-hoc wireless networks. *Proc. 7th Int. Workshop on Security Protocols*, Springer-Verlag, 1999.

12. D. W. Carman, P. S. Kruss, and B. J. Matt. Constraints and approaches for distributed sensor network security. NAI Labs Technical Report #00-010, Sept. 2000.

13. A. Hodjat and I. Verbauwhede. The energy cost of secrets in ad-hoc networks. *Proc. IEEE CAS Workshop on Wireless Communication and Networking*, Sept. 2002.

Appendix

To derive the overlap probability p, one must first consider the number of ways C non-overlapping clusters of width W can be arranged along a key pool of size P. It is clear that one cluster can have its starting point at any of P possible positions, giving P possible arrangements for $C = 1$. For two clusters, the first cluster may start in any of P ways, and the second cluster can start in any of $P - W - (W - 1) = P - 2W + 1$ positions in modulo P arithmetic. The $P - W$ term indicates the number of starting positions available after the first cluster is placed. The $W - 1$ term indicates the positions unavailable at the end of the key space (because a W width cluster cannot fit in a space of $W - 1$ or less, lest the clusters overlap). Thus the total number of ways for two clusters to be arranged is $P \cdot (P - 2W + 1)$. Continuing this principle and performing nested calculations to larger cluster sizes, the total number of ways C clusters of width W can be arranged given a key pool of size P is

$$P \cdot \sum_{A_3=1}^{P-CW+1} \sum_{A_4=1}^{A_3} \cdots \sum_{A_C=1}^{A_{C-1}} A_c \tag{6}$$

for $C \geq 3$. For large values of P, we use the equation

$$\sum_{x=1}^{n} x^r \approx \frac{n^{r+1}}{r+1} \tag{7}$$

to obtain an approximate number of arrangements as:

$$P \cdot \frac{(P-CW+1)^{C-1}}{(C-1)!}. \tag{8}$$

The number must later be multiplied by $1/C$ to account for circular shifting possibilities. The probability p that two neighboring nodes share at least one key is $1 - \Pr[\text{no key shared}]$. The probability that no key is shared is the number of ways $2C$ non-overlapping clusters can be arranged, divided by the number of ways C non-overlapping clusters can be arranged in the first node and second node, respectively, multiplied by the ways $2C$ clusters can be put into two partitions of C clusters each:

$$\Pr[\text{no key shared}] = \frac{\binom{2C}{C} \cdot [\# \text{ ways to arrange } 2C \text{ nodes}]}{[\#\text{ways for node 1}] \cdot [\#\text{ways for node 2}]}. \tag{9}$$

Using the approximation earlier, p reduces to:

$$p \approx 1 - \frac{\binom{2C}{C} [P \cdot \frac{(P-2CW+1)^{2C-1}}{(2C-1)!} \cdot \frac{1}{2C}]}{(P \cdot \frac{(P-CW+1)^{C-1}}{(C-1)!} \cdot \frac{1}{C})(P \cdot \frac{(P-CW+1)^{C-1}}{(C-1)!} \cdot \frac{1}{C})} \tag{10}$$

which simplifies to

$$p \approx 1 - \left(\frac{P-2CW+1}{P-CW+1}\right)^{2C} \cdot \frac{(P-CW+1)^2}{P \cdot (P-2CW+1)}. \tag{11}$$

Utilizing the Uncertainty of Intrusion Detection to Strengthen Security for Ad Hoc Networks

Dennis Dreef[1], Sanaz Ahari[3], Kui Wu[2], and Valerie King[2]

[1] Dept. of Computer Science, University of Victoria, PO Box 3055, STN CSC,
Victoria, BC, Canada, V8W 3P6,
ddreef@csc.uvic.ca
[2] {wkui, val}@cs.uvic.ca
[3] Microsoft Corporation, Redmond, WA, USA, 98052,
sanaza@microsoft.com

Abstract. With Mobile ad hoc networks rapidly approaching practical use by the masses, security has become a main concern. Intrusion detection, as the second line of defense, is an indispensable tool for highly survivable networks. Nevertheless, intrusion detection systems suffer from false alarms, which are extremely hard to control in mobile ad hoc networks. As a result, local intrusion detection engines in mobile ad hoc networks may only be able to claim anomaly with low confidence. Such uncertain knowledge is not very helpful in assisting end users with the final decision for intrusion response. The uncertainty of intrusion detection results, however, can provide a rough guideline on potential hazards and can be utilized to enhance security. In this paper, we present a randomized algorithm to utilize such uncertain knowledge to help routing protocols defend against the rushing attack. By eliminating excess bandwidth usage and adopting an adaptive randomized forwarding mechanism, the presented method is more efficient than previously existing solutions.

1 Introduction

Mobile Ad-hoc NETworks (MANETs) are networks with mobile nodes without the support of fixed infrastructure. The lack of fixed infrastructure and mobility pose many new challenges. The first major challenge was the routing problem which has been solved for the most part by protocols such as AODV [1] and DSR [2]. The second major challenge is security. Potential deployments of MANETs may be in un-trusted environments. Unfortunately existing security solutions for the Internet are inapplicable to MANETs due to mobility and the lack of a centralized network management point. Hence security is an important but hard problem for any realistic applications with MANETs.

The area of MANET security has garnered a lot of attention recently with a host of possible attacks and solutions being published [3,4,5,6]. There are generally two main methods to provide system security: intrusion prevention and intrusion detection. Intrusion prevention usually depends on authentication and cryptographic operations to protect the system. The research in security

I. Nikolaidis et al. (Eds.): ADHOC-NOW 2004, LNCS 3158, pp. 82–95, 2004.
© Springer-Verlag Berlin Heidelberg 2004

for the Internet teaches us that intrusion prevention alone may not be enough to provide a comprehensive solution for highly survivable networks. As a result, intrusion detection becomes an indispensable part of a secure system.

Intrusion detection in MANETs, however, tends to have a high false positive rate [7]. The high false positive rate makes it very difficult for accurate intrusion response. This causes a dilemma when using intrusion detection to enhance system security, if the information obtained from the detection module is potentially incorrect. Information that is unreliable will not increase security without high performance penalties.

Nevertheless, the uncertainty of the intrusion detection results can provide us with more or less hints on current network status and can be utilized to help make proper final decisions. Intrusion detection stimulus should be used with a grain of salt, that is, we know it is inaccurate but it can be used as a parameter in the decision making process. Knowing the inaccuracy and dealing with it appropriately is much better than having no information at all.

The main contribution of this paper is to present a strategy to take advantage of the uncertain knowledge of intrusion detection results. Based on intrusion detection system stimulus, the proposed method randomly forwards a routing request to alleviate some of the negative affects the high false alarm rate could have on performance, and to help routing protocols avoid malicious nodes without resorting to complicated cryptographic operations. To the best of our knowledge, we have not seen any research efforts trying to utilize the uncertainty of IDS in secure routing. We hope our paper will trigger more work in this direction.

The rest of the paper is organized as follows. Section 2 presents the background knowledge, including Ad-hoc On-demand Distance Vector (AODV) routing, threat model, and Intrusion Detection System (IDS) for MANETs. In Section 3, we introduce a randomized message forwarding mechanism to help routing protocols defend against the rushing attack [3] with the uncertain knowledge from IDS. In Section 4, we discuss how the proposed solution addresses the different flavors of the rushing attack. Section 5 presents the simulation evaluation of the proposed solution, and Section 6 concludes the paper.

2   Background

2.1   AODV

The rushing attack [3] is an attack against reactive routing protocols. Reactive routing protocols create routes when they are required. This strategy eliminates much of the table maintenance and other overhead that proactive protocols suffer. DSR [2] and AODV [1] are popular reactive protocols. Although each of the many routing protocols for MANETs have advantages and disadvantages, AODV has become a popular protocol due to its simplicity and better industry suppot.

AODV works by discovering a route by broadcasting a route request message (RREQ) with a unique ID to its neighbors. The neighbors create an entry in their routing table for the first RREQ with the given ID. All proceeding RREQ packets with the same ID are considered duplicates and dropped. Each neighbor subsequently broadcasts the RREQ and this procedure continues until the RREQ reaches the destination or a node with a valid path to the destination.

Once the destination node or a node with a valid route to the destination has received the RREQ, it will create a route reply message (RREP) intended for the source. The RREP will be unicast along the reverse path maintained by a precursor list. A valid path from the source to the destination will be established when the RREP message arrives at the source node.

2.2   Threat Model

In this paper, we use the rushing attack [3] as an example to demonstrate how the inaccurate information from intrusion detection can be used to defend against the attack. The various versions of the rushing attack can be used for man-in-the-middle attacks, by allowing the attackers to position themselves in the route.

Rushing Attacks. In on-demand routing, a node requiring a route to a destination floods the network with RREQ packets in an attempt to find a path to the destination. To limit the overhead of this flood, each node typically forwards only one RREQ originating from any route discovery. In particular, existing on-demand routing protocols, such as AODV, only forward the RREQ that arrives first from each route discovery.

In the rushing attack, the attacker exploits the aforementioned property of forwarding RREQs in a predictable manner. It increases its chances of being included in the path by using various methods of advantageous packet forwarding. What follows are descriptions of three variations of the rushing attack as presented in [3].

MAC Layer Rushing Attack: Medium Access Control (MAC) protocols generally impose delays between the time when the packet is handed to the network interface for transmission and when the packet is actually transmitted. As well collision detection for broadcast packets is difficult, so on demand protocols specify a delay between recieving a request and forwarding it. An attacker ignoring delays at either the MAC or routing layers will generally be preferred over similarly located non-attacking nodes, increasing the chances of establishing a path via the attacker.

Spurious Packet Flooding: Advantage in forwarding speed can also be achieved by keeping the buffers of neighboring nodes full. If each node processes the packets it receives in order and if an inadequate request authentication mechanism is used, a node can be kept busy processing spurious packets, thus deteriorating its ability to forward legitimate requests. Protocols employing public key techniques are particularly subject to these attacks, since they require considerable computation to verify each received request.

Wormhole and High Transmission Power: Faster transit of RREQs can be achieved by transmitting packets at a higher wireless transmission power level and reducing the number of nodes that must forward that RREQ to arrive at the target. A more powerful variant of this attack occurs when a tunnel is formed by two attackers where the first attacker simply forwards all control packets to the other attacker which subsequently broadcasts them. If the tunnel provides significantly faster transit than legitimate forwarders, nodes near one end of the tunnel generally will be unable to discover routes to the other end of the tunnel that do not include the tunnel. In general, a wired tunnel will provide faster transit than native, multi-hop wireless forwarding.

Existing Solutions. The proposed solution to these attacks in [3] is a combination of four mechanisms: random message forwarding, secure route discovery, secure neighbor detection and secure route delegation.

Random Message Forwarding: In traditional RREQ forwarding, the receiving node forwards the first RREQ and suppresses all subsequent RREQs. In the method proposed in [3], the node first collects a number of RREQs, and then selects an RREQ at random to forward. This method requires two parameters, the number of RREQs to be collected, n, and timeout value, t. Each node collects the maximum possible number of requests until either the maximum is collected or the timeout period is reached.

Secure Route Discovery: This mechanism [3] is based on the fact that legitimate nodes forward only one RREQ in any route discovery. The solution has three components:

- Secure Neighbor Detection protocol allows two nodes to detect each other as neighbors only if they can communicate and are within maximum transmission range of each other. This mechanism prevents an attacker from claiming to be a neighbor of another node when it is outside its maximum transmission range. This consists of a three-round mutual authentication protocol that uses delay timing to ensure that the other party is within communication range. The initiator sends a Neighbor Solicitation packet and then recieves a reply. The initiator then sends a Neighbor Verification. The delay between the Neighbor Solicitation and Neighbor Reply provides an upper bound on the distance of the neighbor.
- Secure Route Delegation is used to verify that all the secure Neighbor Detection steps were performed between any adjacent pair of nodes. This can be described with the following example:
  Consider two neighbors A and B. A has received an RREQ from node S destined for node R with sequence number, i. Node A performs the Secure Neighbor Detection protocol and verifies that B is its neighbor, then it delegates the RREQ to B by signing the RREQ and forwarding it to B. B verifies the signature and the process continues.
- Buffered requests need to be duplicate-suppression-unique; that is, if any two route requests contain node A, the route prefix leading up to (and including) A must be the same.

Though these solutions do prevent the rushing attack from being as effective as it could potentially be, it does also require heavy usage of precious bandwidth and processing power. From the three-way handshake required for secure neighbor detection to the message signing used in secure route delegation, this solution requires a lot of resources. Furthermore, it waits for an independently chosen timeout period before randomly forwarding one of the received RREQs, causing significant delay during route discovery.

2.3   Intrusion Detection System (IDS)

MANETs do not have concentration points such as gateways, switches and routers where the traffic can be easily monitored. In addition, the nodes typically have a limited local knowledge of the network topology. These factors along with different communication patterns between the wired and wireless worlds make intrusion detection in MANETs more challenging.

Zhang et al. have proposed an anomaly-based model for intrusion detection in MANETs [8]. The architecture involves an IDS agent for each node. These agents are responsible for data collection, local analysis and detection, cooperative analysis and detection, and lastly intrusion response.

Normal behavior patterns are gathered during a training process. During this process the normal profiles of data that is of interest are recorded. The nodes will continually model what they deem normal behavior based on the environmental factors and topology of the network from their point of view.

The framework adopts the premise that normal activities have observable traits and that when detecting anomalies, the values of those traits stray from the norm hence indicating an attack. The detection can achieved by rule induction and data mining techniques.

Nevertheless due to mobility and the lack of centralized control point, MANET intrusion detection usually has a high false positive rate (a normal behavior is flagged as anomalous) that creates a conundrum for users. It is very hard for end users to justify an intrusion response based on unreliable information.

3   Randomized Message Forwarding Based on Uncertain Knowledge

3.1   Assumptions

We believe that security in MANETs can be achieved with the proper combination of intrusion detection and intrusion response. Our proposed solution implements the intrusion response portion of this combination with a few assumptions as to what information the IDS should provide.

The following are the current suggested trace data to be collected for intrusion detection of the rushing attack. This information can easily be obtained from MANET IDS[8].

- Expected number of RREQ packets: it is the average expected number of RREQ packets received for one route during time t.

– Knowledge of neighbors: each node will maintain a table of neighbors. The IDS should be able to provide a confidence level for each neighbor, indicating whether the neighbor is an attacker or has been potentially attacked. For instance, the confidence value of 0% means the neighbor is definitely not an attacker while the confidence value of 100% means the neighbor is surely an attacker.

3.2 Randomized Algorithm

The randomized algorithm requires the following parameters:

1. expected number of RREQ packets to be received, n
2. timeout value, t
3. confidence value, cv. (Note that this is assumed to come from the IDS)

The randomized algorithm works as follows. Based on n and cv it generates a random number, i. To generate i we first determine the node space. Node space is essentially the upper bound on the range in which the i can exist. It is determined by multiplying n and the confidence value, cv. In order to enforce randomness, the node space will always have a minimum size of 2. This is to provide an attacker with a maximum probability of 50% to become part of a route. We then use the node space as the modulus of the random number generator.

$$nodespace = max(2, (n * cv)) \tag{1}$$

$$i = random()\%nodespace + 1 \tag{2}$$

For example, if the confidence level cv for the prior hop is 0%, the local IDS agent is confident that the previous hop is not attacking or compromised, so the node space will have a size of 2 determined using Formula (1) regardless of the size of n. This indicates one of the first two packets that arrive for a given route discovery will be forwarded. On the other hand, if cv is 50% and n is 10, we will generate a random number i using Formula (2) with a value between 1 and 5.

Similarly, after each RREQ is received, the associated cv value for the sending node is obtained from the IDS. We then use the maximum of the previous cv and the new cv to create the node space and a new random number. If the number of RREQs is greater than or equal to the value randomly chosen, then the packet at that index is forwarded. This allows the randomized forwarding mechanism to adapt the confidence value and node space as the IDS provides new information. The following pseudo code shows how it all works together.

```
if(not is_neighbor(source) && is_neighbor(previous_hop)){
        //add to a queue for its broadcast id
        rreq_queue *q = add_rreq(bcast_id, source, packet)
        //get confidence
        conf = IDS:get_confidence(previous_hop)
        q->cv = max(b->cv, conf)
        //determine node space
```

```
    expected = max(2, n*(q->cv/100))
    // calculate the random value
    randq = (random()%expected)+1
    if(q->size >= randq){
            rreq *r = rids_rreq_get(q, randq)
            //parse the randomly chosen RREQ
            packet = r->packet
    }else{return}
}//proceed to normal AODV processing of RREQ
```

As discussed in [3], the threshold value t should ideally be based on the number of legitimate hops between the initiator and the node forwarding the RREQ: closer nodes to the source should choose shorter timeouts than far-away nodes. For example, the immediate neighbors of the source should only wait for t, where t is the propagation time from the source to its neighbors. The legitimate number of hops, however, may not be known. Hence we suggest an adaptive method for determining the threshold value as follows:

1. t is initially set to the average propagation time required for one hop.
2. If in time t, we received x packets where x is considerably smaller than n, we simply adjust t as follows: $t *= 2$.
3. We can keep on doing this until either we have received all n packets, or t has reached a maximum T. Since if T is too large it will suffer from severe performance overhead. The actual value of T should be determined based on the actual network density and traffic. In a dense network a smaller T would be appropriate while in a sparse network T should be larger.

Remark 1. The above algorithm is just an example, demonstrating how the uncertain knowledge from IDS can be utilized for random forwarding. We remind the readers that there possibly exist various types of algorithms that take advantage of the uncertainty of IDS response. For instance, we could add weight to each RREQ packet based on the path length and the associated cv value of prior hop to help make wise decisions; we could also use accumulated cv values or the maximum cv value along the entire path. Since the main purpose of this paper is to present a framework that integrates the uncertain knowledge in secure routing and thus to trigger more research in this direction, we do not intend to list and analyze all possible algorithms.

3.3   Comparison

Superficially, the above proposal seems very similar with the random message forwarding mechanism in [3]. They actually have a lot of differences. One main difference is that the randomized forwarding suggested in [3] waits time t before selecting which RREQ to forward. While our proposal has the potential to forward the very first RREQ received, in fact it could forward any packet received well before timeout t is reached. This is advantageous over waiting for t as it will lessen the delay while still using randomness to increase security. The adaptive backoff allows the routing protocol to adjust to the network status and wait for a larger sample size to choose a RREQ from.

In contrast to the solution in [3], we do not use any cryptography or additional handshaking. This should significantly lower our overhead with respect to their solution, leaving our overhead levels only slightly higher that those of the original protocol. The lower overhead in terms of energy consumption and bandwidth usage is very important in the wireless world. For example, sensors in sensor networks have limited battery life and thus CPU cycles are precious, and the bandwidth at a conference may be heavily used and transmitting extra routing overhead is not a good use of it.

Our method provides a framework to balance the tradeoff between security and efficiency. The solution proposed here will by no means provide 100% security guarantee, but it does increase security at a low cost computationally, bandwidth wise and in terms of delay. Three is no free lunch in secure systems. It is never the case that strong security enforcement requires little performance penalty. In real applications, we are always faced with such balance. A framework such as ours to help balance the tradeoff is invaluable in wireless networks.

4   Dealing with Different Types of Rushing Attacks

4.1   MAC Layer and Wormhole and High Transmission Power Rushing Attack

The attacks as previously described in Section 2.2 use some advantage to get RREQs to their destinations quicker by forwarding packets without honoring MAC layer timing constraints, using increased transmission power level, or creating a tunnel with another attacker. The proposed solution protects against these attacks as follows. Assume the scenario in Figure 1 where the sender, S, forwards a packet to nodes A and B, where A is the attacker employing one of the above methods. The attacker will immediately forward the packet to node C, faster than node B.

To protect against the above scenario, we rely on the randomized forwarding mechanism. With an appropriate confidence value from the IDS, C randomly chooses which RREQ to forward. In this case the RREQ from A is likely discarded and the RREQ from B is forwarded on since the cv value associated with A is high.

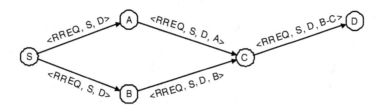

Fig. 1. Sample attack scenario

With the wormhole and high transmission power attacks, it can often be the case that no other RREQs will be received within threshold time, t, thereby forcing a node to forward the attacker's RREQ despite the adaptation. Note that the attacker may be selected as the forwarding node with probability of $1/node\ space$ as defined in Formula (1). Hence the chances of a successful attack are decreased, and get smaller as the network density increases. The denser the network, the higher the number of neighbors a node has. Since neighbors affect the node space and the size of n, there is a greater chance the attacker will not be selected.

4.2   Spurious Packet Flooding

The spurious packet flooding attack, as described in Section 2.2, is typically only effective when cryptography is involved, because of the increased amount of time required to process each RREQ. In normal cases, an attacked node should be able to handle received packets much faster than the speed of wireless transmission from the malicious node. Since we do not require any cryptographic operations, the randomized forwarding mechanism we proposed is not susceptible to this variation of the attack.

5   Simulation and Analysis

5.1   Simulation Model

We modify the original AODV protocol to include our proposed randomized forwarding, namely Randomized AODV (RAODV). In this section, we evaluate the performance of the modified AODV protocol via simulation. To truly determine the effectiveness, an adversarial environment was used with four attackers. A normal environment without attackers was also used for head to head comparison of RAODV and AODV.

We use the ns-2 simulator [9] to evaluate the randomized RREQ forwarding protocol described in Section 3.2. To simulate RAODV, the functionality of processing RREQ messages in AODV was rewritten. We assume the required information is available from an existing IDS.

```
get_confidence ( previous_hop , hops )
    if ( is_attacker ( previous_hop ))
        cv = Θ // where Θ is a high value
    else
        cv = f ( hops ) // where f () is an exponential function
    return cv ;
```

A proper IDS would have generated a high confidence value in its analysis of the attackers behavior. In our simulation, the cv value associated with an attacker is above 80%, which we believe achievable for an effective IDS. The RREQs from non attackers were given varying confidence values based on the hop count. This decision is justified because a higher hop count increases the likelihood that an attacker slipped into the path, much like when a message is passed by word of mouth, the further from the source you hear it, the more likely it becomes the message has been altered. It is believed that this assumption is realistic because as the length of the path grows, the number of nodes increases and the probability that a node in the path has been compromised also increases.

The attack nodes were modified so they no longer forward anything but routing related packets, acting as a black hole for data. Also, the transmission power was increased to double that of the normal nodes and the RREQ forwarding delay was eliminated. This combination gives the attack nodes a great advantage to successfully perform attacks.

A 1500m by 300m rectangular area with 50 nodes was chosen. A random waypoint mobility model was used with a minimum speed of 5 m/s and a maximum speed of 10 m/s. These parameters were chosen to avoid the problems stated in [10]. The pause times used were 0, 30, 60, 120, 300, 600 and 900 seconds, while the simulation time is 900 seconds in duration [11]. Each simulation scenario was run 10 times. When confidence intervals were calculated for the simulation results, the confidence level is set to 90%.

The channel capacity is set to 2 Mbps. Constant Bit Rate (CBR) traffic was used. There were 32 connections with 37 of the 50 nodes involved using 22 different sources and 25 different destinations. The 32 connections started at random times during the simulation time period. This traffic pattern with the large portion of nodes involved in communication poses heavy traffic in the networks. We intend to evaluate RAODV under these harsh conditions in the following simulation.

5.2 Evaluation Metrics

We evaluated control overhead, route creation delay, delivery ratio, the percentage of RREPs forwarded by attackers, and packet drop ratio of the attackers.

Control overhead with and with out the HELLO packets were both calculated for clarity. To get the overhead statistic, the total number of packets sent during simulation was counted and divided by the total number of routing packets sent during the same simulation. This was calculated twice for each simulation, which delivered results with and without the HELLO packets.

Delivery ratio is defined as ratio of the number of data packets sent by the sources over the number of data packets received by the destinations.

The infiltration ratio is defined as the number of times attackers forward RREPs divided by the number of RREPs sent from the sources. This ratio is adequate; however, it does not serve as an accurate measure because it does not take into consideration the amount of time an attacker remains part of a route. So we also looked at the attackers' packet drop ratio.

Attacker drop ratio is defined as the ratio of the number of packets dropped by attackers over the total number of packets sent from sources. This metric indicates how many packets were swallowed by the attacking nodes. This metric will let us see the effectiveness of the attackers. The infiltration ratio as described above shows us how many times routes were infiltrated but it does not show how effective that infiltration was.

Route creation delay is very important in on-demand routing. Determining if RAODV drastically increases the route creation delay when compared to standard AODV, is of great importance. For this metric, two routes were randomly chosen: one long (6 hops) and one short (3 hops). The route creation delay times were calculated by taking the difference between the time of the first request sent and the time of the first reply received.

5.3 Simulation Results

The results of the simulation are positive with respect to performance. In Table 1, it can be observed that the average AODV response time is very low for the short route but the average RAODV response time is also low. There is only approximately 0.3 seconds difference. The 6 hop route is much different as the difference is approximately 10 seconds, though the median time is lower than that of AODV. From Table 1, RAODV does introduce some delay in route discovery phase. Nevertheless, compared with the results stated in [3] in which the security solution proposed incurs significant increase in route discovery delay, our method is quite acceptable.

Table 1. Route Creation Delays. Note all delays are in seconds

Route	AODV	RAODV			
Length	Mean	Mean	Median	Standard Deviation	90% Confidence Interval
3 hops	0.4050	0.7485	0.4050	1.0392	0.5405
6 hops	5.0481	15.7657	2.8803	20.1670	10.4898

The routing overhead difference between AODV and RADOV is trivial as shown in Figure 2(a) and 2(b). The overhead levels of both RAODV and AODV follow nearly the same path while both counting the HELLO messages and not counting them. The overhead of RAODV is 15.8% without HELLO messages, while it is 29.1% with them. AODV has an average overhead of 15.0% without HELLO messages, and an average of 28.4% with. From this it could be generalized that RAODV increases the overhead by less than 1% over AODV. This is a big advantage of our proposed solution over other existing solutions in [3].

From Figure 2(c) and Figure 2(d), it becomes obvious that the difference in data delivery ratio is small, just as with overhead. AODV averaged 53.0% delivery while RAODV achieved 53.4% delivery, giving a 0.4% difference in delivery ratio in favor of RAODV. Intuitively, RAODV should have much higher packet delivery ratio than AODV since RAODV avoids the packet drop by attackers. The similar packet delivery ratios for AODV and RAODV are mainly because we intended to pose very harsh traffic condition in the network in the

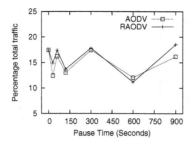

(a) Routing Overhead excluding HELLO

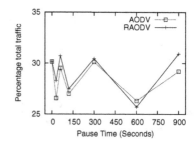

(b) Routing Overhead including HELLO

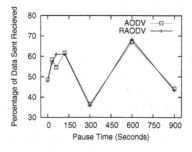

(c) Data Delivery ratio under non-adversarial conditions

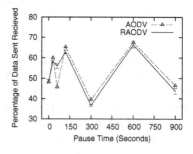

(d) Data Delivery ratio while under attack

Fig. 2. (a) and (b) Routing overhead graphs displaying AODV and RAODV performance at different pause times. (c) and (d) Data delivery ratio showing the performance of AODV and RAODV at different pause times.

simulation. That is, if RAODV selects alternative paths, the penalty of using the possibly longer length will be enlarged because of heavy end-to-end traffic. Our simulation results actually approximate the worst situation for RAODV or the lowest bound for packet delivery ration for RAODV. The exaggerated penalty of using long length is confirmed again in Figure 3(b), from which we observe that attackers drop more packets with AODV. The similar delivery ratios of AODV and RAODV imply that more packets are dropped from normal nodes with RAODV because of channel contention caused by heavy traffic through longer paths. The performance of RAODV in light traffic or other "good" network topology (e.g. there are several alternate paths with similar length from a source to a destination) will certainly be better.

For effectiveness with respect to security, route compromise is an important metric to look at. Figure 3(a) shows that in all but one scenario RAODV included attackers less often. For the simulations with lower pause times, we see that both

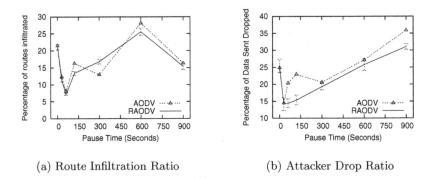

(a) Route Infiltration Ratio   (b) Attacker Drop Ratio

Fig. 3. (a) shows route infiltration results of AODV and RAODV (b) shows the effectiveness of the attackers against AODV and RAODV

protocols perform similarly. When pause time is long, such as larger than 600 seconds in the simulation, RAODV has smaller infiltration ratios.

This trend is mirrored in Figure 3(b) as we see with RAODV the attackers are dropping fewer packets. Despite the higher infiltration with pause time 300s, we still see RAODV outperforms AODV, indicating that though more routes were infiltrated, the infiltrated paths did not last long. Also, by our definition, the infiltration may not necessarily mean the attacker will be finally included in a valid path. While the attackers exist, AODV dropped 23.8% of the packets sent by the sources, while RAODV only dropped 20.7%, meaning the attackers were 12.4% less effective against RAODV.

To summarize, the randomized forwarding mechanism as simulated only marginally increases control overhead and has similar packet delivery ratio with original AODV even under harsh traffic condition. It does increase security and does limit the effectiveness of the attackers. The results show that a randomized forwarding mechanism removes the predictability that attackers rely on, while only slightly affecting performance.

6   Conclusion and Future Work

In this paper, we present a scheme to utilize the uncertainty of IDS results to enhance network security. This work is not intended to provide 100% security guarantee, a fundamental design principle for other existing solutions [3]. Instead, our proposal is to present a framework to effectively balance the tradeoff between security and performance efficiency, a case we are actually faced in realistic applications. Such tradeoff mechanisms are extremely important in wireless networks because of stringent constraints of bandwidth and energy resources.

From recent research activities in intrusion detection for MANETs [7,8], IDS tends to raise high false positive alarms. Controlling false positive alarms is very challenging for MANETs, and this difficulty is unlikely to be solved in a short time. It is a discouraging fact that very few researchers are willing to "waste" energy in such an elusive problem. This paper tries to put some fresh air in this direction by showing that even the uncertain knowledge of IDS can still be utilized to enhance security. We have observed that predictability is an enemy of security and inaccurate intrusion detection is an enemy of correct intrusion response. We contend that the randomization we proposed provides a possible solution to both problems, giving increased security with performance comparable to that of AODV as specified in [1].

In the future, we plan to propose other mechanisms in utilizing the uncertainty of IDS and perform analysis on the bounds of randomized algorithms in making correct decisions.

References

1. C. Perkins, E. Belding-Royer, and S. Das: AODV, RFC 3561, IETF, July 2003.
2. D. Johnson, D. Maltz, and Y. Hu: DSR, draft-ietf-manet-dsr-09, IETF, April 2003.
3. Y. Hu, A. Perrig, and D. B. Johnson: Rushing Attacks and Defense in Wireless Ad Hoc Network Routing Protocols. Proceedings of the 2003 ACM Workshop on Wireless Security (WiSe 2003), pp. 30-40, San Diego, California, September 2003.
4. P. Ning and K Sun: How to Misuse AODV: A Case Study of Insider Attacks against Mobile Ad-hoc Routing Protocols, Technical Report: TR-2003-07, 2003.
5. Y. Hu, A. Perrig, and D. B. Johnson: Packet Leashes: A Defense against Wormhole Attacks in Wireless Ad Hoc Networks, Proceedings of the Twenty-Second Annual Joint Conference of the IEEE Computer and Communications Societies (INFOCOM 2003), vol. 3, pp. 1976-1986, San Francisco, California, April 2003.
6. Y. Hu, A. Perrig, and D. B. Johnson: SEAD: Secure Efficient Distance Vector Routing for Mobile Wireless Ad Hoc Networks, Proceedings of the 4th IEEE Workshop on Mobile Computing Systems & Applications (WMCSA 2002), pp. 3-13, Calicoon, New York, June 2002.
7. K. Wu and B. Sun: Intrusion detection for wireless mobile ad hoc networks, book chapter in Ad Hoc and Sensor Networks, Y. Pan, and Y. Xiao (Eds.), Nova Science Publishers, hardbound, to appear, 2004.
8. Y. Zhang, W. Lee, and Y. Huang: Intrusion Detection Techniques for Mobile Wireless Networks, ACM/ Kluwer Wireless Networks Journal Vol. 9, No. 5, pp. 545-556, September 2003.
9. The Network Simulator ns-2 version 2.27, March 2004.
10. J. Yoo, M. Liu, and B. Noble: Random Waypoint Considered Harmful, Proceedings of the IEEE Conference on Computer Communications (INFOCOM), pp. 1312-1321, San Fransico, California, April 2003.
11. S. R. Das, C. E. Perkins, and E. M. Royer: Performance comparison of two on-demand routing protocols for ad hoc networks, Proceedings of the IEEE Conference on Computer Communications (INFOCOM), pp. 3-12, Tel Aviv, Israel, March 2000.

Weathering the Storm: Managing Redundancy and Security in Ad Hoc Networks*

Mike Burmester, Tri Van Le, and Alec Yasinsac

Department of Computer Science, Florida State University
Tallahassee, Florida 323204-4530
{burmester,levan,yasinsac}@cs.fsu.edu

Abstract. Many ad hoc routing algorithms rely on broadcast flooding for location discovery or more generally for secure routing applications, particularly when dealing with Byzantine threats. Flooding is a robust algorithm but, because of its extreme redundancy, it is impractical in dense networks. Indeed in large wireless networks, the use of flooding algorithms may lead to a *broadcast storm* in which the number of collisions is so large that we get system failure. Further reducing unnecessary transmissions greatly improves *energy efficiency* of such networks. Several variants have been proposed to reduce the relay overhead either deterministically or probabilistically. Gossip is a probabilistic algorithm, in which packet relaying is based on the outcome of coin tosses. The relay probability can be *fixed*, *dynamic* or *adaptive*. With dynamic Gossip, local information (local connectivity) is used. With adaptive Gossip, the decision to relay is adjusted adaptively based on the outcome of coin tosses, the local network structure and the local response to the flooding call. The goal of gossiping is to minimize the number of relays, while retaining the main benefits of flooding, i.e., effective distance.

In this paper we consider ways to reduce the number of redundant transmissions in broadcast flooding while guaranteeing security. We present several gossip type protocols, which exploit local connectivity and adaptively correct local relay failures. These use a (geodesic) cell based approach and preserve *cell-distance*. Our last two protocols are non probabilistic and guarantee delivery, the first such protocols to the best of our knowledge.

Keywords: Ad hoc networks, secure MANETS, flooding, Gossip, broadcast redundancy, broadcast storms, secure routing.

1   Introduction

Ad Hoc networks are self-organizing wireless networks, absent of any fixed infrastructure [9,18,11]. Nodes in such networks communicate through wireless transmissions of limited range, sometimes requiring the use of intermediate nodes to

---

* This material is based on work supported in part by the U.S. Army Research Laboratory and the U.S. Research Office under grant number DAAD19-02-1-0235.

I. Nikolaidis et al. (Eds.): ADHOC-NOW 2004, LNCS 3158, pp. 96–107, 2004.

reach a destination. Nodes in ad hoc networks are limited in their power supply and bandwidth. Node mobility further complicates the situation.

Two primary issues in ad hoc network research are efficiency and security. Because of their nature, efficiency is essential in ad hoc networks. Also naturally, ad hoc networks are more vulnerable to security risks than fixed, wired networks. Unfortunately, efficiency and security are also competing properties, in that improving efficiency is likely to reduce security and efforts to increase security are likely to negatively impact efficiency. The security, efficiency trade-off in ad hoc networks is the focus of this paper.

Routing in ad hoc networks is an active area of research [4,5,6,19,17,9,18]. The de facto route discovery algorithm for such networks is flooding [1,2,3,17]. With broadcast flooding, each node that receives a message retransmits that message exactly once. Flooding [17] has many positive properties for ad hoc networks including maximal coverage, distance preservation, and redundancy. Maximal coverage means that if a time-relevant path[1] exists between a source and any destination, flooding will discover that path. Distance preservation is the property that, since it discovers every path, flooding will always find the shortest path between the source and destination. Redundancy is a positive attribute in ad hoc networks because ad hoc networks are naturally less reliable, and less secure, than their static counterparts.

Conversely, many seek to replace flooding as the ad hoc routing algorithm of choice because of its inefficiency that is directly related to redundancy [13, 10,20,20]. In fact, in dense networks, the redundancy may be catastrophic if a broadcast storm [15] is triggered. A solution to the broadcast storm problem is to reduce message propagation. This is the approach taken in probabilistic retransmission protocols, also referred to as gossip [9,20,8] protocols.

Gossip is similar to flooding, with one important distinction. In gossip, when a node receives a message for the first time, rather than immediately rebroadcast as in flooding, it engages a probabilistic process to determine whether or not to retransmit. Essentially, it rebroadcasts each message with probability p.

From a security point of view, the gossip approach has undesirable properties. Chief among them is that gossip gives non-faulty nodes the chance to forego participation, while allowing faulty, or Byzantine, nodes undue influence in the routing process. Thus, nodes that may be highly reliable and efficient in a fair environment, are prone to failure in the face of Byzantine attacks. Fortunately, mild adaptations of gossip can offer substantial security enhancement at limited efficiency cost.

The rest of this paper is organized as follows. In Section 2 we explain our approach and present a basic gossip protocol. We define our cell-grid and the concept of cell-to-cell propagation. In Section 3 we present four basic gossip protocols which correct failures of the basic protocol adaptively by using local

---

[1] Since ad hoc networks are dynamic, a path may form or dissolve during the flooding process. Whether the flooded message finds nodes involved is time dependent. Nodes that are connected before the "flood passes" will receive the messages, nodes disconnected at "flood passage" will not. Heretofore, we do not address dynamic path routing algorithms optimizations, per se.

neighbor information. In Section 5 we present an adaptive gossip protocol that uses cell-grid information. Finally, in Section 6 we discuss security issues. We conclude in Section 7.

2   The Cell-Grid and Basic Gossip Protocols

Our goal in this section is to find gossip protocols that minimize message propagation while retaining some of the basic features of flooding: distance information, maximal coverage and redundancy.

We first define a cell-grid and show how it can be used for gossiping. We are only concerned with large dense networks for which the redundancy in flooding may cause a broadcast storm. We shall assume that the density is evenly distributed, in particular that no parts of the network are sparce. Finally, for simplicity, we assume that all nodes of the ad hoc network have the same range: one hop. This will be our unit of measurement.

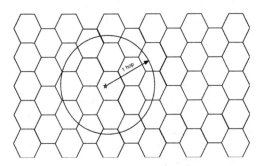

Fig. 1. The cell-grid and a node with its broadcast range.

A *cell-grid* is a covering (or tiling) of the Euclidean plane with regular hexagons, or *cells* – as shown in Figure 1. The cells are the basis for message propagation in our protocols. In these protocols, at least one node from each cell will be active and propagate the message, resulting in cell-to-cell propagation. Effectively, our approach reduces node-to-node flooding to cell-to-cell flooding. To minimize the number of propagations we must choose the size of each cell to be maximal subject to faid-out. Therefore, for cell-to-cell propagation, we choose the grid size so that the maximum distance between any two points of two adjacent cells must be no more than 1 hop, since in the worst case, there may only be nodes on the boundary of the grid.

Let ℓ be the length of an edge of the regular hexagon cell in hops. The maximum distance between two adjacent cells is:

$$\ell \sqrt{(2\sqrt{3})^2 + 1} = \ell\sqrt{13} = 1 \text{ hop,}$$

as shown in Figure 2. Since we want this distance to be bounded by one hop, we have $\ell = 1/\sqrt{13}$. Then the area of a cell is: $2\ell\sqrt{3} = 2\sqrt{3/13}$ hop squared, which

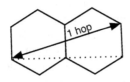

Fig. 2. The maximum distance between points of an adjacent pair of cells is 1 hop.

is roughly $1/5$ of a hop squared, or $1/5\pi$ of a hop circle.[2] We will apply this observation shortly. We shall also assume that the set of nonempty cells form a connected region.

2.1   Cell-Based Gossip

Our first gossip protocol is non-adaptive. It aims at ensuring that at least one node from each cell retransmits message. If there are r nodes in a cell this would be achieved, on average, if each cell node were to broadcast the message with probability $p = 1/r$. Obviously if there are no nodes in a cell then the network is locally sparse and our protocols will fail. Since the selection is probabilistic, there is a possibility of *propagation failure*. This is roughly: $(1 - \frac{1}{r})^r \sim e^{-1}$, for large r. To reduce this we can use a larger message propagation probability, say $p = k/r$, where k is a small integer. In this case the probability of failure will be approximately e^{-k}. Observe that for a cell of the cell-grid to be excluded, there must be propagation failure in all six of its neighbor cells. This reduces even further the propagation failure to e^{-6k}. While this may appear to restrict the applicability of our protocols, we contend that the primary target of ad hoc network applications is dense networks, for which our protocols are designed.

The easiest way to approximate r is to assume a lower bound for the density of the network. Suppose that the minimum degree is n_0 nodes per hop squared. We must have at least one node per cell, so in this minimal configuration, $r = n_0/5$, and therefore $p = 5/n_0$. The first gossip protocol we describe is that presented by Haas-Halpern-Li [10]. The input is: p, s, m, where p is the (fixed) propagation probability, s is the source and m is the message.

Gossip1. $(n_0, k; s, m)$

Node s broadcasts m.

FOR EACH node x that receives m for the first time DO
    broadcast m with probability $p = 5k\pi/n_0$.

2.2   Dynamic Cell-Based Restransmission

A more dynamic way to approximate r is to compute it based on the number of neighbors of a node (its degree). In this case, for node x, $r_x = d_x/5\pi$, where d_x

---

[2] Hop circle is the area of a circle with one hop radius.

is the degree of x (the area of a cell is $1/5\pi$ of a hop circle). Of course now nodes need to now the number of their neighbors. This can be achieved by having all nodes make short "hello" calls at regular intervals.

We now describe our second gossip protocol, which is an extension of the protocol in [10]. Let N be the node set of the ad hoc network, p_0, $0 < p_0 \leq 1$, be a constant and id_x an identifier for node x.

Gossip2. $(N, k, p_0; s, m)$

[FOR EACH node x DO:
  periodically broadcast "hello, id_x" with probability p_0
  compute $d_x = p_0^{-1} \times \#\{\text{hello's received during one time-period}\}$][3].
Node s broadcasts m.
FOR EACH node x that receives m for the first time DO
  broadcast m with probability $p_x = 5k\pi/d_x$.

In this protocol the number of hello calls has been reduced by using the probability p_0. This protocol takes into account the local density and therefore will reduce the probability of propagation failure in networks whose density is not uniform, or for which the given lower density bound is too low. However we may not get adequate coverage, particularly if the network diameter is large. For this goal, failures must be adaptively corrected.

Theorem 1. *Let c be the number of nonempty cells in network.*

1. *The success probability of Gossip1 is at least $1 - ce^{-k}$.*
2. *The success probability of Gossip2 is at least $1 - (1 + c)e^{-k/(1+\epsilon)}$, where $\epsilon = \sqrt[3]{k/2n_0p_0^2}$.*

Proof. Let n_x be the number of neighbor nodes of x. Because nodes are distributed evenly in the local area, the number of nodes in the same cell with x is approximately $n_x/5\pi$. Further, the probability that a neighbor node y of x will rebroadcast a gossip of m is $p = 5k\pi/n_0$, so the probability that no neighbor node y in the same cell with x will gossip in the first protocol is:

$$(1 - p)^{\frac{n_x}{5\pi}} = \left(1 - \frac{5k\pi}{n_0}\right)^{\frac{n_0}{5k\pi}\frac{kn_x}{n_0}} < \left(\frac{1}{e}\right)^{\frac{kn_x}{n_0}} < e^{-k}.$$

Therefore the probability that there exists a cell with no gossipy node inside is at most $ce^{-k} = e^{-k+\ln c}$. Consequently, with probability at least $1 - e^{-k+\ln c}$, every cell has a gossipy node. By our assumption, all the cells are connected. Thus, the message m will be gossiped to all cells. That means Gossip1 will succeed with probability at least $1 - e^{-k+\ln c}$, or the probability that it will fail is negligible.

In Gossip2, the probability that $d_x > n_x(1 + \epsilon)$ is at most $e^{-2n_x(p_0\epsilon)^2}$ for all $\epsilon > 0$ (Okamoto bound [16]). Therefore, using the same arguments as above, we can show that Gossip2 will succeed with probability at least $1 - e^{-k/(1+\epsilon)+\ln c} -$

---
[3] Hello calls are sent and received in the background.

$e^{-2n_0(p_0\epsilon)^2}$. Taking $\epsilon_0 > 0$ such that $\epsilon_0^2(1+\epsilon_0) = k/2n_0p_0^2$, we obtain the success probability of Gossip2 is at least

$$1 - (1+c)e^{-k/(1+\epsilon_0)} < 1 - (1+c)e^{-k/(1+\sqrt[3]{k/2n_0p_0^2})},$$

where $\epsilon_0 < \sqrt[3]{k/2n_0p_0^2}$, and the probability of failure is negligible.

3   Adaptive Gossiping

In this section, we present three gossiping protocols that adaptively correct propagation failures in Gossip2 by using readily available information. In particular, we consider a variation of Gossip2, in which we avoid "Hello" calls by using random broadcast delay. In the second protocol, we assume that nodes can measure the signal strength of the received signals. This information is used to decide whether a node should rebroadcast the message in order to maximize the coverage. In the third protocol, in addition to strength, nodes can also obtain the relative direction of the broadcasting source. This additional information helps reduce the failure propagation probability and network congestion while maintaining coverage. Notice that each of these approaches is essentially stateless, which is a primary feature of flooding.

3.1   Adaptive Gossip

Our first adaptive Gossip variation relies on probabilistic delay generation to serialize message retransmissions and extends the protocols in [15,10]. In this protocol, when each node hears a message for the first time, it generates a wait period, randomized within an a priori selected given limit, e.g. between one and five seconds. The node then waits for the the selected amount of time, counting the number of times that it hears the subject message retransmitted. If the counter meets the retransmission threshold before the wait period ends, the node does not retransmit and discards all further copies of the message. If the desired number of retransmissions has not been received, the node retransmits the message and discards all further copies.

AdaptiveGossip $(k; s, m)$

Node s broadcasts m.
FOR EACH node x that receives m for the first time DO
    delay at random within the contention time.
    IF number of received gossips of m is less than $5k\pi$ THEN broadcast m.

Theorem 2. *The success probability of AdaptiveGossip is at least $1 - ce^{-k}$.*

Proof. From the protocol, it is clear that in any neighborhood of a node x, there will be at least $5k\pi$ gossipy nodes (provided $5k\pi \leq n_x$). Since nodes in a local neighborhood are distributed randomly, the probability that a given neighbor y

of x will be in the same cell with x is $\frac{1}{5\pi}$. Therefore the probability that there are no gossipy nodes in the same cell with x is at most

$$(1 - \frac{1}{5\pi})^{5k\pi} = ((1 - \frac{1}{5\pi})^{5\pi})^k < e^{-k}.$$

Consequently, the success probability of AdaptiveGossip is at least $1 - ce^{-k}$.

3.2   Signal Strength

We would like to ensure that at least one node in every cell will propagate the message. In this protocol, the signal strength information is used to estimate whether the sender and receiver are in the same cell. If this is the case then propagation is not needed. The estimation is obtained by calculating the probability that the sender is in the receiver's cell given his signal strength. The protocol is given below.

Let g_m be a received gossip of m, and $sigstrength(g_m)$ be the signal strength of g_m, taking values in the range $[0, 1]$, and $S_0 \in [0, 1]$ is a signal strength.

AdaptiveSignalStrengthGossip $(S_0, k; s, m)$

Node s broadcasts m.
FOR EACH node x that receives m for the first time DO
    delay at random within the contention time.
    IF number of g_m such that $sigstrength(g_m) < S_0$ is less than k THEN
broadcast m.

Lemma 1. *Let $h \in [0, l]$ be a positive real number. Then the probability $p(h)$ that "a randomly chosen node y of distance h hops to x is not in the same cell with x" is at most $\frac{2}{3}\frac{h}{l}(2 - \frac{h}{l})$.*

Proof. We consider two cases. First, if x has distance at most h from the boundary of the cell, then the probability $p(h)$ is at most $\frac{2}{3}$, where the maximum achieves when x is at one of the six corners of the cell. The first case happen with probability $1 - (1 - \frac{h}{l})^2 = \frac{h}{l}(2 - \frac{h}{l})$.

Second, if x has distance more than h from the boundary of the cell, then $p(h) = 0$. Therefore overall, the probability that y is not in the same cell with x is at most $\frac{2}{3}\frac{h}{l}(2 - \frac{h}{l})$.

Theorem 3. *The success probability of AdaptiveSignalStrengthGossip is at least*

$$1 - (\frac{2}{3}f_0(2 - f_0))^k, \quad where \ f_0 = \frac{h(S_0)}{l}.$$

In particular, the failure probability is at most $(\frac{2}{3}f_0(2 - f_0))^k \leq (\frac{2}{3})^k$.

Proof. From the protocol, for each *non*-gossipy node x, there will be at least $5k\pi$ gossipy nodes y that are of distance at most h_0 from x. From Lemma 1, the probability that each such node y is not in the same cell with x is at most $\frac{2}{3}\frac{h_0}{l}(2 - \frac{h_0}{l})$. In total, the probability that all these nodes y are not in the same cell with x is at most $(\frac{2}{3}\frac{h_0}{l}(2 - \frac{h_0}{l}))^k$.

In the following adaptation, we would like to ensure that at least one node in every cell will propagate the message. Signal strength information is used to estimate whether the sender and receiver are in the same cell. If this is the case then propagation is not needed. The estimation is obtained by calculating the *probability* that the sender is in the receiver's cell given his signal strength. The protocol is given below.

AdaptiveVariableSignalStrengthGossip $(k; s, m)$

Node s broadcasts m.
FOR EACH node x that receives m for the first time DO
    delay at random within the contention time.
    let $v_x = \sum_{g_m} -\ln p * (sigstrength(g_m))$, where g_m is any received gossip of m.
    IF $v_x < k$ THEN broadcast m.

Theorem 4. *The success probability of AdaptiveVariableSignalStrengthGossip is at least* $1 - ce^{-k}$.

Proof. For each node x, the probability that a received g_m was broadcast from outside the cell of x is $p(sigstrength(g_m))$. So the probability that no no in the cell of x retransmits the message is at most $\prod_{g_m} p(sigstrength(g_m)) = e^{-v_x}$. If x does not retransmit m itself then $v_x \geq k$, thus the failure probability is at most $ce^{-v_x} \leq ce^{-k}$.

3.3   Summary

Adaptive gossip allows nodes to determine retransmision non-deterministically based on local observations. This allows wireless ad hoc networks to avoid failure results from broadcast storms, and to improve their energy efficiency greatly. Instead of using probabilistic coin flippings to reduce collisions as done in many other gossip protocols [9,20,8], our protocols use random contention time to avoid them. Additionally, by taking advantage of signal strength information readily available at almost no cost to the local hosts, our protocols guarantee full network coverage in each broadcasts with exponentially small chance of failure. In the next sections, we will see how to further eliminate this chance effectively.

4   Signal Direction

In the preceeding section, gossiping may fail to reach some nodes. While this small probability of failure is acceptable in certain cases, for other scenarios, such as those in route finding and broadcasting of control information, a guaranteed broadcast protocol is often desired. In the next sections, we analyze possible ways to address this problem, at the same time avoid broadcast storms caused by flooding.

We now augment our signal strength protocol with direction information to assure a higher success rate for our broadcast protocol. Essentially, the nodes listen to see if the target message has sufficiently propagated without their par-

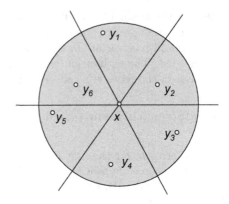

Fig. 3. The broadcast area of x is completely covered by $y_1, \ldots, y_6$.

ticipation. If not, they retransmit. More specifically, each intermediate node will gossip if and only if, after a random time period, it has not received gossips from all directions. Let $dir(m)$ be the direction of the source of a received message m, taking integer values in range $[1..6]$. The nodes perform the following protocol.

AdaptiveSignalDirectionGossip $(k; s, m)$

> Node s broadcasts m.
> FOR EACH node x that receives m for the first time DO
> > delay at random within contention time.
> > FOR EACH direction t in $\{1, \ldots, 6\}$ DO
> > > IF there is no g_m received from direction t THEN
> > > broadcast m and stop.

Theorem 5. *Protocol AdaptiveSignalDirectionGossip always succeeds.*

Proof. Divide the area around node x into 6 angles, 60° each. Let A_0 be such an angle. If x does not retransmit then there is at least one node y in A_0 that retransmits. Let $A_0' \subset A_0$ be the part of A_0 that is within distance one hop from x, i.e. A_0' is one-sixth of the unit circle around x. We see that every node in A_0', when broadcasting, would cover A_0' completely; and y is such a node in A_0'. Thus x does not need to retransmit: its broadcast area is already covered. The broadcast area of x is illustrated in Figure 3. Therefore, if the flooding covers the whole network area then the reduced flooding also covers the whole network area completely.

5   Geodesic Gossiping

In this section, we will look at geodesic based gossiping protocol. In contrast to location-aware networks [7,14], Our protocol is lightweight. It uses only local location information and still preserves the privacy of mobile nodes.

In the following protocol, we assume that each node x can obtain its cell identifier cid_x, where the cell is a regular hexagon of radius l, which is also the edge length. To avoid malicious attacks, we also require that the location cid_x is obtained from a *tamper-proof* blackbox given to x that can sign the location information cid_x.[4]

AdaptiveGeodesicGossip (s, m)

Node s broadcasts (cid_s, m).
FOR EACH node x DO
    delay at random within contention time.
    IF no valid gossip $g_m = (cid_x, m)$ is received THEN
    broadcast (cid_x, m).

Theorem 6. *Protocol AdaptiveGeodesicGossip always succeeds.*

Proof. We see that:

1. By assumption, all cells are connected.
2. Each cell will have at least one transmission since all nodes will retransmit unless another node within their cell transmits.
3. The message will propagate since the hop length spans 2 cells.

6   Security and Efficiency Issues

We have proposed five new adaptive broadcasting protocols, namely

- AdaptiveGossip
- AdaptiveSignalStrengthGossip
- AdaptiveVariableSignalStrengthGossip
- AdaptiveSignalDirectionGossip, and
- AdaptiveGeodesicGossip.

The first three protocols use redundancy to probabilistically guarantee message transmission. These offer exponentially small failure probability in k, by using linear redundancy in k. In the last two protocols, message delivery is always *guaranteed* while minimizing redundancy. In Figure 4, λ is the relative frequency of hellos versus broadcast requests. The transmission rate is the fraction of nodes in the neighbor of x that retransmit.

The figures for the adaptive protocols are accurate even when there are *malicious* nodes. Note that while our protocols are designed to prevent broadcast storms under reasonable circumstances, we do not consider all-out denial of service attacks [12] in this paper. Rather, we assume that denial of service is handled by choke points at the physical level.

---

[4] To avoid replay attacks, the location information should include precise time of measurement. A signed cid_y should be regarded as *unsigned* if it was signed τ time in the pass, where τ is the upper bound on transmission time of a message in one hop. If privacy is a concern then encryption should be used. All these requirements can be done, for example, by using a one-way keyed hash function with a common secret key. For example, $cid_x := AES_{sk}(cell - location(x), time)$ where $time$ is properly rounded.

Protocol	Failure Rate	Transmission Rate	Notes
Gossip1 [10]	ce^{-k}	$\frac{5k\pi}{n_0}$	$n_0 = \inf_x n_x$
Gossip2	$(c+1)e^{-k/(1+\epsilon)}$	$\frac{5k\pi}{n_x} + \lambda p_0$	$\epsilon = \sqrt[3]{k/2n_0 p_0^2}$
AdaptiveGossip	ce^{-k}	$\frac{5k\pi}{n_x}$	
SignalStrengthGossip	$1 - (\frac{2}{3}f_0(2-f_0))^k$	$\frac{k}{f_0^2 n_x}$	$0 < f_0 < 1$
VariableSignalStrengthGossip	ce^{-k}	$\frac{4k}{n_x}$	
SignalDirectionGossip	0	$\frac{6}{n_x}$	
GeodesicGossip	0	$\frac{5\pi}{n_x}$	

Fig. 4. A comparison of the security features of the proposed gossip protocols.

7 Conclusion

In this paper, we have identified an approach for controlling redundancy in transmission protocols in ad hoc networks. We have mathematically shown the negative impacts of redundancy on ad hoc network bandwidth and how the redundancy can be controlled. Specifically, we give mechanisms that allow network managers the ability to trade off redundancy and its resulting overhead to provide delivery reliability and we show how security issues are addresses with this controlled redundancy.

Our approach is founded on the concept of grid-to-grid message passing. Essentially, if we construct a comprehensive tiling of the network area, message redundancy and volume can be tuned to meet the demands for reliability and security. We give protocols to accomplish these objectives and give proofs of theorems related to the security properties of those protocols.

Finally, we show how density is the dominant factor in controlling redundancy in dynamic networks.

References

1. B. Awerbuch, D. Holmer, C. Nita-Rotaru and H. Rubens, *An On-Demand Secure Routing Protocol Resilient to Byzantine Failures*, ACM Workshop on Wireless Security (WiSe'02), September 2002.
2. E.M. Belding-Royer and C.-K. Toh, *A review of current routing protocols for ad-hoc mobile wireless networks*, IEEE Personal Communications Magazine, pages 46-55, April 1999.
3. J. Broch et al, *A performance comparison of multi-hop wireless ad hoc network routing protocols*, Proc. ACM MOBICOM, pp. 85-97, 1998.
4. Mike Burmester and Tri van Le. *Tracing Byzantine faults in ad hoc networks*, Proceedings, Computer, Network and Information Security 2003, New York, December 10-12, 2003.
5. Mike Burmester and Tri van Le. *Secure Multipath Communication in Mobile Ad hoc Networks*, International Conference on Information Technology, Coding and Computing, Las Vegas Nevada, April 5-7, 2004.
6. Mike Burmester and Alec Yasinsac, *Trust Infrastructures for Wireless, Mobile Networks*, WSEAS Transactions on Telecommunications, January 2004.

7. S. Capkun, M. Hambdi and J. Hubaux, *Gps-free positioning in mobile ad hoc networks*, Proceedings of Hawaii Int. Conf. on System Sciences, Jan 2001.

8. J. Cartigny and D. Simplot, *Border Node Retransmission Based Probabilistic Broadcast Protocols in Ad-Hoc Networks*. In Proc. 36th International Hawaii International Conference on System Sciences (HICSS?03), Hawaii, USA. 2003.

9. S. Corson and J. Macker, Mobile Ad hoc Networking (MANET): *Routing Protocol Performance Issues and Evaluation Considerations*, RFC2501, January 1999.

10. Z.J. Haas, J.Y. Halpern and L. Li. *Gossip-based ad hoc routing*, in Proc. of INFOCOM'02, 2002, pp. 1707–1716.

11. D.B. Johnson and D.A. Maltz, *Dynamic Source Routing in Ad-Hoc Wireless Networks*, Mobile Computing, ed. T. Imielinski and H. Korth, Kluwer Academic Publisher, pp. 152–181, 1996.

12. Karpijoki, V. *Signalling and Routing Security in Mobile Ad Hoc Networks*, Proceedings of the Helsinki University of Technology, Seminar on Internetworking - Ad Hoc Networks, Spring 2000.

13. B. Karp and H. Kung, *Greedy Perimeter Stateless Routing for Wireless Networks*, Proceedings of the 6th International Conference on Mobile Computing and Networking, Boston, USA, 2000, 243-254.

14. Y.B. Ko and N.H. Vaidya, *Location-Aided Routing in Mobile Ad Hoc Networks*, Proceedings of ACM/IEEE MOBICOM '98, October 1998.

15. S-Y. Ni, Y-C. Tseng, Y-S. Chen, and J-P. Sheu. *The broadcast storm problem in a mobile ad hoc network*. In Proceedings of the Fifth Annual ACM/IEEE International Conference on Mobile Computing and Networking, pages 151-162, Aug 1999

16. M. Okamoto, *Some inequalities relating to the partial sum of binomial probabilities*. Annals of the Institute of Statistical Mathematics, 10:29–35, 1958.

17. C.E. Perkins, E.M. Royer and S.R. Das. *IP Flooding in ad hoc networks*. Internet draft (draft-ietf-manet-bcast-00.txt), Nov 2001. Work in progress.

18. C.E. Perkins and E.M. Royer. *Ad hoc on-demand distance vector routing*, IEEE Workshop on Mobile Computing Systems and Applications, pp. 90-100, Feb. 1999.

19. E. Royer and C-K. Toh. *A Review of Current Routing Protocols for Ad-Hoc Mobile Wireless Networks*. IEEE Personal Communications Magazine, April 1999, 46-55.

20. Y. Sasson, D. Cavin and A. Schiper. *Probabilistic Broadcast for Flooding in Wireless Mobile Ad hoc Networks*. In Proc. of IEEE WCNC 2003.

A Secure Autoconfiguration Protocol for MANET Nodes

Fabio Buiati, Ricardo Puttini, and Rafael De Sousa

Department of Electrical Engineering
University of Brasília
Campus Universitário Darcy Ribeiro – Asa Norte – Brasília – DF BRAZIL
fabio@redes.unb.br, {puttini,desousa}@unb.br

Abstract. In this paper we propose a secure autoconfiguration model for Manet. Our design is based in a distributed and self-organization certification service, which provides node identification and authentication for the autoconfiguration protocol. We have defined some modifications in the Dynamic Configuration and Distribution Protocol (DCDP) in order to extend the protocol functionalities to include security-aware node identification and authentication services. The overall security is also enforced with intrusion detection techniques.

1 Introduction

Mobile ad hoc networks (Manet) provide a flexible way of developing ubiquitous broadband wireless access, allowing mobile networks to be readily deployed without using any previous network infrastructure. Such networks, also called spontaneous networks, are multi-hop wireless networks where a node may, at any time, disappear from, appear into or move within the network. Considering this very definition of Manet, two basic network services may be identified: routing and autoconfiguration [1].

While considerable work has been done, in the later years, in designing and standardization of Manet routing protocols, the design of Manet autoconfiguration protocols is yet in early stage. As a consequence, proposals for security enhancements to Manet routing protocols are appearing rapidly e.g. [2-6], but to our best knowledge, the literature about secure Manet autoconfiguration is rare or even inexistent. In this paper, we intend to fill in this gap, proposing a secure autoconfiguration protocol for Manet.

The design of an autoconfiguration protocol for Manet must take the self-organized and distributed approach. Following these premises, there are presently two basic approaches in the conception of Manet autoconfiguration service:

Conflict-detection allocation scheme adopts a trial and error policy. A node picks a random address and performs duplicate address detection (DAD), which requires positive response from all configured nodes in Manet. Perkins et al. [7] have taken this approach.

Conflict-free allocation scheme uses the concept of binary split, which means that configured nodes have disjoint address pools. Each configured node can independently assign addresses to new nodes arriving in the Manet. This is the case

I. Nikolaidis et al. (Eds.): ADHOC-NOW 2004, LNCS 3158, pp. 108–121, 2004.

for the Dynamic Configuration and Distribution Protocol (DCDP), proposed by A. Misra et al. [8].

Our proposition is based on the DCDP protocol [8], with the enhancements proposed by M. Mohsin and R. Prakash [9]. More precisely, in this paper, we analyze the security requirements of the basic conflict-free allocation protocol and we define some modifications in the protocol proposition [9], in order to secure it.

A central problem in the definition of secure services for Manet relates to the adequate specification of membership, allowing differentiation of the nodes that are trustable from those that are not. In our design, the definition of secure autoconfiguration relates to the specification and enforcement of Manet membership.

There are two important aspects in the above membership definition. The first one is the very definition of trust for some particular Manet. Different scenarios can arise here, going from quite open "trust everybody" policies to very restrictive "trust nobody" policies. As nodes in a Manet must rely on each other for the provision of the network service, the definition of trust can also follow this same collaborative approach. The second aspect consists in defining how the nodes could be uniquely identified in the Manet. The IP address clearly is not a good choice, especially if we are dealing with the design of the autoconfiguration service, whose main goal relates to allocation of IP addresses to uninitialized nodes. Some authors have claimed that MAC address could serve for this purpose, arguing that commercial wireless devices do not allow the modification of their built-in MAC address [10]. We do not assume that hypothesis, given that special designed wireless devices or even firmware modified on-the-shelf devices do not meet this condition, allowing MAC addresses to be tampered. Of course, IP and MAC addresses are normally used for interface identification on the behalf of network and data link layer services, respectively. However, these are not good choices for the purpose of node identification, from the security point of view.

We propose a single solution to accomplish with both aspects. Collaborative trust is achieved by adopting a "K-out-of-N" trust model (N is the non-fixed total number of nodes currently in the Manet), which means that any node is trusted by all other N nodes in the Manet if, and only if, any previously trusted K-nodes trust on it. For example, an arriving node may not be trusted in the Manet until it is trusted by some coalition of K-nodes that are already trusted in the network. In our proposal, this model is realized by means of a distributed certification service [11]. The rationale behind such system is simple: whenever a node begins to trust on some other node, it issues a partial certificate on the behalf of that node. The later can recovery his completely trusted certificate (e.g. becomes a Manet member) as it collects partial certificates from any K nodes in the Manet. The certificate itself serves for the purpose of identifying uniquely the nodes, from the security point of view.

Security is achieved in our autoconfiguration protocol by requiring that a node will only have access to and/or participate on the service if it is already trusted in the Manet, e.g. if it possesses a completely trusted certificate. Thus, distributed certificate issuing must be considered in the autoconfiguration protocol, as it must proceeds before successful completion of the autoconfiguration process. Once certified, a node must use its certificate to authenticate all autoconfiguration protocol messages. Such security solution has the following features: (1) an untrusted node is not able to attack the network by maliciously requesting autoconfiguration services to other nodes; (2)

an untrusted node is not able to disrupt the autoconfiguration service by answering maliciously to autoconfiguration services requests; and (3) a compromised node that begins acting maliciously may be identified by means of some intrusion detection mechanism, provided that the authentication of the autoconfiguration protocol messages has non-repudiation property.

The reminder of the paper is organized as follows. Section 2 presents a brief review of conflict-free allocation autoconfiguration. In section 3, we discuss our adversary and security models. In section 4, we explain our Manet certification service. Section 5 is dedicated to the design of a secure autoconfiguration protocol, using a certificate based Manet Authentication Extension (MAE). Section 6 points out some possible interaction with intrusion detection service. Related works are discussed in section 7. Finally, section 8 concludes the paper.

2   MANET Autoconfiguration

DCDP [8] is a protocol for distribution of IP configuration information, which is based on the binary buddy system model [12]. In DCDP, nodes are either requestors of or responders to individual configuration requests. All MANET nodes that are already initialized keep disjoint blocks of available IP addresses. An arriving node broadcasts an autoconfiguration request in his 1-hop neighborhood asking for network configuration information to nearby nodes. Any node receiving the request may serve it by leasing half of one of its IP addresses blocks to the requester node. This binary splitting method assures that IP addresses blocks have sizes expressed as a power of two provided the nodes are initialized accordingly.

The basic DCDP protocol is enhanced in [9] to accomplish with several concerns related to node mobility that aren't discussed in the original protocol proposal, such as node departure, crash and synchronization. The proposal in [9] also describes network partitioning and merging. Our secure autoconfiguration protocol design is based on the original DCDP protocol with the improvements proposed in [9].

We define an "arriving node" as any node wishing to join the Manet and "configured node" as any Manet node that is already initialized with at least one IP addresses block[1]. A configured node uses the first IP address of its IP addresses block for his IP address.

2.1   Node Arriving

An arriving node needs to be initialized with an IP addresses block, from where it allocate his own IP address. In DCDP such operation proceeds in five steps, as shown in Fig.1:

---

[1] As discussed further ahead in the paper, due to the K-out-of-N trust model adopted in our design, we require, at least, K nodes to be initialized in the Manet before having full availability of autoconfiguration and certification services. This bootstrap procedure may include out-of-band initialization, which is not discussed in this paper.

1. The arrived node broadcasts an *addr_request* message.
2. Upon receiving of this request, the configured node replies with an *addr_reply* message, informing the size of its biggest IP address block. It is possible to have more than one configured node replying the request.
3. The arriving node selects the configured node having the biggest IP address block size and unicasts back to it a *server_poll* message. The other replies are discarded.
4. When the configured node receives the *server_poll* message, it splits the biggest IP address block into two disjoint sets with the same size, allocating half of this IP addresses block to the arriving node.
5. Upon receiving the new allocated IP addresses block, the arriving node picks up first address to itself. It sends an IP_assignment_ok message indicating that it has been configured. The new configured node is said to be a "buddy node" for the node serving the request.

Once the process is finished, every node in Manet keeps a table with node identification and assigned IP addresses for all nodes in the network, for synchronization purposes. Buddy nodes are also marked in this table.

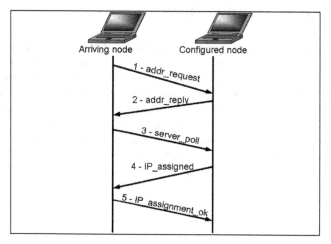

Fig. 1. IP Address assignment process

2.2 Node Departure

There are two ways for a node to leave the network:

Graceful departure: When a node wishes to leave the Manet, it broadcasts a departure_request message. On receiving this message, a configured node sends a departure_reply granting permission for this node to leave the network. Departure is acknowledged by the node leaving the network with a graceful_departure message, freeing all the IP addresses blocks that were allocated to the leaving node.

Abrupt departure: A nodes may occasionally leave the network abruptly without passing by the process above. This is the case for node crashes or failures or even if a node moves away. To avoid IP address leaks, the nodes resynchronize periodically by broadcasting its local table in hello messages. After each synchronization, the nodes must scan its local table for its buddies. If a node discovers that one of his buddies is missing in the table, it reclaims the IP addresses blocks allocated to the missing node for itself.

2.3   Network Partitioning and Merging

Network partition proceeds naturally, as the IP addresses blocks allocated to the MANET nodes forming each network partition are disjoint. However, merging of two partitions may result in address conflicts, if the partitions have been originated from different MANETs. Anytime a merging is detected (by use of *hello* messages), conflict detection and resolution is initiated. This is simply accomplished by requiring the nodes from the partition with the larger IP addresses blocks to give up their IP addresses blocks and executes autoconfiguration again.

3   Adversary and Security Model

In this paper, we focus on vulnerabilities related to fabrication (generation of false messages) and impersonation (masquerading as another node) of Manet autoconfiguration protocol messages. We do not elaborate in modification (malicious modification of messages) attacks because these attacks are likely to have minor importance in the case of the autoconfiguration protocol being considered (section 2), as all the communications take places in the 1-hop neighborhood (e.g. messages are not forwarded). There are also some cases where passive eavesdropping vulnerabilities may be considered. Additionally, trivial attacks based in resource consumption and non-cooperation are possible too. We do not elaborate in these kinds of attacks either.

3.1   Adversary Model

We define an "adversary" as any node announcing erroneous information in fabricated Manet autoconfiguration protocol messages. Also, a "target" is any node accepting and using this erroneous information. We admit that fabricated messages have valid syntax.

Adversaries may exploit any message defined as mandatory for the Manet autoconfiguration protocol. There are two basic categories of fabrication attacks against the autoconfiguration protocol:

1. Requester Attacks: In this case, the adversary fabricates messages requesting autoconfiguration services. As example, an adversary may either request address allocations, making unavailable to allocation for correct nodes all IP addresses in the address pools that are being assigned; or request liberation of some address pools (possibly impersonating some other nodes), resulting in future relocation of those addresses that are not actually free to other nodes.
2. Server Attacks: In this case, the adversary fabricates messages answering requests from other nodes in the Manet. As examples, an adversary can answer an

addr_request message (see Fig.1) telling that he is the node with the biggest addresses block, forcing the requester to select it as a server for the address allocation and, then he provides no answer to *server_poll* messages, denying the autoconfiguration service to the arriving node. Alternatively, the adversary can serve a request with addresses that are already assigned to other Manet nodes, resulting in conflicting allocation of some addresses.

Whenever launching these attacks mentioned above, an adversary may exploit some vulnerabilities related to the particular features of Manets:

1. promiscuous nature of the wireless link – an adversary is able to promiscuously listen to wireless transmissions coming from its neighbors;
2. peer-to-peer communication model – an adversary may communicate directly with any node within the transmission range of its wireless interface; and
3. mobility – an adversary can move away, with limited speed, to gather information about other far away nodes.

Moreover, Manet nodes have a non-negligible probability of compromise due to vulnerabilities related to OS, software bugs, backdoors, viruses, etc. Also, a mobile node without adequate physical protection is also prone to being captured [13]. Although we do not elaborate on such vulnerabilities, we admit that an adversary may be able to compromise or capture a mobile node. We do not restrict the consequences of a node break-in. Thus, during break-in, any secret information (including private or shared keys) stored locally may be exposed to the intruder. Any broken node may be either used to launch attacks or impersonated. As there is no practical way to distinguish between these situations, we do not differentiate compromised nodes from adversaries, from the security point of view. Finally, we admit that multiple attackers can coexist in the network and may collaborate on the purpose of system break-ins.

3.2 Security Requirements

The first line of defense for the autoconfiguration protocol is related to differentiating between trusted and untrusted nodes. We require that only trusted nodes are able to participate in the autoconfiguration service. Thus, untrusted nodes won't be able to generate neither Requester nor Server Attacks.

Also, given the self-organized nature of Manets, we also require the trust relations to be collaboratively established and maintained.

Finally, as stated in the later section, we do not neglect the probability of node break-ins. This means that we must deal with such occurrences, in the security model. Indeed, whenever a node is compromised it becomes untrusted and must be excluded from the autoconfiguration service. Actually, it may be difficult to detect when a node has been compromised.

However, concerning the security of the autoconfiguration service, a compromised node must actively fabricate fake autoconfiguration messages, in order to disrupt the service (e.g. requester and server attacks). If we are able to detect such misbehaving actions and the nodes originating the fake messages, we can use such mechanism to declare the misbehaving node untrusted. Thus, we require:

1. autoconfiguration protocol messages to be authenticated with non-repudiation, which binds the messages with the node originating them;

2. correct nodes to be able to detect misbehaving actions (message generation) against the autoconfiguration service;
3. correct nodes to combine message non-repudiation and misbehavior detection to generate accusations against adversary nodes.

Finally, we must also require this detection and accusation mechanism to be collaboratively executed, avoiding a single compromised node from generating accusations against correct nodes.

3.3 Security Model

In this section we present a security model used to accomplish with the security requirements defined above. The designed security model is shown in Fig.2.

As briefly discussed in section 1, differentiating between trusted and untrusted nodes is basically a matter of membership specification and enforcement, which must be collaboratively defined. We adopt in our security model a distributed certification service in order to deal with both requirements. Such service imposes a "K-out-of-N" trust model, where a node is trusted by all other N nodes in the Manet if and only if it is trusted by at least K different nodes that are already trusted in the Manet. Each node runs an instance of a Local Certification Service (L-Cert), which collaborates with L-Certs placed in other Manet nodes in order to provide the distributed certification service.

All autoconfiguration protocol messages are authenticated with non-repudiation by a Manet Authentication Extension (MAE) [11], which is attached to each message, allowing unique and undeniable identification of the message originator. Authentication can be readily verified, provided the certificate of the message originator is locally available or supplied with the message.

Finally, a Local Intrusion Detection and Response System (L-IDS) provide misbehavior identification. The detection tasks are also collaboratively executed. Whenever an intrusion is detected, an alert is generated to the Local Certification Service, which executes collaborative certificate revocation, completing the collaborative detection and accusation mechanism. As we discuss in section 4, certificate revocation is done by counter-certificates, which are signed against the

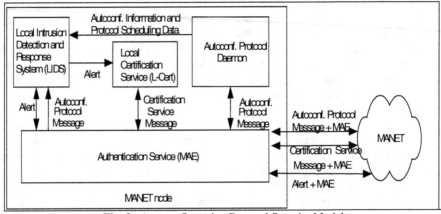

Fig. 2. Autoconfiguration Protocol Security Model

certificate being revoked. Certificate revocation also follows the "K-out-of-N" model, the certificate being effectively revoked whenever K different Manet nodes have signed an accusation (e.g. a partial counter-certificate) against it.

4 MANET Certification Service

The design of self-organized certification services for Manet has been discussed in a few recent papers [13,14,15], which are based in a distributed certification authority (DCA) trust model. The distribution of the CA capabilities is achieved by sharing the CA private key (KCA) among the network nodes by means of threshold cryptography [16]. Each Manet node has an active instance of a Local Certification Service (L-Cert) that holds a CA private-key-share (SKCA). Any K of these L-Cert may collaboratively assure the CA functions. The KCA, however, is not recoverable by any node.

Coalitions of K L-Cert are dynamically established to provide certification services, which include certification issuing, renewal and revocation.

Our certification service is adapted from previous work [13,14,15]. Certificate issuing, renewal and revocation, as well as secret share issuing and updating are directly taken from them. We have proposed policy based certificate issuing and renewal, local certificate cache and CRL, and usage with multiple DCA [11].

4.1 Certificate Issuing and Renewal

A node without certificate or needing to renew his certificate must ask to other nodes in the Manet for a certificate issuing. The authentication policy must specify how the nodes receiving a certificate request serve such request. Different criteria can be specified for certificate issuing and renewal policy. Possible policy options are: (1) serve according to some policy-specified identity check; (2) serve manually (some MANET user/manager must be prompted to decide if certificate request can be served); (3) deny (certificate request are rejected and an error message is returned to requester); (4) reject (certificate request is silently discarded); and (5) other (user defined).

Technically, there is no difference between certificate issuing and renewal, but the policies for such services can be specified separately. Whenever a node needs to receive a new certificate, he locally prepares a certificate request, which should contain the required identity information (as specified in the certification policy) and his public key. If the node is requesting a certificate renewal, the old (but not expired) certificate is also sent with the certification request. Any private-key-share holder receiving the certification request may answer the request. If a valid certificate is found along with the certification request, the request is treated as a certificate renewal. Otherwise, the request is viewed as a new certificate-issuing request. The appropriate policy is applied. After applying the certification policy, the node can decide about signing the certification request with his private-key-share, generating a partial certificate request (as the signed certificate is not yet signed with K_{CA}), which should be unicasted back to the requester (Fig.3). Whenever receiving K valid partial certificates, the requesting node can compute his new certificate.

Fig.3 illustrates the three basic phases in certification issuing/renewal, which are fully executed in the local 1-hop neighborhood, provided that the requester have, at least, K 1-hop neighbors [11]:

1. the L-Cert locally broadcasts a service request message;
2. each L-Cert receiving the request applies the security policy to decide if the request can be served and answer the request accordingly, unicasting a service response message back to the requester;
3. L-Certs combine any K responses to complete the process.

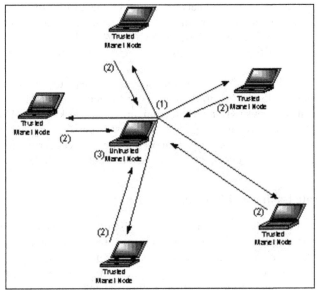

Fig. 3. Basic Certification Service Protocol

4.2   Certificate Revocation and CRL

Certificate revocation is done by signing a counter-certificate with KCA. The decision for signing a counter-certificate should be taken in two different cases: self-revocation and intrusion detection revocation. Self-revocation is done when any node decides to revoke his own certificate (e.g. due to the exposure of its private key). To do that, the node broadcasts a self-signed counter-certificate in his neighborhood. Any node receiving a self-signed counter-certificate generates a partial counter-certificate by signing the original counter-certificate with their own SKCA. Intrusion detection revocation is more complex because in this case one node asks for the revocation of other node's certificate. When a node detects other misbehaving/compromised node, he creates a partial counter-certificate for the compromised node by signing a counter-certification with his SKCA. Whenever K nodes detect the same node in the network as a compromised one, K partial counter-certificates should have been generated. In both self-revocation and intrusion detection revocation, the partial counter-certificates being generated are immediately flooded into the network. Any node receiving K partial counter certificates can recover the KCA signed counter-certificate, which is also flooded in the network. Any node receiving/recovering a signed counter-certificate must store the certificate in his local CRL. Counter certificates are maintained in the CRL while the current (t_now) time is lesser than the revoked certificate expiration time (t_exp), e.g. t_now < t_exp.

5   MAE for MANET Autoconfiguration Protocol

Our proposal for the autoconfiguration protocol has been derived from [9], with the following modifications:
1. a node needs to have a trusted certificate before having access to the autoconfiguration service. We use this certificate for the purpose of node identification; and
2. all messages have an attached Manet Authentication Extension (MAE), which carries the authentication information and the certificate of the node originating the message.

The original syntax of the protocol messages [9] have been modified as there is no need to have an "ID" field included in the message, the meaning of this filed being replaced by the sender's certificate, which is included in the MAE. The new syntaxes for the protocol messages are presented in the Appendix A.

We assume that the node has been successfully configured with its certificate before requesting and participating in the autoconfiguration service. The certificate can be either setup off-line or it can be dynamically requested. We present in section 5.2 a proposal for execution of certificate request by nodes that are not initialized (e.g. a node that does not have an IP address).

5.1   Authentication of Autoconfiguration Protocol

Authentication of the autoconfiguration service is provided by a Manet authentication extension (MAE), which is appended to each protocol message. This MAE contains all the authentication information required to correctly assure authenticity, integrity and non-repudiation to the message being protected. Actually, the MAE used for the autoconfiguration protocol follows the same format of the MAE that we have previously defined for securing of the Manet routing service [11]. MAE syntax is shown in Appendix B.

MAE is composed by authentication objects. At least one (mandatory) authentication object should be present in the MAE and contains a digital signature (DS) object, which authenticates all non-mutable fields of an autoconfiguration message. The message originator must sign the DS with its private key. The corresponding public key is bond with the originator identity in the certificate, which must available for verification of the signature in the nodes that receives the message. If node certificates are not locally available (certificate distribution is discussed the next section) MAE may also contains a certificate object (CERT), carrying the certificate of the MAE signer along with the message.

5.2 Obtaining a Certificate Before Autoconfiguration

We have shown in section 4.1 that certificate issuing (and renewal) occurs in a two-way handshake protocol, usually executed in the local 1-hop neighborhood. An uninitialized Manet node without valid a certificate must execute this certificate issuing process before being able to apply for autoconfiguration services. In this section, we propose a simple procedure for allowing request for certificate issuing without needing previous access to autoconfiguration service.

The procedure solution consists of four steps:

1. The uninitialized node randomly chooses a temporary IP address and makes the certificate issuing request using this address. The request is broadcasted in the local 1-hop neighborhood only.
2. The nodes receiving the request (neighbor nodes) decides to serve the request according to the certificate issuing policy and normally unicast the partial certificate signed with SK_{CA} back to the requester (using the IP address provided with the certificate issuing request message). The TTL of the IP packet may be set to 1, assuring that the packet containing the response message won't be forwarded any further from the local neighborhood.
3. The requester promiscuously listens to certificate issuing responses from its neighbors (as the responses may not have the requester MAC address as destination, since it is possible that the address randomly chosen by the request (step 1) is duplicated in the Manet.
4. After collecting any different K certificate issuing response messages, the node recovers its certificate and starts the autoconfiguration process. The temporary IP address isn't used anymore.

Duplicate address allocation is possible, as the uninitialized node chooses a temporary randomly. However, as long as the uninitialized node broadcasts the service request only in its neighborhood, there will be a problem only if the duplicated address in the same neighborhood from the requester. If some kind of proactive routing protocol is used, this could be avoided by simple inspection of hello messages exchanged in the neighborhood, which identifies all neighbors' IP addresses.

Another solution could be reserving an special range of IP addresses for temporary allocation, reducing the probability of duplicated allocation to the case where more then one uninitialized node request the certificate issuing at the same time, using the same IP address.

As long as there is no binding between certificates and IP addresses (see section 1), we may also have a timeout timer for collecting the response messages (step 3). If this timer expires before the reception of K responses, another request could be made, with a different choice for the temporary IP address.

5.3 Secure MANET Merging

Merging of two MANETs that have been bootstrapped by different DCA (e.g. different K_{CA} are used for signing node certificates) requires the establishment of bi-directional cross trust relation between the DCA. A cross trust relation between the nodes trusting in two different DCAs (e.g. DCA1 and DCA2) can be established if there are, at least, K nodes trusting both DCAs. This is simply done by generating a certificate for DCA1 and a certificate for DCA2 that are signed by K private-key-shares from DCA2 and DCA1, respectively. The CAs certificates generated in this processes should be flooded in the network, finalizing the cross trust relation establishing. Establishing of cross trust relation is also controlled by the certification policy.

6   Intrusion Detection

We have defined a distributed and collaborative intrusion detection system architecture specially designed for Manet [17], based on the Local IDS concept. This system was successfully implemented for detection of common attacks against Manet routing protocols. Presently, we are adapting the same system to detect attacks against the designed autoconfiguration protocol, such those described in the section 3.1. Preliminary results in this direction are very promising. We hope to publish the results concerning intrusion detection of active attacks against autoconfiguration protocol in a companion paper in the near future.

7   Related Works

Perkins et al. proposed [7] a "trial and error" protocol for IP address autoconfiguration. A node picks a random address and performs duplicate address detection (DAD) requests for a positive response of all the configured nodes in the MANET. Does not work with network partition and merging and uses flooding to obtain an approval of all nodes configured. Autoconfiguration can fail when occur unbounded delays.

The mechanism proposed at [18] is similar the used in [7] using "trial and error" method. So, the duplication address detection (DAD) proposed not only checks address duplication during the initialization of address, but also during ad hoc routing by intermediate nodes. It utilizes a hybrid scheme consisting of two phases named Strong DAD and Weak DAD that checks if there is any addresses duplication during the assigning IP address and when occur merging of partitions respectively. In contrast of [7], the partitioning and merging of ad hoc networks is considered.

Boleng [19] utilizes a mechanism that consists of variable length addresses to save significant storage and transmission overhead by using only the minimum number of bits to uniquely address a MANET node. The address length can be expanded to accommodate new nodes as required. Partitioning and merging of networks are considered as well.

In [20], Nesargi and Prakash present a scheme that is not scalable because assigning a new IP address depends on the approval of all known the MANET nodes.

We believe that extension of our autoconfiguration security model to other autoconfiguration protocols such as [7,18,19,20] is quite straightforward, as those protocols shares the same security requirements with our proposal.

We have been working with security aspects of Manet routing protocol, which resulted in the design of a Manet Certification Service and MAE [11] and in the design of distributed and collaborative IDS for Manet [17]. The MAE and IDS designed for securing routing protocols are completely reused in this work, proving the extensibility of the techniques employed in our model.

8   Conclusion

In this paper we have described a general authentication service along with a conjugate certification service to secure the MANET autoconfiguration protocol.

Concerning the trust model used "K-out-of-N"; the most critical parameter is K. As the authentication and autoconfiguration services are provided by any coalition of K nodes, implementation with large K can tolerate more adversaries, however the service availability is degrades. With a small K, the system become high availability, but more vulnerable to attacks from adversaries. Robustness may be found adopting policies so that nodes beyond the one-hop neighborhood can use autoconfiguration service.

We have implemented the DCDP protocol allowing operation with or without MAE service (e.g. with or without security). Execution of all autoconfiguration tasks (e.g. node arrival, departure and network synchronization were tested. In our test-bed, 6 nodes running Linux system were loaded with an IEEE 802.11b card, which was configured for ad hoc operation. Four of those nodes were previously setup with certificates and IP addresses blocks, one was used to create tampering attacks (as described in section 3) and the other was used to perform node arrival and departure operations. Both graceful and abrupt depart were tested. If the MAE service was not included and processed along with the autoconfiguration protocol messages, the attacker could successfully disrupt the autoconfiguration protocol in all the tested tasks. Alternatively, if the MAE service was enabled, the attacker was not able to disrupt the autoconfiguration service, as it was not trusted by other nodes. Thus, arrival and departure could be executed normally.

We are currently working in intrusion detection aspects of the autoconfiguration protocol. For the moment, we are evaluating a misuse detection engine with distributed and cooperative features, similar to [17], presented for securing routing protocols. Also, we are investigating the possibility of having single MAE processing and intrusion detection for both routing protocol and autoconfiguration service.

References

1. S. Corson and J. Marker – Mobile ad hoc networking (MANET): Routing protocol performance issues and evaluation consideration. RFC 2501 (informational), IETF, 1999.
2. Y. C. Hu, D. Johnson, and A. Perrig. SEAD: Secure eficient distance vector routing for mobile wireless ad hoc networks. In Fourth IEEE Workshop on Mobile Computing Systems and Applications (WMCSA '02), June 2002, pages 3--13, June 2002.
3. Y. C. Hu, A. Perrig, and D. Johnson, "Ariadne: A secure On-demand routing protocol for ad hoc networks", in the Proceedings of ACM MobiCom 2002, Sep. 2002.

4. B. Dahill, K. Sanzgiri, B. N. Levine, C. Shields and E. Royer, "A secure routing protocol for ad hoc networks". In the Proceedings of the 2002 IEEE International Conference on Network Protocols (INCP 2002), Nov. 2002.

5. M. Guerrero and N. Asokan, "Securing Ad Hoc Routing Protocols", in the Proceedings of 2002 ACM Workshop on Wireless Security (WiSe'2002), in conjunction with the ACM MOBICOM2002, September, 2002.

6. P. Papadimitratos and Z. J. Haas. Secure routing for mobile ad hoc networks. SCS Communication Networks and Distributed Systems Modeling and Simulation Conference (CNDS 2002), Jan 2002.

7. Charles E. Perkins, Jari T. Malinen, Ryuji Wakikawa, Elizabeth M. Belding-Royer and Yuan Sun, "IP Address Autoconfiguration for Ad hoc Networks", draft-ietf-manet-autoconf-01.txt, Internet Engineering Task Force, MANET Working Group, November 2001.

8. A. Misra, S. Das, A. McAuley, and S. K. Das. Sun. Autoconfiguration, Registration and Mobility Management for Pervasive Computing. IEEE Personal Communications, vol. 08, Issue 04, Aug. 2001.

9. M. Mohsin and R. Prakash. IP Address Assignment in a Mobile Ad Hoc Network. IEEE Milcom 2002.

10. H. Yang, X. Meng and S. Lu, "Self-Organized Network Layer Security in Mobile Ad Hoc Networks", in the Proceedings of ACM Workshop on Wireless Security – 2002 (WiSe'2002), September, 2002.

11. R. Puttini, L. Me, R. de Sousa, "MAE – MANET Authentication Extension for Securing Routing Protocols", 5th IEEE International Conference on Mobile and Wireless Communications Networks (MWCN2003), Oct. 2003.

12. D.E Knuth, The Art of Computer Programming Vol. I, Fundamental Algorithms 3rd Edition, Addison Wesley, 1997.

13. H. Luo, P. Zerfos, J. Kong, S. Lu and L. Zhang, "Self-securing Ad Hoc Wireless Networks", in the Proceeding of Seventh IEEE International Symposium on Computer and Communications (ISCC'02), 2002.

14. J. Kong, P. Zerfos, H. Luo, S. Lu and L. Zhang, "Providing robust and ubiquitous security support for MANET," IEEE ICNP 2001, 2001.

15. L. Zhou and Z. J. Haas. Securing ad hoc networks. IEEE Network Magazine, 13(6):24--30, November/December 1999.

16. A. Shamir – How to Share a Secret. Communications of the ACM, 22(11):612-613, 1979.

17. Puttini, R; Percher, JM; Me, L, Camp, O; de Sousa, R. "A Modular Architecture for a Distributed IDS for Mobile Ad Hoc Networks". International Conference on Computer Science and Applications (ICCSA - SIAM), Montreal (Canadá) in Lecture Notes on Computer Science vol. 2669, Springer-Verlag, pp. 91-113, 2003.

18. Jae-Hoon Jeong, Hyun-Wook Cha, Jung-Soo Park and Hyoung-Jun Kim, "Ad Hoc IP Address. Autoconfiguration", draft-jeong-adhoc-ip-addr-autoconf-00.txt, Internet Engineering Task Force, MANET Working Group, May 2003.

19. J. Boleng, "Efficient network layer addressing for mobile ad hoc networks", Tech. Rep. MCS-00-09, The Colorado School of Mines, 2000.

20. Sanket Nesargi and Ravi Prakash, "MANETconf: Configuration of hosts in a mobile ad hoc networks" INFOCOM, 2002.

Analysis of the Information Propagation Time Among Mobile Hosts*

Tassos Dimitriou[1], Sotiris Nikoletseas[2], and Paul Spirakis[2]

[1] Athens Institute of Technology, Greece
tdim@ait.gr
[2] Computer Technology Institute and Department of Computer Engineering &
Informatics, University of Patras, Greece
{nikole,spirakis}@cti.gr

Abstract. Consider k particles, 1 red and $k-1$ white, chasing each other on the nodes of a graph G. If the red one catches one of the white, it "infects" it with its color. The newly red particles are now available to infect more white ones. When is it the case that all white will become red? It turns out that this simple question is an instance of information propagation between random walks and has important applications to mobile computing where a set of mobile hosts acts as an intermediary for the spread of information.

In this paper we model this problem by k concurrent random walks, one corresponding to the red particle and $k-1$ to the white ones. The *infection time* T_k of infecting *all* the white particles with red color is then a random variable that depends on k, the initial position of the particles, the number of nodes and edges of the graph, as well as on the structure of the graph.

We easily get that an upper bound on the expected value of T_k is the worst case (over all initial positions) *expected meeting time* m^* of two random walks multiplied by $\Theta(\log k)$. We demonstrate that this is, indeed, a tight bound; i.e. there is a graph G (a special case of the "lollipop" graph), a range of values $k < n$ (such that $\sqrt{n} - k = \Theta(\sqrt{n})$) and an initial position of particles achieving this bound.

When G is a clique or has nice expansion properties, we prove *much smaller bounds* for T_k. We have evaluated and validated all our results by large scale experiments which we also present and discuss here. In particular, the experiments demonstrate that our analytical results for these expander graphs are tight.

1 Introduction, Problem Definition, and Motivation

Properties of *interacting particles* (moving on a finite graph) are of interest within a number of areas of Science, primarily Physics in the early days, but increasingly

---

* This work has been partially supported by the IST/FET Programme of the European Union under contract number IST-2001-33116 (**FLAGS**) and within the 6FP under contract 001907 (DELIS).

I. Nikolaidis et al. (Eds.): ADHOC-NOW 2004, LNCS 3158, pp. 122–134, 2004.

Biology, the Social Sciences and Computer Science today, since they model interesting phenomena like magnetism, spatial competition, tumor growth, spread of infection, economic systems and mobile communication. The book by Liggett ([13]) and Chapter 14 of the Aldous and Fill book ([1]) are good references.

We consider here a special case of interacting particles: The motion of each of the particles is a random walk on the graph. The walks are *concurrent*; for simplicity, the initial positions of the particles are those of the steady-state probability distribution of the walks.

Let k the number of particles. Let $G(V, E)$ the graph of the walks. Let $|V| = n \geq k, |E| = m$. Initially, one particle is *red* and all others are *white*. Let $\pi()$ the steady-state distribution of each of the walks. The particles interact according to the following *Infection Rule:* When a red particle meets (at a graph node) one (or more) white particle(s), the white particle(s) turn red.

Initially, each vertex of G can have at most one particle (thus $k \leq n$). Let ϕ be the initial distribution of particles.

Definition 1: Let T_k^ϕ be the least time instant at which all particles are red.

Note that T_k^ϕ is a random variable. We are interested here in the expected value of T_k^ϕ, $E_\phi T_k^\phi$.

Definition 2: We call *Infection Time*, w.r.t. an initial distribution ϕ, the value of $E_\phi T_k^\phi$.

Definition 3: Let an initial distribution ϕ be called *pure* when it assigns a single position (vertex) to each particle, with probability 1.

Definition 4: We call the *Infection Time* T_k the worst $E_\phi T_k^\phi$ over all pure distributions ϕ.

Note that $\pi()$ always exists when the random walk is of *continuous time* with *transition rates* $q_{vx} = \frac{1}{d_v}$ if $\{v, x\}$ is an edge (d_v is the degree of vertex v), and $q_{vx} = 0$ if not, because, then, the random walk is aperiodic and ergodic.

In fact, we first consider continuous time walks. Note that the discrete-time walks defined by the transition probability $p_{uv} = \frac{1}{d_u}$ if $\{u, v\} \in E$, and $p_{uv} = 0$ else, have the same stationary distribution mean hitting times as in the continuous case ([1]). Our main goal is to study the Infection Times of various (unweighted, undirected, finite) graphs.

Motivation. The problem studied here is an example of (i) the spread of a virus (i.e. the red colour) in computer networks (see e.g. [11,15,16]) (ii) information spreading (such as rumor spreading, gossiping) (see [3,9]) (iii) a considerable fraction of the communication time of stations in ad-hoc mobile networks, under the existence of a (mobile) infrastructure as a virtual intermediary pool for getting messages from senders and delivering them to receivers (see [4,5,6]).

We elaborate in more detail on (iii). Recall that an ad-hoc mobile network is a collection of mobile hosts, with wireless communication capabilities, forming a temporary network without the aid of any established fixed infrastructure. In [4, 5,6] we have studied the problem of basic communication, i.e. to send information from a sender mobile user, MH_S, to a receiver mobile user, MH_R. For such dynamically changing networks of high mobility we have proposed protocols for basic communication which exploit *the accidental meetings* of the mobile hosts and the co-ordinated (by the protocol) motion of a *small part* of the network.

We abstract the 3D network area by a *motion-graph*, whose vertices model cubes of volume close to that of the transmission sphere of hosts, and whose edges connect adjacent cubes.

The protocol works as follows: The nodes of the support move fast enough to visit (in sufficiently short time) the entire motion graph. When some support node is within communication range of a sender, it notifies the sender that it may send its message(s). The messages are then stored "somewhere within the support structure". When a receiver comes within range of a support node, the messages are then forwarded to the receiver.

Clearly the size, k, and the shape of the support affects performance. In [4,5, 6] we study two basic alternatives for the support structure: a) the "snake" support, where the nodes of the support move always remaining pair-wise adjacent (i.e., forming a chain of k nodes). Essentially the support moves as a "snake", with the head doing a random walk on the motion graph and each other node executing the simple protocol "move where the node preceding me was before". b) a different approach is to allow each member of Σ to perform an *independent* random walk on the motion graph, i.e., the members of Σ can be viewed as *"runners"* running on G. *When two runners meet, they exchange any information* given to them by senders encountered.

In [4,5,6] we perform, using Markov Chain techniques, a rigorous average case analysis for the expected communication time of the snake protocol. The infection propagation time of graphs in this paper can be used to estimate the expected communication time in the runners case. We expect that, because of the "parallelism" of runners, *the runners protocol significantly outperforms the snake protocol* (a fact experimentally validated in [4,5,6]).

2   New Results and Related Work

We first show an *upper bound* on the expected infection propagation time T_k for any $k > 2$ and for any undirected graph G, where we consider continuous time walks.

We demonstrate that this bound is *tight* on a Lollipop graph for certain values of k and an initial position of particles.

Then, we turn into discrete time walks and derive *much smaller* bounds on T_k for *the clique and for expander graphs*. Note that our model of continuous time walks is just the "continuization" of the corresponding discrete time chains. Thus the results hold in both cases.

Previous work and Comparison. Coppersmith et al ([7]) evaluated the expected *meeting time* of *two* random walks (i.e. T_2). They showed that in the worst case (and assuming discrete time), $T_2 = (4/27 + o(1))n^3$. Tetali and Winkler ([19]) gave some earlier bounds on T_2. Sunderam and Winkler ([18]) examined a related but different problem: each node of a clique is a processor having some information piece. They estimate techniques to minimize the time when all processors know all pieces of the information. There, processors do not move and messages are exchanged in discrete, synchronized rounds.

Works on other models of interacting particles (e.g. the anti-voter and voter models), can be found in Ch. 14 of [1] and in ([13]). Infection models were studied well in the past, under the direction of population biology. Their targets are similar but the main difference is the missing of the *graph* as space of motions. See [17] and references there.

So, our work *extends* the results of [7,19] to $k > 2$ concurrent random walks on a finite undirected graph.

Roadmap. Our first results on sections 3 and 4 refer to *continuous time* concurrent random walks. The results on special graphs (sections 5, 6) refer to discrete time walks. In section 7, we discuss our experimental findings.

3   A Tool: Exponential Tails of Hitting Times and Meeting Times

Definition 1. Let $A \subseteq V$ and $i \in V$. Define $E_i T_A$ to be the expected value of the first hitting time to A (i.e. the first time to arrive at a node of A) of a random walk on G, starting from vertex i. Let T_A be the corresponding random variable. Let $t_A^* = \max_{i \in V} E_i T_A$.

Definition 2. Let E an event about a random walk on G. We denote by $\Pr_\phi(E)$ the probability of E when the starting position of the walk is according to the distribution $\phi()$.

Lemma 1 (SubExponentiality Lemma 1a). *([1])* For any $t : 0 < t < \infty$ it is

$$\sup_\phi \Pr_\phi\{T_A > t\} \leq exp(-\lfloor t/et_A^* \rfloor)$$

The machinery stated above can be used also for the meeting times:

Definition 3. Let $M_{i,j}$ be the first time that two independent copies of a random walk on G meet given that they start from i, j. Let $m^* = \max_{i,j} M_{i,j}$.

Working exactly as above, we then get:

Lemma 2 (SubExponentiality Lemma 1b). *([1])*

$$\Pr\{M_{i,j} > t\} \leq exp(-\lfloor t/et_m^* \rfloor)$$

Remark: Note that the above Lemmata are useful only when t is large, i.e. above et_m^* or et_A^*. When t is small then the Lemmata hold trivially because they just say that a tail probability is less than or equal to 1.

Definition 4. For a continuous time walk on G let T_j be the *first hitting time* vertex j, i.e. the first time for the walk to be at vertex j. Let $M_{i,j}$ be the meeting time of two independent copies of the walk, started at vertices i and j, and $EM_{i,j}$ its expected value.

Another interesting fact about the relation of hitting times and meeting times, for continuous time walks, is the following:

Lemma 3 (Aldous and Fill [1], Chapter 14, Proposition 5).

$$\max_{i,j} EM_{i,j} \leq \max_{i,j} E_i T_j$$

This can be easily seen when i, j are vertices of degree 1 in the $n + 1$-vertex star graph.

4   A Tight Upper Bound on Infection Time for General Graphs

Consider, instead of our process (call it P_1) of infection, another process of the same walks governed by the rule:

"When particles meet, they coalesce into clusters and the cluster thereafter sticks together and moves as a single random walk. If the cluster contains red particles then it is colored red".

Definition 5. Let $C_{k,n}$ be a random variable, which is the time at which all particles coalesce into one single cluster.

Clearly, $\forall t > 0$, it is $\Pr\{C_{k,n} > t\} \geq \Pr\{T_k > t\}$, and thus

$$E_\phi C_{k,n} \geq E_\phi T_k \tag{1}$$

for all initial position distributions ϕ. But $EC_{k,n}$ can be bounded as follows:

Lemma 4. $EC_{k,n} \leq em^*(2 + \log k)$, where e is the basis of the natural logarithms.

Proof. We proceed exactly as in Aldous and Fill, Chapter 14, Proposition 11 ([1]) with the only difference being that initially we have k particles instead of n. The proof uses the second of the subexponentiality Lemmas. Order the initial positions of the particles arbitrarily as $i_1, \ldots, i_k$. First, let the k particles perform independent random walks forever, with the particles starting at positions i, j first meeting at time $M_{i,j}$ (say). Then, when two particles meet, let them cluster and follow the future path of the lowest-labelled particle. Similarly, when two clusters meet, let them cluster and follow the future path of the lowest-labelled particle in the combined cluster. Then it follows that

$$C_{k,n} \leq \max_j M_{i_1,j}$$

Let $m^* = \max_{i,j} EM_{i,j}$. By the second subexponentiality lemma then

$$\Pr\{M_{i,j} > t\} \leq \exp\left(-\left\lfloor \frac{t}{em^*} \right\rfloor\right) \tag{2}$$

So, by (2), we get

$$EC_{k,n} = \int_0^\infty \Pr\{C_{k,n} > t\}dt \leq \int_0^\infty \min\left\{1, \sum_{j \in \{i_2,\dots,i_k\}} \Pr\{M_{i,j} > t\}\right\}dt$$

and thus

$$EC_{k,n} \leq \int_0^\infty \min\left\{1, k\exp\left(-\frac{t}{em^*}\right)\right\}dt = em^*(2 + \log k)$$

Hence, by (1) we get

Theorem 1. $T_k \leq em^*(2 + \log k)$

Although this bound is crude for a lot of graphs, there are graphs and positions of particles achieving it. Consider a special case of the Lollipop graph, with a clique of size $n - \sqrt{n}$ and a path of length $l = \sqrt{n}$ extending out of it. Initially, there is one red particle in the clique. There are k white particles, all being at the furthest k vertices of the path. Here $k << \sqrt{n}$. In the sequel, take k so that $\sqrt{n} - k = \Theta(\sqrt{n})$. For a graphical representation of an instance of this graph where $n = 16, k = 3$, see Figure 1 in the next page.

Clearly the worst case expected meeting time is $m^* = \Theta(l^2) = \Theta(n)$, i.e. the average time for the random walk of a white particle to enter the clique, by elementary random walk properties (see [10]).

Let the white particles be labelled (from the furthest to the clique inwards) as $i_1, \dots, i_k$. Let their times to enter the clique be the random variables $S_1, \dots, S_k$, respectively. Note that it is very hard for the red particle to exit the clique. This can happen only when it gets to the vertex where the path and the clique meet (i.e. once every $\Theta(n)$ time), but even then the probability of exiting the clique is $\frac{1}{n - \sqrt{n} + 1}$. On the average, thus, the red particle stays inside the clique for $\Theta(n^2)$ expected time. Let $S = \max\{S_1, \dots, S_k\}$. Clearly then, the infection time is $T_{k+1} = S + \Theta(1)$ w.h.p., where $\Theta(1)$ accounts for the last white particle meeting another particle (which will be red by then). Thus,

$$E(T_{k+1}) = E(S) + \Theta(1) \tag{3}$$

From the subexponentiality property of individual hitting times (Lemma 1) we have that $\forall i = 1, \dots, k$ and for $s^* = \max_{i=1,\dots,k} E(S_i) = \Theta(n)$ we get

$$\Pr\{T_i > \lambda es^*\} \leq e^{-\lambda}, \quad \lambda = 1, 2, 3, \dots \tag{4}$$

So,

$$\Pr\{S > \lambda es^*\} \leq ke^{-\lambda}, \quad \lambda = 1, 2, 3, \dots \tag{5}$$

i.e.

$$\Pr\left\{\frac{S}{es^*} - \log(ek) > x\right\} \le e^{-x}, \quad x \ge 0 \tag{6}$$

So $E\left(\frac{S}{es^*} - \log(ek)\right) \le 1$ and hence

$$E(S) \le (2 + \log k)es^* \tag{7}$$

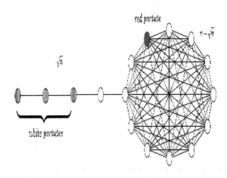

Fig. 1. The Lollipop graph we consider here

In fact, for the one-dimensional random walks this bound is tight! In the continuous time, the number of +1 and -1 steps of a single walk in the line are independent Poisson(t) variables (of mean 1). Then a bound as in (4) holds with equality, and from

$$\Pr\{S < t\} = \Pr\{T_1 < t\} \cdot \Pr\{T_2 < t\} \cdots \Pr\{T_k < t\}$$

(because of the independence of the walks), also relations similar to (5) hold with equality. Thus,

Lemma 5. $E(S) = \Theta(s^* \log k)$

But $s^* = \Theta(m^*)$ for the $\sqrt{n}$ line as we said. So,

Lemma 6. $E(S) = \Theta(m^* \log k)$

Thus, by (3) and (7), we get:

Theorem 2. In the Lollipop graph and for k such that $\sqrt{n} - k = \Theta(\sqrt{n})$, there is a set of initial positions of particles so that $T_{k+1} = \Theta(m^* \log k)$. Thus, the bound of Theorem 1 is tight.

Note however than even in the lollipop graph, for large k, the infection time is much smaller than this upper bound. This is demonstrated in the experiments section.

5   Infection Times on the Clique

As our experiments also indicate (see Section 7), the bound stated in Theorems 1 and 2 is not tight for "close to regular graphs". In the next sections we derive *much better* bounds for such graphs. The experiments we have conducted and which we discuss in Section 7, demonstrate that our analytical results are tight. The case of the clique K_n is instructive:

Theorem 3. *Consider $k \leq n$ particles, one red, $k-1$ white, sitting on the nodes of a K_n. Then the expected time when all particles become red after randomly walking on the K_n is bounded by*

$$\frac{4n}{k-1}(\ln k + O(1)).$$

Denote by R_t and W_t the number of red and white particles at time t, where $R_t + W_t = k$ and $R_0 = 1$. Our goal is to show that for $t^* = O((n/k)\ln k)$ the expected number of red particles is k. It will be helpful in the analysis to consider each round of particle's movement as a two stage process. In the first stage of round t we move all the red particles to a random neighbor of their current location. Then we move the white particles. If a white particle falls into the position of a red one, it becomes infected and changes its color. At the end of the second stage we count all the red particles and we obtain R_{t+1}. Observe that this model is exactly the same as the original one but is simpler to analyze and give us the following intermediate result.

Lemma 7. *Let R_t the number of red particles at the end of round t. Then at the end of round $t+1$ the expected difference $\Delta_t = R_{t+1} - R_t$ satisfies*

$$\Delta_t \geq \frac{R_t(k - R_t)}{n}\left(1 - \frac{R_t}{2n}\right)$$

Proof. Let $r = R_t$ the number of red particles. To simplify the analysis we allow each particle to stay in the current node with probability $1/n$. Thus the next node can be any of the nodes $1, \ldots, n$ with equal probability $1/n$. Now consider what happens when the red particles move.

If r is small then all these particles will land almost surely to different nodes. But when r is large the red particles may form clusters which in effect decrease the population of red particles that may infect the white ones. Thus we need to compute the number r' of different clusters of red nodes and treat each cluster as one red "super-particle" that can propagate its color.

So what is r'? We can compute r' by treating the r particles as balls which have to be thrown in n bins, where each bin corresponds to one of the possible n neighbors. The number of different clusters will simply be the expected number of occupied bins in the balls experiment, which is by definition n minus the number of empty bins. Now the expected number of empty bins is easily found to be $n(1 - \frac{1}{n})^r$, which is bounded from above by $n(1 - \frac{r}{n} + \frac{r^2}{2n^2})$. Thus we get that the number of different super-particles r' is

$$r' \geq n - n\left(1 - \frac{r}{n} + \frac{r^2}{2n^2}\right) = r - \frac{r^2}{2n} = r\left(1 - \frac{r}{2n}\right).$$

All these super-particles are now available in stage two to infect a white particle if it happens to land in the same node occupied by the cluster. The probability that a white particle changes its color to red is at least r'/n, thus on the average we expect $W_t(r'/n) = (k - r)(r'/n)$ of them to become infected. We conclude that at the end of round $t + 1$, the expected number of red particles will be

$$R_{t+1} \geq R_t + \frac{(k - R_t)R_t}{n}\left(1 - \frac{R_t}{2n}\right)$$

or equivalently since $R_t \leq k \leq n$

$$\Delta_t \geq \frac{R_t(k - R_t)}{2n}$$

and the proof of the lemma is complete.

This lemma will help us estimate the number of rounds before all particles become red. An easy but not so accurate bound follows by observing that Δ_t is always at least $(k - 1)/2n$. Thus during successive rounds we make at least $(k - 1)/2n$ progress towards infecting all particles, therefore the total number of rounds required is at most $2n$. A more accurate counting however gives the desired result.

Define T_i to be the time when $R_{T_i} = i$, that is the moment when the number of red particles is exactly i. Thus in the time interval $[T_i, T_{i+1})$, the expected number of red particles is at least i and smaller than $i + 1$. In this time interval the expected progress Δ_i in *all* rounds between T_i and T_{i+1} is therefore at least

$$\Delta_i \geq \frac{i(k - i - 1)}{2n}.$$

But then the expected number of rounds between T_i and T_{i+1} is at most $1/\Delta_i$ since the number of red particles increases by one. Thus overall, the number of required rounds T_k is

$$T_k \leq \sum_{i=1}^{k-2} \frac{2n}{i(k - i - 1)} + \frac{2n}{k - 1} < \frac{4n}{k - 1}(\ln k + O(1))$$

thus completing the proof of the theorem.

One immediate generalization of the random walk on a clique is when p denotes the probability of visiting a random neighbor from the current node. Clearly, Theorem 3 corresponds to $p = 1/n$. So:

Theorem 4. *Consider $k \leq n$ particles, one red, $k - 1$ white, sitting on the nodes of a K_n, where the probability of visiting any node is p. Then the expected time when all particles become red is*

$$O\left(\frac{(1/p)^2}{nk}\ln k\right).$$

6   The Case of Graphs with Good Expansion

A graph $G = (V, E)$ informally is an expander if the number of neighbors of any given set of nodes S is not "too small". More formally

Definition 6. *The expansion v of a graph $G = (V, E)$ is*

$$v = \min_{|S| \leq |V|/2} \left\{ \frac{N(S)}{S} \right\}$$

where $N(S)$ is the set of vertices in $V - S$ which are adjacent to some vertex in S.

Expander graphs have many useful properties but perhaps the most important one is that a random walk on the nodes of the graph is unlikely to remain trapped for a long time in a small portion of the state space. So if $P = [p_{u,v}]_{n \times n}$ denotes the one-step transition matrix of the walk and π_0 the initial distribution, then the distribution at time t, π_t, quickly converges to a unique stationary distribution π.

An instructive case is when the graph is d-regular. For random d-regular graphs or for explicit constructions of families of d-regular graphs it is known that the expansion is a small constant. Thus the time to approximate the stationary distribution, which in this case is $1/n$, by a small amount $\epsilon = n^{-2}$ is at most $O(d \log n)$ (see e.g. [2]).

This observation will help us get a good upper bound for the infection time of k particles on a d-regular graph G with good expansion properties. In particular we will reduce this case to the clique case presented in Section 5 by "simulating" each step of the walk on a clique with $O(d \log n)$ steps of a walk on the d-regular graph.

More precisely we move the k particles on G by breaking their walks into epochs of duration $O(d \log n)$. We let each particle perform a $O(d \log n)$ steps walk inside each epoch and we consider what happens at the end of this walk. In particular, we allow white particles to be infected only if they meet a red one at the end of the epoch. This gives an upper bound on the infection time because white particles may truly be infected in some intermediate time.

Each epoch has the effect of simulating one step at the clique, since by the corollary to expander mixing lemma, the probability of visiting any particular neighbor is equal (by an n^{-2} amount) to the stationary distribution $1/n$.

Applying Theorem 4, we conclude that the expected number of epochs required to infect all the red particles is at most $O((n/k) \log k)$. Combined with the $O(d \log n)$ length of each epoch we find that the infection time of a d-regular graph with good expansion properties is at most $O\left(\frac{m}{k} \log k \log n\right)$ where m is the number of edges of the graph.

Observe that the same bound (to within a polylogaritmic factor) applies for more general expander graphs. For such graphs the expansion v is bounded below by $1/\log^{O(1)} n$, where n is the number of nodes of the graph, thus each epoch never requires more than $O(d \log^{O(1)} n)$ steps to reach the stationary distribution.

7   Experiments

We implemented the experiments in C++ using classes and data structures of the LEDA Platform ([14]). We investigated four classes of graphs: lollipops, cliques, 4-regular graphs and $G_{n,p}$ graphs. For each graph type we conducted experiments for various graph sizes (n, the number of vertices) and numbers of particles (k). For each instance we repeated the experiment enough times to get good average results and smooth time curves. For the expander graphs we tested $n \in \{100, 200, 400, 1600, 3200\}$, $k \in \{$ 2, 10, 20, 25, 30, 40, 50, 70, 100 $\}$, carrying out 200 repetitions in each case, while for the lollipop we tested $n \in \{100, 200\}$, $k \in \{2, 10, 20, 25, 30, 40, 50, 70, 100\}$, conducting 100 repetitions. In the figures (that, due to limited space, are included in full paper [8]) we display the findings for the largest graph size (n) in each case.

The infection stochastic process has been implemented evolving in discrete time rounds. Initially, each particle is placed on a vertex of the graph studied. In the beginning of each round every particle moves randomly and uniformly to one of the neighbouring vertices of the vertex it currently resides. If during a round a red particle resides on the same node with some white particles, then the white particles become red. The process ends when all particles become red. We measure in each case the expected number of rounds needed for all particles to become red.

In the "lollipop" graph case, we had to experimentally evaluate m^*, the maximum of the expected meeting times of two particles, over all possible pairs of starting vertices of the particles. To obtain a realistic value of m^*, we took the average of the meeting times of the two particles in 80 repetitions, for every vertex pair, and then the maximum of all these mean values.

The main experimental findings are the following (for figures, see [8]): (i) As shown in Fig. 2, in the lollipop the crude upper bound is indeed tight for small k, while it is trivial for larger values of k. (ii) The bound we prove for the clique is very tight (see Fig. 3 where the analytical and the experimental curves almost coincide). The same is happening in the regular graphs (the distance between the two curves in Fig. 4 is only due to the fact that the degree (and thus the expansion) is very small. (iii) A similar behaviour is exhibited in the $G_{n,p}$ case, as shown in Fig. 5 This similarity in cliques, regular graphs and $G_{n,p}$ graphs, shown in Fig. 7, is due to their common expansion properties. iv) As shown in Fig. 6, the infection time in lollipops is much bigger compared to that of the expander graphs (note that in this Figure we display the lollipop infection time divided by 100, for its curve to fit in the Figure).

Acknowledgement. We warmly thank H. Euthimiou, whose contribution to the experimental part has been valuable.

References

1. D. Aldous and J. Fill: *Reversible Markov Chains and Random Walks on Graphs*. Unpublished manuscript. http://stat- www.berkeley.edu/users/aldous/book.html, 1999.
2. N. Alon. Eigenvalues and Expanders. *Combinatorica*, 6:83–96, 1986.
3. J. Aspnes and W. Hurwood: Spreading rumors rapidly despite an adversary. In *Proc. 15th ACM Symposium on Principles of Distributed Computing* ? PODC?96, (1996).
4. I. Chatzigiannakis, S. Nikoletseas and P. Spirakis: Distributed Communication and Control Algorithms for Ad- hoc Mobile Networks. In the Journal of Parallel and Distributed Computing (JPDC), Special Issue on Mobile Ad-hoc Networking and Computing, 63 (2003) 58-74, 2003.
5. I. Chatzigiannakis and S. Nikoletseas: Design and Analysis of an Efficient Communication Strategy for Hierarchical and Highly Changing Ad-hoc Mobile Networks. In ACM/Baltzer Mobile Networks Journal (MONET), Special Issue on Parallel Processing Issues in Mobile Computing, Guest Editors: A. Zomaya and M. Kumar, accepted, to appear in 2004.
6. I. Chatzigiannakis, S. Nikoletseas, and P. Spirakis: An Efficient Communication Strategy for Ad-hoc Mobile Networks In *Proc. 15th International Symposium on Distributed Computing* – DISC'01. Lecture Notes in Computer Science, Volume 2180 (Springer-Verlag, 2001), pp. 285–299. Also Brief Announcement in *Proc. 20th Annual Symposium on Principles of Distributed Computing* – PODC'01, pp. 320–322, ACM Press 2001.
7. D. Coppersmith, P. Tetali and P. Winkler: Collisions among random walks on a graph. In SIAM J. Disc. Math. 6:3 (1993), pp. 363-374.
8. T. Dimitriou, S. Nikoletseas and P. Spirakis: Analysis of the Information Propagation Time among Mobile Hosts. (http://www.cti.gr/RD1/nikole/english/psfiles/infection.ps) In the Proceedings of the 3rd International Conference on Ad-Hoc Networks & Wireless (AD HOC NOW), 2004.
9. S. Even and B. Monien: On the number of rounds needed to disseminate information. In *Proc. 1st ACM Symposium on Parallel Algorithms and Architectures*, SPAA 89, (1989).
10. W. Feller. An Introduction to Probability Theory and its Applications. *John Wiley*, New York, 1957.
11. J. Kephard and S. White: Directed-Graph Epidemiological Models of Computer Viruses In *Proc. IEEE Symposium on Security and Privacy*, 1991. Also IBM Technical Report.
12. M. R. Jerrum and A. Sinclair: Conductance and the rapid mixing property of Markov chains: The approximation of the permanent resolved. In *Proc. 20th ACM Symposium on Theory of Computing (STOC)*, pages 235–244, 1988.
13. T. Liggett: Stochastic Interacting Systems: Contact, Voter and Exclusion Processes, Springer Verlag 1999.
14. K. Mehlhorn and S. Naher: LEDA: A Platform for Combinatorial and Geometric Computing. Cambridge University Press. (1999).
15. S. Nikoletseas, G. Prasinos, P. Spirakis and C. Zaroliagis, "Attack Propagation in Networks", in the Theory of Computing Systems (TOCS) Journal, Special Issue on the Thirteenth (13th) Annual ACM Symposium on Parallel Algorithms and Architectures (SPAA 2001), accepted.

16. R. Ostrofsky and M. Yung: How to Withstand Mobile Virus Attacks In *Proc. 10th ACM Symposium on Principles of Distributed Computing* PODC 91, (1991), pp. 51-59.
17. J. R. Norris: Markov Chains. Cambridge University Press, 1997.
18. V.S. Sunderam and P. Winkler, In Disc. Appl. Math. 42 (1993), pp. 75-86.
19. P. Tetali and P. Winkler: Simultaneous reversible Markov chains. In Combinatorics, Paul Erdōs is Eighty Vol. 1, D. Miklos, V.T. Sos and T. Szonyi (editors), Janos Bolyai Mathematics Institute, Budapest, 1993, pp. 433-452.

Consensus with Unknown Participants or Fundamental Self-Organization*

David Cavin, Yoav Sasson, and André Schiper

École Polytechnique Fédérale de Lausanne (EPFL), CH-1015 Switzerland
{david.cavin,yoav.sasson,andre.schiper}@epfl.ch

Abstract. We consider the problem of bootstrapping self-organized mobile ad hoc networks (MANET), i.e. reliably determining in a distributed and self-organized manner the services to be offered by each node when neither the identity nor the number of the nodes in the network is initially available. To this means we define a variant of the traditional consensus problem, by relaxing the requirement for the set of participating processes to be known by all at the beginning of the computation. This assumption captures the nature of self-organized networks, where there is no central authority that initializes each process with some context information. We consider asynchronous networks with reliable communication channels and no process crashes and provide necessary and sufficient conditions under which the problem admits a solution. These conditions are routing and mobility independent. Our results are relevant for agreement-related problems in general within self-organized networks.

1 Introduction

The paper addresses the problem of bootstrapping a self-organized MANET. More precisely, the paper addresses the following question. Consider some geographical region R that is initially empty. At some point, one or more mobile nodes enter the region and want to deploy one or more services. However, to deploy the service(s), it is necessary for the first nodes that enter R to agree on an initial set of nodes, in order for these nodes to decide which node is going to provide what service. Let us call these nodes *I-nodes* (Infrastructure nodes).

To decide which node is going to provide which service, we need to solve an agreement problem that decides on a set *I-nodes*, and outputs *I-nodes* at each node in *I-nodes*. Once this problem is solved, each node in *I-nodes*, based on the knowledge of *I-nodes*, can locally determine which node is responsible for providing what service(s). For example, assume that the agreement is on *I-nodes* = $\{n_1, n_2, n_3\}$, and consider that there are five services s_1 to s_5 to provide. Based on the knowledge of *I-nodes*, n_1 will provide the services s_1 and s_2, n_2 will provide the services s_3 and s_4, and n_3 will provide the service s_5. Or, if there is

---

* The work presented in this paper was supported by the National Competence Center in Research on Mobile Information and Communication Systems (NCCR-MICS), a center supported by the Swiss National Science Foundation under grant 5005-67322.

I. Nikolaidis et al. (Eds.): ADHOC-NOW 2004, LNCS 3158, pp. 135–148, 2004.

only one service to provide, n_1 will provide it, and n_2, n_3 know that they have no service to provide.

The problem is easily solved if there is some fixed node fn that is always in R: fn can act as a centralized decision point. However, this solution is not self-organized, since it relies on some preexisting infrastructure. This leads to the following question: is it possible to solve the problem without any preexisting centralized infrastructure, i.e., in a fully self-organized way?

Deciding on the set *I-nodes* can be modeled as a consensus problem [1]. In the consensus problem, a set Π of processes have to agree on a common value (called the decision value) that is the initial value of one of the processes. Consensus has been extensively studied in traditional networks with process failures, and algorithms based on various system models have been developed [2,3,4]. The importance of consensus is due to the fact that it is a basic building block for solving several other important fault-tolerant distributed problems. However, there is a fundamental difference between the classical consensus problem, and the problem addressed in the paper: in the paper *the set Π is unknown* (and Π is precisely the information we want to obtain). This makes it a new problem that we call *Consensus with Unknown Participants* or simply *CUP*. Note that the notion of consensus with *uncertain* participants appears in [5]. However, the specification is different, and the context is also different (it is used as a building block for implementing a dynamic atomic broadcast service in a wired synchronous network). Thus, the results in [5] are unrelated to the results established in this paper.

The CUP problem is formally defined in Section 2, which also defines the model in which the problem is solved. The classical consensus problem is hard to solve because processes may crash. With CUP, the difficulty of the problem is due to the *unknown* participants. So, for simplification, we assume in the paper that processes (i.e., mobile nodes) do not crash. We also assume that the nodes in R always form a connected network, and we assume the existence of an underlying multihop routing protocol: if some node n knows the existence of a node n' in R, then n can reliably send a message to n'. *Moreover, n can only send reliably a message to nodes that it knows.* Given these assumptions, the results obtained in the paper are independent of the underlying routing algorithm or mobility pattern of the nodes.

Clearly if all nodes in R do not know any other nodes, then no node can send any message to any other node, and CUP cannot be solved. This leads us to add to our model the notion of *participant detectors*, which are distributed oracles attached to each node n. The participant detector of n provides to n a (possibly small) subset of the nodes in R, e.g. by receiving messages or beacons. In Section 3 we introduce various classes of participant detectors, and we compare them based on the classical notion of reduction. In Section 4 we identify necessary and sufficient conditions for solving CUP, and we solve CUP using the participant detector named *one sink reducibility*. In Section 5 we illustrate how CUP can be used for solving the bootstrapping problem described in this section. Finally, we conclude and present future work in Section 6.

2   Consensus with Unknown Participants

We consider a finite set Π of processes drawn from a (finite or infinite) universe $\mathcal{U}$. The processes in Π have to solve the traditional consensus problem, but contrary to the usual model for consensus, the processes in Π do not necessarily know each other. This assumption captures the *self-organization* nature of the type of system that we consider: there is no central authority that initializes each process with some context information.

To solve consensus, processes in Π communicate by message passing. However, *process $p \in \mathcal{U}$ can send a message to process $q \in \mathcal{U}$ iff p knows the existence of q*. Similarly, process q can send a message to p iff q knows the existence of p. So if p knows q, but q does not know p, the communication is asymmetric. If q does not know p and receives a message from p, from there on q knows p, i.e., q can send a message to p. Communication channels are reliable and the system is asynchronous: we do not assume any bound on the transmission delay of messages nor on the process relative speeds. We finally make the strong assumption that processes never crash. Indeed, the necessary conditions established in the paper are still true in models with process crashes.

The specification of consensus with unknown participants (CUP) is close to the classical specification of consensus [6]: an instance of the CUP problem is defined by the primitives $propose(v_i)$ by which each process $p_i \in \Pi$ proposes its initial value v_i, and $decide(v)$, by which a process decides on a value v. The decision must satisfy the following conditions:

Validity. If a process decides v, then v is the initial value of some process.
Agreement. Two processes cannot decide differently.
Termination. Every process eventually decides.

In the classical consensus problem processes know each other, i.e., processes in Π know Π. This is not the case here. Moreover, in the classical consensus problem processes may crash. The crash of processes makes the problem difficult. Here processes do not crash; what makes the problem difficult is the ignorance of the set of other processes having to solve an instance of the consensus problem.

3   Participant Detectors

3.1   Presentation and Specification

If each process $p \in \Pi$ knows only itself, then p cannot communicate with any other process in Π, which clearly makes it impossible to solve consensus. We are interested in identifying the minimal information that processes must have about the other participants to make consensus solvable. We capture the information that process p has about other processes by the notion of *participant detectors*. Similarly to failure detectors [4], participant detectors are distributed oracles associated with each process. In the setting of the classical consensus problem, where the set of participants is *known*, the task of failure detectors is to maintain a set of processes which are suspected to have crashed. In our context however, the set of participants is *unknown*. By querying their local participant detector, processes may obtain an approximation of Π, the set of processes participating in consensus.

We denote by PD_p the participant detector of process p. Process p can query its participant detector PD_p, which returns a set of processes. We denote by $PD_p(t)$ the query of p at time t. The information returned by PD_p can evolve between queries, but verifies the following two properties.

Property 1 (Information Inclusion). The information returned by the participant detectors is non-decreasing over time:

$$\forall p \in \Pi, \forall t' \geq t : \; PD_p(t) \subseteq PD_p(t')$$

Property 2 (Information Accuracy). The participant detectors do not make mistakes in the sense that they do not return a process that does not belong to Π:

$$\forall p \in \Pi, \forall t : \; PD_p(t) \subseteq \Pi$$

PD_p gives p an initial context, i.e., an approximation of Π. This context will allow p to start sending messages to other processes. Moreover, messages received by p will allow p to increase its knowledge of Π. We define below participant detectors reflecting different levels of accuracy of participant estimation. To define these detectors, we consider

1. The (undirected) graph $\mathcal{G} = (\mathcal{V}, \mathcal{E})$, where the vertices $\mathcal{V} = \Pi$ and the (undirected) edge $(p, q) \in \mathcal{E}$ iff $q \in PD_p$ or $p \in PD_q$.
2. The directed graph $\mathcal{G}_{di} = (\mathcal{V}, \mathcal{E})$, where the vertices $\mathcal{V} = \Pi$ and the directed edge $(p, q) \in \mathcal{E}$ iff $q \in PD_p$.

Participant Detector 1 (Connectivity CO). *A participant detector satisfies the* connectivity *property iff the (undirected) graph $\mathcal{G}$ is connected.*

Participant Detector 2 (Strong Connectivity SCO). *A participant detector satisfies the* strong connectivity *property iff the directed graph $\mathcal{G}_{di}$ is strongly connected.*

Participant Detector 3 (Full Connectivity FCO). *A participant detector satisfies the* full connectivity *property iff the directed graph $\mathcal{G}_{di} = (\Pi, \mathcal{E})$ is such that for all $p, q \in \Pi$, we have $(p, q) \in \mathcal{E}$.*

Finally, we introduce the last detector, satisfying the *one sink reducibility* property. The motivation behind this particular participant detector will become clear to the reader in Section 4.

Participant Detector 4 (One Sink Reducibility *OSR*). *A participant detector satisfies the* one sink reducibility *property iff the graph $\mathcal{G}$ is connected and the directed acyclic graph obtained by reducing $\mathcal{G}_{di}$ to its strongly connected components has one and only one sink[1].*

Note that Properties 1 and 2 allow processes to query their participant detectors at different times. It is easy to see that if some participant detector PD satisfies the the property of CO, SCO, FCO or OSR if queried at some time t, it also satisfies the property when queried at some time $t' > t$. Consider for example OSR and assume that PD satisfies OSR at time t. If the graph $\mathcal{G}_{di}$ reduced to its strongly connected components contains at most one sink, adding edges to $\mathcal{G}_{di}$ at time t' cannot increase the number of sinks.

3.2   Comparing Participant Detectors

Given the system model of Section 2, we are interested in establishing relationships between participant detectors. To this means we use the notion of *reducibility*, borrowed from the concepts introduced in [4] for comparing failure detectors.

Definition 1 (Reducibility). *A participant detector PD is reducible to a participant detector PD', noted $PD \succeq PD'$, if there exists an algorithm $A_{PD \to PD'}$ that transforms the participant detector PD into the participant detector PD'.*

In other words, if a participant detector PD is reducible to PD', it is possible to use PD in order to emulate PD'. PD' is said to be weaker than PD: any problem that can be solved using PD' can be solved using PD instead. Figure 1 illustrates the reduction.

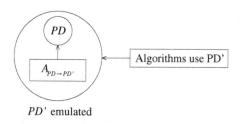

Fig. 1. Transforming Participant Detectors

Definition 2 (Equivalence). *If PD is reducible to PD' ($PD \succeq PD'$) and PD' is reducible to PD ($PD' \succeq PD$), then we say that PD and PD' are equivalent, noted by $PD \cong PD'$.*

---

[1] A *sink* in a directed graph is a vertex with out-degree 0, i.e. there are no edges leaving the vertex. An example graph representing $PD \in OSR$ is provided in Figure 2, which depicts a graph with four strongly connected components A, B, C, D and one sink, namely C.

If two participant detectors are equivalent, they belong to the same *equivalence class*. Any result that applies to one (e.g. problem that can be solved, impossibility result), applies to the second as well.

4   Solving CUP

4.1   Preliminary Results

The goal of this section is to identify the necessary and sufficient requirements for solving CUP by exploring the participant detectors presented in Section 3. We begin with the following intuitive elementary result.

Proposition 1. *The Connectivity participant detector is necessary but not sufficient to solve CUP.*

Proof.
i) Necessary: Assume that the resulting graph $\mathcal{G}$ returned by the participant detectors is disconnected, i.e. there exists two components C_1 and C_2 of processes that cannot communicate with each other and independently execute consensus. Let v_1 be the initial value of the processes in C_1 and v_2 of those in C_2 (with $v_2 \neq v_1$). Consensus terminates in both components, with processes in C_1 deciding on v_1 and processes in C_2 on v_2, leading to a violation of the agreement property.

ii) Not sufficient: Consider $\Pi = \{p_1, p_2, p_3\}$ such that $PD_{p_1} = \{p_1, p_2, p_3\}$, $PD_{p_2} = \{p_2\}$ and $PD_{p_3} = \{p_3\}$. Let v_2 be the initial value of p_2, v_3 the initial value of p_3, with $v_2 \neq v_3$. Clearly, $PD \in CO$. Processes p_2 and p_3, unaware of any other process besides themselves, execute consensus and decide on values v_2 and v_3 respectively, violating the agreement property.   □

The Full Connectivity Oracle is a trivial oracle in the sense that it fully compensates for the uncertainty of the participating processes, leading to the following proposition:

Proposition 2. *The Full Connectivity participant detector is sufficient but not necessary to solve CUP.*

Proof.
Sufficient: It is easy to see that Algorithm 1 (on page 141) meets the consensus specification given our model. Indeed, the Full Connectivity participant detector provides the full set Π of participants to all processes. Therefore, each process can deterministically choose a common leader, e.g. the process with the lowest identifier. The leader process sends its initial value v to all processes in Π, which decide on v upon reception.

Not necessary: Consider $\Pi = \{p_1, p_2, p_3\}$, with $PD_{p_1} = \{p_1, p_2, p_3\}$, $PD_{p_2} = \{p_1, p_2\}$ and $PD_{p_3} = \{p_2, p_3\}$. We have that $PD \notin FCO$. Nevertheless, the following algorithm solves consensus. Let p_i^{min} denote the smallest process returned by the participant detector of p_i. If $p_i = p_i^{min}$ then p_i decides on its own initial value and sends the value to PD_{p_i}. If $p_i \neq p_i^{min}$, then p_i waits for a value and decides upon reception.   □

Algorithm 1 Solving consensus with $PD \in FCO$ for a process $p_i \in \Pi$

1: **propose**(v_i):
2:   $participants_i \leftarrow PD_i$;
3:   $leader_i \leftarrow min(participants_i)$;
4:   **if** $p_i = leader_i$ **then**
5:     $decision_i \leftarrow v_i$;
6:     **send** $decision_i$ to all $p_j \in participants_i$;
7:   **end if**
8:
9:   {Upon **receive**($decision$):}
10: **decide**($decision$);

We now establish the equivalence between the Strong Connectivity and Full Connectivity participant detectors.

Proposition 3. *The Strong Connectivity and Full Connectivity participant detectors are equivalent.*

Proof. With a Full Connectivity participant detector, each process can communicate with every other process in Π, hence $\mathcal{G}_{di}$ is strongly connected and $PD \in FCO$ trivially implies $PD \in SCO$, i.e., $FCO \succeq SCO$.

We prove $SCO \succeq FCO$ with Algorithm 2, a token-based depth-search algorithm that discovers all members of Π assuming $PD \in SCO$. Every process $p_i \in \Pi$ initiates the participant discovery by invoking discover_participants() after having queried its participant detector $PD \in SCO$ (l.4). Notice that processes query their participant detector only once. Every token carries the following information :

$token_i.issuer$: The identity of the process having issued the token.
$token_i.visited$: A set containing the processes visited by the token.
$token_i.tovisit$: A set containing the known processes that have not yet been visited by the token.

Prior to forwarding the token, p_i adds in $token_i.tovisit$ the processes it can communicate with, i.e. processes returned by its local strong connectivity participant detector (l.8). The token is then forwarded to any process present in $token_i.tovisit$ (l.9). Upon reception of $token_i$ by a process p_j, p_j checks whether it is the issuer of $token_i$. If yes, the algorithm terminates and returns the set of visited processes (l.13). If not, it updates the data structures stored in $token_i$ as follows. First, it adds to $token_i.tovisit$ all processes in PD_{p_j} that have not yet been visited (l.15). p_j then removes its own ID from $token_i.tovisit$ (l.16), adds it in $token_i.visited$ (l.17). If there are no more processes to visit, p_j sends the token back to $token_i.issuer$ (l.18-19). Otherwise, it simply forwards $token_i$ to any one of them (l.21).

In order to prove the correctness of the algorithm, we need to show that it terminates and returns all processes. Consider the token issued by p_i. To prove that the algorithm terminates, we must show that p_i eventually executes line 13.

This happens when $token_i$ (the token issued by p_i) has returned to p_i. Assume that $token_i$ is located at some process p_j. Either $token_i.tovisit$ contains some process not visited by the token, in which case p_j forwards the token to one of them or p_j sends back $token_i$ to $token_i.issuer$. Since the number of processes is finite, eventually all processes are visited, in which case line 19 is executed and eventually p_i executes line 13.

It remains to prove that when process p_i executes line 13, $token_i.visited$ contains all the processes. We prove the result by contradiction. Assume that $token_i$ is located at some process p_j, and p_j sends back the token by executing line 19 while $token_i.visited$ does not contain some processes X. This means that for all processes $p_j \in token_i.visited$, PD_{p_j} does not contain any of the processes in the set X. A contradiction with SCO. □

Algorithm 2 Participant discovery algorithm for process $p_i \in \Pi$

1: $token_i.issuer \leftarrow p_i$;
2: $token_i.visited \leftarrow \emptyset$;
3: $token_i.tovisit \leftarrow \emptyset$;
4: $neighbors_i \leftarrow PD_i$; {Participant detector invocation}
5:
6: **discover_participants()**:
7:    $token_i.visited \leftarrow \{p_i\}$;
8:    $token_i.tovisit \leftarrow neighbors_i \setminus \{p_i\}$;
9: **send** $token_i$ to any $p_j \in token_i.tovisit$;
10:
11: {Upon **receive**($token_j$) from p_k :}
12: **if** $token_j.issuer = p_i$ **then**
13:    **return** $token_j.visited$; {Algorithm terminates}
14: **else**
15:    $token_j.tovisit \leftarrow token_j.tovisit \cup (neighbors_i \setminus token_j.visited)$;
16:    $token_j.tovisit \leftarrow token_j.tovisit \setminus \{p_i\}$;
17:    $token_j.visited \leftarrow token_j.visited \cup \{p_i\}$;
18:    **if** $token_j.tovisit = \emptyset$ **then**
19:       **send** $token_j$ to $token_j.issuer$;
20:    **else**
21:       **send** $token_j$ to any $p_l \in token_j.tovisit$;
22:    **end if**
23: **end if**

Since the FCO and SCO participant detectors are equivalent, it is straightforward to obtain an algorithm that solves CUP with SCO: execute Algorithm 1 with FCO, where FCO is obtained from SCO using Algorithm 2.

It follows from Proposition 2 and Proposition 3 that the Strong Connectivity detector is also sufficient but not necessary for solving CUP. Since by Proposition 1, CO is necessary but not sufficient, we must look for a participant detector somewhere in between CO and SCO for it to be both necessary and sufficient for solving CUP. The OSR participant detector, introduced in the next section, is such a participant detector.

4.2   OSR Is Necessary and Sufficient

The following proposition is the main result of the paper and serves as a basis for designing algorithms that solve CUP.

Proposition 4. *The One Sink Reducibility participant detector is necessary and sufficient to solve CUP.*

Proof.
i) Necessary: Suppose that $PD \notin OSR$, i.e. the directed acyclic graph obtained by reduction of $\mathcal{G}_{di}$ to its strongly connected components has at least two sinks S_1 and S_2. Since there is no outgoing path leaving components S_1 and S_2, processes in S_1 and S_2 can be unaware of the existence of the other sink. We prove the result by contradiction.

Assume there exists an algorithm $\mathcal{A}$ that solves CUP. Let all initial values of processes in S_1 be different from all initial values of processes in S_2. By the termination property of consensus, processes in S_1 and processes in S_2 must eventually decide. Let us assume that the first process in S_1 that decides, say p, does so at t_1, and the first process in S_2 that decides, say q, does so at t_2. Delay all messages sent to S_1 and S_2 such that they are received after $max(t_1, t_2)$. So the decision of p is on the initial value of some process in S_1, and the decision of q is on the initial value of some process in S_2. Since these initial values are different, the agreement property of consensus is violated. A contradiction with the assumption that $\mathcal{A}$ solves CUP.

ii) Sufficient: The proof is by giving Algorithm 3 (see page 144) that solves CUP with $PD \in OSR$ (see next section).   □

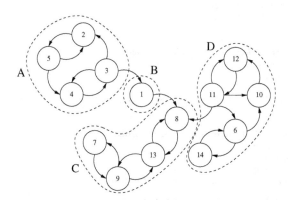

Fig. 2. An example graph $\mathcal{G}$ representing $PD \in OSR$, along with its strongly connected components (arrows represent the information provided by the participant detectors).

Algorithm 3 Solving consensus with $PD \in OSR$ for a process $p_i \in \Pi$

1:
2: $leader_i \leftarrow \perp$; {leader estimate}
3: $leaders_i \leftarrow \emptyset$; {set of leaders}
4: $decisionRequestors_i \leftarrow \emptyset$; {set of processes requiring a notification of the decision}
5: $proposal_i \leftarrow false$; {initial value}
6: $decision_i \leftarrow \perp$; {decision value}
7: $participants_i \leftarrow \emptyset$; {set of processes returned by Algorithm 2}
8:
9: **propose**(v_i):
10: $participants_i \leftarrow discover\_participants()$; {execute and store result from Algorithm 2 using $PD \in OSR$}
11: $leader_i \leftarrow min(participants_i)$;
12: **if** $p_i = leader_i$ **then**
13:    {the process may be the sink leader, send a message}
14:    $proposal_i \leftarrow v_i$;
15:    send *am I the sink leader?* to all $p \in participants_i$;
16: **else**
17:    send *decisionRequest* to $leader_i$;
18: **end if**
19: {the procedure exits and the process waits for a decision}
20:
21: **Upon reception of** *am I the sink leader?* **from process** p_j:
22: **if** $p_j \neq leader_i$ **then**
23:    {disagreement on leader identity}
24:    send $(nack, leader_i)$ to p_j;
25: **else**
26:    send *ack* to p_j;
27: **end if**
28:
29: **Upon reception of** *ack* **from process** p_j:
30: **if** *ack* received from $\forall p \in participants_i$ **then**
31:    {the process is indeed the sink leader; propagate $proposal_i$ as the decision}
32:    $decision_i \leftarrow proposal_i$;
33:    send $(decision_i, leader_i)$ to all $p \in participants_i$;
34: **end if**
35:
36: **Upon reception of** $(nack, leader_j)$ **from process** p_j:
37: {the process is a local leader}
38: **if** $leader_j \notin leaders_i$ **then**
39:    {request a decision from $leader_j$}
40:    $leaders_i \leftarrow leaders_i \cup \{leader_j\}$;
41:    send *decisionRequest* to $leader_j$;
42: **end if**
43:
44: **Upon reception of** *decisionRequest* **from process** p_j:
45: **if** $decision_i \neq \perp$ **then**
46:    send $decision_i$ to p_j;
47: **else**
48:    $decisionRequestors_i \leftarrow decisionRequestors_i \cup \{p_j\}$;
49: **end if**
50:
51: **Upon reception of** $(decision_j, leader_j)$ **from process** p_j:
52: **if** $decision_i = \perp$ **then**
53:    $decision_i \leftarrow decision_j$;
54:    **decide**$(decision_j)$;
55:    send $(decision_j, leader_j)$ to all $p \in decisionRequestors_i$;
56: **end if**

4.3 Solving CUP with OSR

Intuition of the Algorithm

Algorithm 3 (on Page 144) solves CUP with $PD \in OSR$. We present the general intuition with help of the example graph of Figure 2.

The key strategy is to ensure that the nodes belonging to the sink of the graph obtained by reduction to its strongly connected components (component C in the figure) decide before other strongly connected components. The decision can then be appropriately propagated to the rest of the participants. Processes query their $PD \in OSR$ participant detector *only once*, at the beginning of the consensus execution. Initially, processes have no other knowledge besides that returned by the participant detector and are unable to discern whether or not they belong to the sink. To augment their knowledge, processes execute the token discovery algorithm presented in Algorithm 2. Although participant detectors are queried only once, a process can still discover and communicate with nodes discovered later in the course of the computation, e.g. by receiving a message from a process not returned by its PD. In Figure 2, processes in the strongly connected component A will, by executing the token discovery algorithm of Figure 2, discover processes in A, B and C. Similarly, processes in B will discover those in B and C; processes in D will discover those in D and C. Processes in C however will only discover the processes in the same component C. Let $discovered(p_i)$ denote the processes discovered by process p_i.

After the execution of Algorithm 2, every process p_i elects as a leader the process in $discovered(p_i)$ with the lowest identifier (line 11). In our example, process 1 will be the leader in components A and B, process 7 in C and process 6 in D. Non leader processes will send a *decisionRequest* message to their leader to get the decision value (line 17). The message is received at line 44; upon reception of this message the leader registers the decision request in the set $decisionRequestors_i$ (line 48).

The leaders then identify among themselves the leader of the sink component. In order to do so, each leader sends at line 15 the *am I the sink leader?* message to all the processes it has previously discovered (the set $participants_i$ in Algorithm 2), and waits for acknowledgments (*acks*) or negative acknowledgments (*nacks*) from *all* these processes. Upon receiving *am I the sink leader?* from a leader l, a process p either responds with *ack* if p's leader is l (line 24), or otherwise with $(nack, leader)$, where *leader* is p's leader (line 26).

The sink leader (process 7 in our example) will be the only leader to receive only *acks*. Other leaders will receive one or more $(nack, leader_j)$ messages and will send a *decisionRequest* message to $leader_j$ to get the decision value (line 41). The message is received at line 44; upon reception of this message the leader either sends the decision if available (line 46) or registers the request using the set $decisionRequestors_i$ (line 48).

Finally, the sink leader sends its initial value to all the processes in the set $participants_i$ (line 33). The sink leader itself receives this message at line 51, decides at line 54 and then, using the $decisionRequestors_i$ set, sends the decision to all non sink leaders. The non sink leaders propagate the decision to the non leaders registered in their *decisionRequestors$_i$* set. In our example, the sink leader (process 7) sends the decision to all processes in C, as well as to local leaders 1 and 6. Process 1 will propagate the decision to the processes in component A and process 6 to the processes in component D.

Note that the nodes in component C have been able to decide without information about participants outside of their component. The correctness of the algorithm is discussed below.

Correctness of Algorithm 3 (sketch)

We only give a sketch of the proof. We start with three lemmas.

Lemma 1. *If $PD \in OSR$, the set participants$_i$ returned by Algorithm 2 at line 10 contains (at least) every node of the sink component.*

Proof. Algorithm 2 ensures that the token created at a process p_i will visit every process reachable from p_i. Assuming $PD \in OSR$, there exists a path from p_i to every process in the sink, i.e., every process in the sink is reachable from p_i. □

Lemma 2. *If $PD \in OSR$, for any process in the sink component, the set returned by Algorithm 2 contains exactly every node of the sink component.*

Proof. By definition a sink is a strongly connected component without any outgoing link. So the discovery Algorithm 2 executed by a process in the sink component will return all processes in the sink component and no other. □

Lemma 3. *The leader of the sink component is the only process to impose its initial value as the decision.*

Proof. The leader of the sink component is the only leader that will receive only *acks*, from every process of its *participants* set (l.30). So it will be the only process to send its proposal (l.33). □

With help of the above lemmas, we can show that Algorithm 3 meets the specifications of CUP (see Section 2).

Validity. By line 32 the decision is the initial value of some process. So the validity property is trivially satisfied. □

Agreement. By Lemma 3 the leader of the sink component is the only process to send its initial value as the decision. The agreement property trivially follows from this. □

Termination. To prove that Algorithm 3 terminates for every process, we must show that every process executes line 55 (after reception of the message $(decision_j, leader_j)$, line 51). We distinguish three cases :

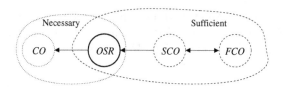

Fig. 3. Relationships between participant detectors

- *Processes in the sink.* By Lemma 2 the set *participants$_i$* of the sink leader contains all processes in the sink. So, by line 33 (sending of the decision by the sink leader to *participants$_i$*) all processes in the sink component eventually receive the decision message.
- *Local leaders.* Since there is only one leader (the sink leader) that receives only acks, all other leaders will receive nacks. Specifically, by Lemma 1 they will receive (*nack, leader*) from the processes in the sink component, with *leader* = *sink leader*. So the local leaders will send *decisionRequest* to the sink leader (l.41), and the sink leader will eventually send them the decision (l.55).
- *All other processes.* Every non leader process registers itself with its leader by sending the message *decisionRequest* (l.17). Since every local leader eventually decides (previous case), every local leader will send the decision to all these processes (l.55).   □

4.4   Necessary and Sufficient Participant Detectors for CUP

Figure 3 summarizes the relationships between the participant detectors and CUP.

5   Bootstrapping a Self-Organized MANET

We now illustrate how CUP can be applied to solve the problem presented in the introduction, i.e. for bootstrapping a MANET in a self-organized manner. Let R be some geographical region that is initially empty. The mobile nodes have no prior knowledge of the identity nor of the number of peers in R (or in the entire network). At some point, one or more mobile nodes enter the region and want to deploy one or more services. However, to reliably deploy the service(s), it is necessary for the first nodes that enter R to agree on an initial set of infrastructure nodes, *I-nodes*, in order to determine which node is going to provide what service.

To obtain the set of *I-nodes*, each node in R executes the CUP algorithm described in Section 4.3, with the *participants* set of Algorithm 3 as the initial value. Since the decision is the initial value of the sink leader, the decision of CUP is the sink component. Let the set *I-nodes* be the decision of the CUP algorithm (i.e. the sink nodes). By the CUP specification (Section 2), this set will be the same for each node in R. Based on the knowledge of *I-nodes*, nodes in

R can locally decide which node is responsible for providing what service(s). The unequivocal determination of the *I-nodes* through CUP leads to a unequivocal attribution of the services to be offered by each node.

6   Summary and Future Work

The paper has addressed the question of bootstrapping a self-organized MANET, i.e. for nodes that have no prior knowledge about their peers in the network to reliably agree on the set of services to be offered by each node. To this means we have introduced CUP, a variant of the consensus problem where neither the identity nor the number of participating processes is known. We have identified (mobility and routing independent) necessary and sufficient conditions, based on the notion of *participant detectors*, for which CUP admits a solution. We have also presented an algorithm, using the participant detectors called OSR, that enables nodes to reliably solve the aforementioned bootstrapping problem in a self-organized manner. Since consensus is a basic building block for solving other distributed algorithms, our results concerning CUP are in particular relevant for other agreement related problems (e.g. total-ordered broadcast, leader election) in self-organized settings such as MANET.

For future work, we intend to investigate the additional requirements necessary for solving CUP when processes may crash.

References

1. Fischer, M.J., Lynch, N.A., Paterson, M.: Impossibility of distributed consensus with one faulty process. JACM **32** (1985) 374–382
2. Lynch, N.: Distributed Algorithms. Morgan Kaufmann, San Francisco, CS (1996)
3. Dwork, C., Lynch, N., Stockmeyer, L.: Consensus in the presence of partial synchrony. J. ACM **35** (1988) 288–323
4. Chandra, T.D., Toueg, S.: Unreliable failure detectors for reliable distributed systems. Journal of the ACM **43** (1996) 225–267
5. Bar-Joseph, Z., Keidar, I., Lynch, N.: Early-delivery dynamic atomic broadcast. In Malkhi, D., ed.: Proceedings of the 16th International Symposium on Distributed Computing (DISC'02). Volume 2508 of Lecture Notes in Computer Science., Toulouse, France, Springer (2002) 1–16
6. Fischer, M.J.: The consensus problem in unreliable distributed systems (a brief survey). In: Proceedings of the 1983 International FCT-Conference on Fundamentals of Computation Theory, Springer-Verlag (1983) 127–140

Connectivity of Wireless Sensor Networks with Constant Density

Sarah Carruthers and Valerie King

University of Victoria, Victoria, BC, Canada

Abstract. We consider a wireless sensor network in which each sensor is able to transmit within a disk of radius one. We show with elementary techniques that there exists a constant c such that if we throw cN sensors into an $n \times n$ square (of area N) independently at random, then with high probability there will be exactly one connected component which reaches all sides of the square. Its expected size is a constant fraction of the total number of sensors, and it covers a constant fraction of the area of the square. Furthermore, the other connected components of sensors and uncovered contiguous portions of the square are each very small ($O(\log^2 n)$ and $O(\log n)$ respectively), so that their relative size grows smaller as N increases. We also discuss some algorithmic implications.

1 Introduction

In wireless sensor networks, the connectivity of the network is established via radio transmission between sensors. For two sensors to be able to communicate, they must be within some critical range of each other, as transmission capability is finite. The connectivity of the entire network is composed of these sensor-to-sensor links. Depending on the disbursement of the sensors, it is possible for islands of sensors to be isolated from the rest of the network; there need only exist a gap that isolates them. We will model the wireless sensor network as a set of sensors placed independently at random in a $n \times n$ square.

We show that there exists a c such that if cN (hereafter $N = n^2$) sensors are placed independently at random in a $n \times n$ square, then with high probability there exists a large connected component of size $O(N)$ which covers at least one point on every side of the square. Furthermore, with high probability, we can bound the number of sensors of any other connected component by $O(\log^2 n)$ and any uncovered contiguous area by $O(\log n)$. Thus, if the density (number of sensors per area) is a constant, the size of each contiguous portion of the square which is uncovered or unconnected to the main connected component diminishes relative to the area of the square, as the square grows larger.

Our technique is to decompose the square into a grid of small squares (which we call *boxes*) and then use simple counting arguments to prove the necessary bounds. The boxes are small enough to ensure that any disconnected set of sensors must be surrounded by a path of empty boxes. We first prove a bound on the number of empty boxes. In Lemma 3, we bound the probability that there

I. Nikolaidis et al. (Eds.): ADHOC-NOW 2004, LNCS 3158, pp. 149–157, 2004.
© Springer-Verlag Berlin Heidelberg 2004

exists a long path of empty boxes. This implies that exactly one component touches all four sides of the square.

We then reapply Lemma 3 to bound the number of sensors which are surrounded by sensors in the large component, but are themselves cut off from transmitting to these sensors by a large encircling gap. Imagine the large component as a giant piece of Swiss cheese or a donut. There are holes in its interior, and these holes may in fact contain connected components of sensors which can transmit to each other, but not to the large component. For the remainder of the paper, we will refer to such isolated connected components as *timbits*.

2   Related Work

Recently, there has been much work done in the area of connectivity of wireless networks. In 1998, Gupta and Kumar [2] showed that with limited transmission strength, a wireless network achieves asymptotic connectivity. Similarly, in 2003 Shakkottai, Srikant and Shroff [4] show this is also true for networks in which some sensors may fail. Xue and Kumar showed in 1998 that a sensor need be connected to $\Theta(\log n)$ nearest neighbours in order for the network to be asymptotically connected [3], i.e., $\Theta(N \log N)$ sensors are necessary for every sensor to be connected in a square of area N.

In 2003, Jennings and Okino [1] showed bounds on the probability that a network with n sensors in a unit square, each with a fixed radius of transmission, is connected. What they fail to prove, however, is the size of the largest connected component. They corroborate their results with simulations. In these simulations, up to 100 sensors were placed randomly in a square of area $A = 600^2$. The transmission radii of the sensors were chosen as a proportion of the length of the square, with a fixed radii picked for a particular graph. These simulations show that the size of the largest connected component is large $(.9A)$, when the radius of transmission is $.3\sqrt{A}$. In contrast, our results show that in a square of area A, for any constant $d < 1$, a *constant* radius of transmission suffices for the largest connected component to cover an expected area of dA, in a network of constant density.

Very recently it has come to our attention that in 1995, Penrose considered a more general problem in a sophisticated analysis using percolation theory [5]. This work is cited in another paper written in 2000 by Diaz, Penrose, Petit and Seina [6]. See also [7].

3   Preliminaries

In our model, we say two sensors are *neighbours*, i.e., there is a transmission link between them, if they are within distance one of each other. Two sensors x and y are *connected* if there is a path of sensors $z_1, ..., z_k$ such that x is a neighbour of z_1, y is a neighbour of z_k, and for each $i < k$, z_i is a neighbour of z_{i+1}. A *connected component* is a maximal set of sensors which are connected. A point is *uncovered* if it is of distance greater than one from any sensor; otherwise the point is covered. A set of points is *uncovered* (*covered*) if every element of the set is uncovered (covered).

We toss cN sensors independently at random into an $n \times n$ square of area N. We partition the $n \times n$ square into $2\sqrt{2}n$ equally spaced rows and columns so that each small square (hereafter *box*) has size $1/(2\sqrt{2}) \times 1/(2\sqrt{2})$, for a total of $8N$ such boxes. This size of box is chosen so that if a box A contains a sensor, then this sensor has a link to any sensor in any box which is adjacent or diagonal to A. A box is empty if it contains no sensors; otherwise it is nonempty.

The first question we ask is: if we throw cN sensors into the $n \times n$ square, how many boxes will remain empty?

Lemma 1. *For any constant c_3, there exists a constant c such that if we throw cN sensors into an $n \times n$ square, the expected number of empty boxes is $N/5c_3$.*

Proof. We observe that the problem of determining the number of empty boxes is the same as the well-known balls in bins occupancy problem: Given r bins and s balls thrown in at random, how many bins are empty? Here, $r = 8N$ and $s = cN$. The expected number of empty bins is $r(1 - 1/r)^s \leq 8Ne^{-c/8}$ which is less than $N/5c_3$, for some $c \geq 24 + \ln c_3/8$.   □

Lemma 2. *Let c_1, d be any constants, where $c_1, d > 1$. There exists a constant c such that if cN sensors are thrown independently at random into the square, then the probability that the number of empty boxes is at least N/c_1 is less than $1/e^{dN}$.*

Proof. Fix $\lfloor N/c_1 \rfloor$ empty boxes. The probability that a sensor doesn't fall into one of these empty boxes is less than $(1 - 1/c_1)$. The probability that no sensors fall into these boxes is $(1 - 1/c_1)^{cN}$. Since there are $\binom{N}{\lfloor N/c_1 \rfloor}$ ways to choose the set of empty boxes, the probability of this occurring for any such set of boxes is bounded by:

$$\binom{N}{\lfloor N/c_1 \rfloor}(1 - 1/c_1)^{cN} \leq (c_1 e)^{N/c_1}(1/e)^{(c/c_1)N} \leq (1/e)^{dN} \, , \qquad (1)$$

when $c > c_1 d + \ln c_1 + 1$. We derive these inequalities from the following two facts: (1) $\binom{s}{t} \leq (se/t)^t$ and (2) $(1 - 1/t) \leq (1/e)^{1/t}$, where $1/t < 1$, which we will use repeatedly in this paper.   □

4   A Unique Largest Component

By a *path of empty boxes* we mean a "manhattan" path in the sense that it consists of up, down, left and right turns. We define a square of size r to be a square composed of r boxes, i.e. smaller squares of size $1/(2\sqrt{2}) \times 1/(2\sqrt{2})$, such that $\sqrt{r}$ is an integer value. We define a timbit as a connected component which covers portions of no more than two sides of the $n \times n$ square.

Observation 1. *If a set of sensors is unconnected to another set, there must be a path of empty boxes separating that set from the other. In particular, every timbit must be surrounded by a path composed of empty boxes, and possibly the boundary of the square.*

We now characterize the paths that a random set of less than $\lfloor r/c_1 \rfloor$ empty boxes can form in a square of size r. We show:

Lemma 3. *Given a square of size r, with $\lfloor r/c_1 \rfloor$ empty boxes occurring in it randomly, then the probability that there is a path of empty boxes of length $l \geq 3\log_2 r$ is less than $(r/3)(3/(c_1 - 1))^l$.*

Proof. We give an upper bound on the probability by (over) counting the number of possible ways to place the empty boxes so that a path of length l is formed and then divide by the possible number of ways of placing the $\lfloor r/c_1 \rfloor$ empty boxes in the square. There are r ways to choose the start of the path and 3^{l-1} ways to continue it (we don't go back the way we came). The number of ways to pick the remaining sensors is $\binom{r-l}{\lfloor r/c_1 \rfloor - l}$. The total number of ways of placing r/c_1 empty boxes in the square is $\binom{r}{\lfloor r/c_1 \rfloor}$. Thus we have:

$$r3^{l-1} \frac{\binom{r-l}{\lfloor r/c_1 \rfloor - l}}{\binom{r}{\lfloor r/c_1 \rfloor}} , \tag{2}$$

which is

$$\leq (r/3)(3/(c_1 - 1))^l . \tag{3}$$

This expression is less than $1/r$ when $c_1 > 7$ and $l > 2\log_2(r/3))$. □

From the lemma above, we see that with high probability there is no path of empty boxes in the square which goes from one side of the square to its opposite. All other connected components in the square either share a border on the outer boundary of the square, or are *timbits*.

Theorem 1. *(1) With high probability, there is a single connected component which touches all four sides of the $n \times n$ square, (2) all other contiguous connected components have area $O(\log^2 n)$ and (3) all contiguous uncovered portions have area at most $O(\log n)$.*

Proof. Proof of (1) is immediate from Lemma 3.

Proof of (2) follows from Lemma 3 as well, however, if the perimeter of all such other contiguous connected components is $O(\log n)$, then the area of these components is $O(\log^2 n)$.

Proof of (3) follows from a slight modification to Lemma 3. In order to bound the area of the uncovered portion, we will treat it as a tree of length l embedded in the grid. This tree can be described by walk of length $2l$. We modify Lemma 3 as follows. Rather than having 3^{l-1} ways to continue that path, we have 4^{2l}

ways, i.e., we may have to back-track. The result is that the probability that
there is an uncovered portion of the $n \times n$ square is:

$$r 4^{2l} \frac{\binom{r-l}{\lfloor r/c_1 \rfloor - l}}{\binom{r}{\lfloor r/c_1 \rfloor}} , \tag{4}$$

which is

$$\leq (r 4^l)(4/(c_1 - 1))^l . \tag{5}$$

This expression is less than $1/r$ when $c_1 \geq 33$, and $l > 2 \log r$.   □

It remains to show that this single connected component contains $\Omega(N)$
sensors. The problem is that it may contain holes. We next bound the expected
area which is not covered by the largest component.

5 The Largest Connected Component

We now show the expected size of the largest connected component. We prove
an upper bound on the expected portion of all areas of the $n \times n$ square which
may not be part of the largest component, i.e. all boxes containing elements of
timbits and all empty boxes. These parts are bounded as follows:

- Lemma 5 bounds the expected area of all boxes which haven't too many
  empty boxes, and which contain timbits.
- Lemma 7 bounds the expected area of squares which have too many empty
  boxes, even though they have a sufficient number of sensors.
- Lemma 8 bounds the expected area of squares which have too many empty
  boxes because they receive too few sensors.
- Lemma 1 bounds the expected area of empty boxes.
- Observation 5 bounds the area which contains rectangles of abnormal size,
  when the $n \times n$ square is partitioned into r-squares.

We add this all up, and subtract this amount from N to get a lower bound on
the expected area covered by the largest component (and the number of sensors
in the largest component).

We do this by partitioning the $n \times n$ square into smaller squares. We first
bound the expected area of squares which have long paths of empty boxes, then
bound the expected area of squares with too many empty boxes.

5.1 Timbits

In order to bound the area of the $n \times n$ square which is occupied by timbits,
we will decompose the $n \times n$ square into grids of squares of size $3^2, ..., \log^2 n$.
The size of one of these squares is equal to the number of boxes it contains. An
r-square is a square of size r. Note that we need only consider squares up to size
$\log^2 n$ as we have already shown that with high probability, there are no paths
of empty boxes of length $O(\log n)$. We then use these grids of squares to bound

the number of timbits that can be contained (along with the path that encircles them) in a square of a particular size. Counting over all possible sizes, we can bound the total area occupied by timbits. In Lemma 5, we end up over counting this area, because we can't say how a timbit may fall within a fixed $\sqrt{r} \times \sqrt{r}$ grid. What we can say however, is that any timbit will reside in at most four r-squares, for some r.

Note however, that when partitioning the $n \times n$ square into these grids, the last row or column of this grid may contain rectangles of area less than r. To account for this, we bound the size of the $n \times n$ square which can contain these *deviant* rectangles.

Observation 2. *The total area of these deviant rectangles is $O(r \log n)$, which is less than $N/5c_3$ for sufficiently large N.*

Lemma 4. *For every timbit contained in the $n \times n$ square, there exists an s such that the timbit is contained in up to four grid squares of size s, one of which contains a path of length at least $(\sqrt{s} + 1)/2$.*

Proof. We define the length of a timbit as follows. Let x_1 (x_m) be the column number of the left-most (right-most) box in the timbit which contains a sensor. Similarly, let y_1 (y_m) be the row number of the top-most (bottom-most) box of the timbit which contains a sensor. Let the length of the timbit be: $max\{x_m - x_1, y_n - y_1\}$.

The empty path which surrounds the timbit must be at least $2t + 6$ in length to completely enclose the timbit. Place around the timbit the smallest square, s, so that it completely contains both the timbit and the path surrounding it. Note that size of $s = (t + 2)^2$. Regardless of how the timbit sits in this any grid (as described earlier), it will only reside in at most four of these squares. This path must be at least $2t + 6$ in length, which in terms of s is $2\sqrt{s} + 2$.

Note that for any timbit, this implies that there are at most four squares of size s that contain an empty path of length greater than or equal to $(1/4)(2\sqrt{s} + 2) = (\sqrt{s} + 1)/2$.

We must also consider the case in which the timbit covers one or more of the borders of the $n \times n$ square. It is easy to see, however, that the statement of the lemma still holds, as any timbit covering a border, must still be cut off from the rest of the components by a path, this path must be at least half the length of the timbit, and this timbit must lie in at least one square of size s which contains at least half of this path of empty boxes. Thus the path must be at least $(\sqrt{s} + 1)/2$, in at least one square of size s, as required.   □

Lemma 5. *For any constant c_3, there exists a constant c_1 such that the expected sum over r of all areas of r-squares which have at most r/c_1 empty boxes and contain timbits is less than $N/5c_3$.*

Proof. This follows from Lemma 3. To account for all possible timbits, for a particular r, we multiply the probability of the path occurring $((r/3)(3/(c_1 -$

$1))^l$), by the number of boxes in the square $(4r)$, times the area that squares of this size take up in the $n \times n$ square $(8N/r)$. Summed over all possible r-values, this gives us the expected size of the portion of the $n \times n$ square which is occupied by these timbits. So for a fixed r, the expected area of these timbits is:

$$\leq 4r(8N/r)(r/3)(e/c_1 - 1)^{(\sqrt{r}+1)/2} . \tag{6}$$

And over all possible r-values, the expected area of the $n \times n$ square occupied by these timbits is:

$$\leq \sum_{i=3}^{(\log n)} 4i^2(8N/i^2)(i^2/3)(3/(c_1 - 1))^{(\sqrt{i^2}+1)/2}$$

$$= 4N/3 \sum_{i=3}^{(\log n)} (i^2)(3/(c_1 - 1))^{(i+1)/2} \leq N/5c_3 , \tag{7}$$

for some c_1. □

Lemma 6. *For any constants $c_1, d > 1$, there exists a constant c, such that the probability that an r-square receives cr sensors and ends up with greater than r/c_1 empty boxes is $1/e^{dr}$.*

Proof. This follows from Lemma 2. □

Lemma 7. *For any constants c_1, c_3, there exists a constant c such that for all r, the expected sum of area of all boxes in the $n \times n$ square which are contained in an r-square which (1) receives cr sensors and (2) ends up with more than r/c_1 empty boxes, is less than $N/5c_3$.*

Proof. For a fixed r, the area of r-squares which have too many empty boxes is just equal to the area of all r-squares in the $n \times n$ square, times the probability that an r-square receives cr sensors and ends up with more than r/c_1 empty boxes. This follows from Lemma 6, and is: $(8N/r)(r)(1/e^{dr}) = 8N/e^{dr}$. Summed over all possible r values, the total area of the $n \times n$ square which receives cr sensors, but ends up with more than r/c_1 empty squares is:

$$\sum_{i=3}^{\log n} 8N(1/e^{di^2})$$

$$= 8N \sum_{i=3}^{\log n} (1/e^{di^2}) \leq 8N/e^{d+2} \leq N/5c_3 . \tag{8}$$

□

Lemma 8. *For any constant c_3, there exists a constant c such that if cN sensors are thrown independently at random into a square of area N, then the expected sum of area of all boxes contained in any r-square which receives fewer than c_2r sensors, where $c_2 \geq (1 + \ln(c_3)/2)$, is less than $N/5c_3$.*

Proof. Note that this is the same as the ball and bins problem. Dividing the $n \times n$ square into r-squares, we have $8N/r$ bins, into which we throw, independently at random, cN balls (sensors). We bound the probability (for a fixed r) that an r-square receives fewer than c_2r sensors. The expected number of sensors received by an r-square is $\mu = \frac{cN}{8N/r} = cr/8$. Setting $c = 16c_2$. We use the following Chernoff bound [8]:

$$Pr(X \leq (1 - \delta)\mu) \leq e^{-\mu\delta^2/2} . \tag{9}$$

Setting $\delta = 1/2$, and $(1 - \delta)\mu = c_2r$. This gives:

$$Pr(X \leq c_2r) \leq e^{-c_2r/4} . \tag{10}$$

Let $\delta = 1/2$. The expected number of boxes in the $n \times n$ square, for fixed r, with too few sensors is $(r)(8N/r)(e^{-c_2r/4}) = 8Ne^{-c_2r/4}$. Summed over all feasible r, this area is then:

$$\sum_{i=3}^{\log n} \frac{8N}{e^{i^2 c_2/4}}$$

$$= 8N \sum_{i=3}^{\log n} \frac{1}{e^{i^2 c_2/4}} \leq 8N/e^{c_2+2} \leq N/5c_3 , \tag{11}$$

where $c_2 \geq (1 + \ln(c_3)/2)$. □

Lemma 9. *For any constants c_1, c_3, there exists a constant c such that if cN sensors are thrown independently at random into a square of area N, then the expected area of all boxes contained in any r-square with more than r/c_1 empty boxes, for any r is less than $2N/5c_3$.*

Proof. Let a be the constant c from Lemma 7. Let $b = max\{a, c_2\}$, where c_2 is the constant from Lemma 8. Now, from Lemma 8, there exists a constant c such that if cN sensors are thrown into a square of area N, the expected area of all boxes contained in any r-square which receives fewer than br sensors is less than $N/5c_3$.

From Lemma 7, the expected area of all boxes contained in any r-square receiving br sensors with more than r/c_1 empty boxes is less than $N/5c_3$. Since a box in an r-square is either in an r-square with fewer than br sensors, or in a r-square with more than br sensors, the lemma follows. □

Theorem 2. *For any constant c_3, there exists a constant c such that if cN sensors are thrown into a square of area N independently at random, then the expected area covered by the largest component is $N - N/c_3$.*

Proof. Every point which is not covered by the largest component is contained in either a box containing an element of a timbit, or an empty box. We choose c to be the maximum of the c's in Lemmas 1, 5 and 9. If we subtract out those areas, and the area of the deviant rectangles described in Observation 5, then we are left with the area covered by the largest component, which is $N - N/c_3$, for any constant c_3. □

6   Discussion

This result has algorithmic implications. For example, since the small components contain $O(\log^2 n)$ sensors, a sensor can detect if it is contained in the large connected component by a flooding transmission which takes $c \log^2 n$ steps. If after $c \log^2 n$ steps there continue to be new sensors discovered, then the sensor "knows" that it is in the largest component. Another implication is that the shortest path between any two connected sensors is at most $O(n \log n)$, with high probability.

Further analytical and/or empirical work is needed to determine the bounds on the values of the constants.

Acknowledgements. We would like to thank Dr. Kui Wu for suggesting this problem. This research was funded by NSERC and a Collaborative Research Experience for Women grant from the Computing Research Association.

References

1. E.H. Jennings, C.M. Okino. On the Diameter of Sensor Networks. *Proceedings of the 2002 IEEE Aerospace Conference*, Big Sky, Montana, Mar. 2002.
2. P. Gupta, P. R. Kumar. Critical Power for Asymptotic Connectivity in Wireless Networks. in *Stochastic Analysis, Control, Optimization and Applications: A Volume in Honor of W.H. Fleming* (W. M. McEneany, G. Yin, and Q. Zhang, eds), pp. 547-566, Boston, MA: Birkhauser, March 1998.
3. F. Xue and P. R. Kumar. The number of neighbours needed for connectivity of wireless networks. Manuscript, 2002.
4. S. Shakkottai, R. Srikant, N. Shroff. Unreliable Sensor Grids: Coverage, Connectivity and Diameter. In *IEEE INFOCOM 2003*.
5. M. D. Penrose. Single Linkage Clustering and Continuum Percolation. *Journal of Multivariate Analysis 53*, pp. 94-109, 1995.
6. J. Diaz, M. D. Penrose, J. Petit, M. Seina. Convergence Theorems for Some Layout Measures on Random Lattice and Random Geometric Graphs. *Combinatorics, Probability and Computing 9*, pp. 489-511, 2000.
7. M. D. Penrose. Random Geometric Graphs. Oxford University Press, 2003.
8. R. Motwani, P. Raghavan. *Randomized Algorithms*, p 70, Cambridge Press, 1995.

Range-Free Ranking in Sensors Networks and Its Applications to Localization

Zvi Lotker*, Marc Martinez de Albeniz*, and Stéphane Pérénnes*

CNRS-INRIA-Sophia Antipolis-I3S, 06902, France
zvilo@eng.tau.ac.il, marcalbeniz@hotmail.com, stephane.perennes@inria.fr

Abstract. We address the question of finding sensors' coordinates, or at least an approximation of them, when the sensors' abilities are very weak. In a d dimensional space, we define an extremely relaxed notion of coordinates along dimension i. The *rank_i* of a sensor s is the number of sensors with ith-coordinate less than the i-coordinate of s. In this paper we provide a theoretical foundation for sensor ranking, when one assumes that a few anchor sensors know their locations and that the others determine their rank only by exchanging information. We show that the rank problem can be solved in linear time in $\mathbb{R}$ and that it is NP-Hard in $\mathbb{R}^2$. We also study the usual localization problem and show that in general one cannot solve it; unless one knows a priori information on the sensors distribution.

1 Introduction

One of the fundamental problems in sensor networks is the *localization problem*, i.e., each sensor wishes to know its exact location. In the last two years several variations of this problem had been studied [6,13,11,7,16]. For an application to localization we refer the reader to [17]

Since sensors are supposed to be cheap the idea of using a GPS devices in every sensor is not practical. One possibility to reduce the cost is to divide the sensors into two sets. One set will serve as base stations and will be equipped with a GPS. The sensors in the second set will use the base stations in order to compute their location. Many localization algorithms for sensor networks have been proposed. There are mainly two groups of algorithms: *range based* and *range free*. In the range based case it is assumed that the sensors have an estimation of the point-to-point distance or angle. In the range free case no such assumption is made. Because of the hardware limitations of sensors networks devices, solutions in range-free localization are being pursued as a cost-effective alternative to more expensive range-based approaches.

In this paper we study the range free localization and ranking problems. The main questions that arise are: *Uniqueness* do we have enough information so that all the solutions that are consistent with the sensors observations are "the same"? *Practical feasibility*: Can we find at least one solution in reasonable time? *Distributed efficiency*: can we provide a fast distributed algorithm that solve the problem ? *Best solution*: Can we find the solution which is the most consistent with the observations?

---

* Project Mascotte CNRS-INRIA-Sophia Antipolis-I3S, 06902, France, this work was supported by the European project Aracne.

I. Nikolaidis et al. (Eds.): ADHOC-NOW 2004, LNCS 3158, pp. 158–171, 2004.

Our Results. We define the ranking problem for the d dimensional space ($\mathbb{R}^d$). For the one dimension case, we show that with uniform power (resp. arbitrary power), one can solve the range free ranking problem in linear time (resp. polynomial time). We also show an instances on which any algorithm that solves the range free localization can make large errors due to the lack of information. We also show that the best possible approximation on the coordinate value is a 2-approximation. Last, we show that when the sensors are identically and independently distributed (*i.i.d*), one can solve range free localization with high probability and good precision.

In the 2 dimensional case we show that range free ranking is not only $\mathcal{NP}$ hard but also APX hard. Our proof also implies that the range free localization is APX hard. As a corollary we prove that finding an approximated 2 dimensional realization of a unit disk graph is $\mathcal{NP}$ hard, thus answering an opened question in [13]

Related Work. Localization in Ad-hoc wireless networks and sensor networks have been the subject of extensive research in recent years. A detailed survey of the applications which describes a spectrum of current products and explores the latest research in that field is provided in [8]. The average estimation was proposed in [4] and was tested by simulation. The main differences between the average algorithm and our algorithm is that we compute the average location on all the sensors while the algorithm in [4] uses only the base stations (anchors). Another difference is that we use the average in order to compute the ranking, and then we use the ranking to get a more accurate estimation. In a later publication [5], improves his work by suggesting a novel density adaptive algorithm (HEAP) for placing additional anchors to reduce estimation error. Another approach for solving the localization proposed in [10] consists in sensing nodes and base stations. Instead of single hop broadcasts, anchors flood their location throughout the network maintaining a running hop-count at each node along the way. Nodes calculate their position based on the received anchor locations, the hop-count from the corresponding anchor, and the average-distance per hop; a value obtained through anchor communication again the algorithm was tasted by simulation. In the paper [7] another alternate solution performs best when an irregular radio pattern and random node placement are considered, and low communication overhead is desired this claim was demonstrated by simulation.

In [6] it is shown that for dimension 2 the range based localization problem is $\mathcal{NP}$ hard (mainly because finding a 2 dimensional realization of a network under distance constraints is $\mathcal{NP}$ hard). The authors also show how to solve the localization if the neighborhood graph is rigid.

Paper Organization. In Section 2 we give the basic definitions and notations. In Section 3 we prove our results for ranking in one dimension, we use the ranking problem to get estimation for the location when the sensors are uniformly distributed. In section 4 we study ranking and localization in dimension 2 and higher.

2   Model and Notation

Let the sensor set be S, sensors will be denoted s. In $\mathbb{R}^d$ we denote the real location of a sensor s by $(s(1), \ldots, s(d))$.

We assume that each sensor has a unique label $name(s)$. In the one dimensional case the rank of sensor s is the number of sensors that have a smaller coordinate than $s : rank_1(s) = |\{s' \in S : s'(1) < s(1)\}|$, and in the d dimensional case we have one ranking function per dimension : $rank_i(s) = |\{s' : s'(i) < s(i)\}|$.

The digraph associated to the sensor network has as vertices the sensors and contains an arc (s, s') whenever s can transmit to s', this digraph will be denoted G and we will assume that it is connected. The out-neighbors set of a sensor s will be denoted $N(s)$. The number of hops $h(s, s')$ from s to s' is the distance from s to s' in the graph G. We assume, unless specifically stated, that all sensors can transmit in a ball of radius r, according to some norm (usually the l_2 one), i.e. a sensor u can transmit to all the sensors v such that $|u - v| \leq r$. According to our hypothesis, the associated digraph G is a Unit Disk Graph [1] (UDG) of $\mathbb{R}^d$ for the chosen norm (usually l_2). This implies that G is a symmetric digraph, and we will identify it with its underlying graph.

We say that two sensors s, s' are equivalent if $N(s) = N(s')$. Since in the range free model two equivalent sensors cannot be distinguished, we assume that we do not have equivalent sensors (this is not restrictive since equivalence can be detected and one can elect one sensor per equivalence class). We also assume that communication is carried out in a synchronous manner, i.e., all the vertices are driven by a global clock. Messages are sent at the beginning of each round, and are received at the end of the round. (Clearly, our lower bounds hold for asynchronous networks as well). Since we work above the MAC layer we allow any set of communication to be performed during a round. The algorithms can be modified for asynchronous using standard techniques.

To conclude note that according to our model the only information that an algorithm can extract from the sensors observation is the graph G and the location of the base stations.

3   One Dimensional Case

In many practical cases i.e., in industrial line, road net, narrow street, along a fence or border we can assume that the sensors are placed in a one dimension environment. As we will see, the one dimensional case is much easier than the two dimensional one.

Without loss of generality, we assume in this section that the sensors are placed in the unit interval and that there are only two base stations placed at 0 and 1; we denote them respectively B_0, B_1.

Consider first the more general case, in which the sensor range may vary. Each sensor s knows its transmission range $r(s)$. The graph G associated to the sensor network is then almost like an *interval graph* (see [9] for a definition). Theoretically to solve ranking or localization one would gather all the sensor observations: that is the adjacency matrix of the network G and the sensor ranges and then try to find 1 dimensional representation of G in which vertex B_0 is mapped on 0 and vertex B_1 on 1 and where the interval of transmission associated to sensor s has length $r(s)$. Using the same ideas that allows

---

[1] Unit disk graphs are intersection graphs: each vertex is represented by a disk and two vertices are adjacent if the disks overlap. Note that in our case 2 nodes are connected if the center of one disk is included in the other disk, and vice versa. However, there is a simple reduction between the 2 models: if we divide the radius by 2, we get the unit disk graph model.

to recognize interval graphs (see [9]), one can easily find a solution to the localization problem (note that in section 3.4 we will see that finding a solution reduces to solving a simple linear program). This implies that ranking is unique and can be computed in polynomial time, still the actual localization of the sensors remains very uncertain.

In the next subsections we use the following definitions. Let $\mathcal{A}_i$ be an algorithm that computes the estimated location of a sensor s_i. For each sensor i we denote by $\mathcal{A}_i^t(x)$ the result of the algorithm $\mathcal{A}_i$ in the t iteration. In two dimensions we use the $\mathcal{A}_i^t(x)$ and $\mathcal{A}_i^t(y)$ for the x, y coordinates that are computed by the algorithm $\mathcal{A}_i$ in the t iteration.

3.1 Ranking

The fact that one can solve the ranking problem is mainly due to the fact that there is a natural ordering in $\mathbb{R}$ (and not in $\mathbb{R}^2$). Indeed the 1 dimensional representations of the graph G can be partitioned into 2 classes, in the first one a sensor s gets rank $r(s)$ and in the other $n - r(s)$. We break this left-right symmetry since we know that station B_0 has rank 0.

We now describe a fast distributed algorithm that solves the ranking problem when the sensors range is uniform in $O(h(s_1, B_1))$ time steps.

The general idea of the algorithm is to use the fact that different sensors see different environments. Furthermore, if all the sensors start with being in the neighborhood of the base station in B_0 eventually sensors that are near B_1 will have a bigger average. Our algorithm works in $\lceil \frac{1}{r} \rceil$ phases.

Lemma 3.1. *If* $0 \leq a_1 \leq a_2 \leq a_3 \leq \ldots \leq a_i \leq \ldots \leq a_{i+j} \leq \ldots \leq a_{i+j+k}$ *then* $\frac{\sum_{n=1}^{i+j} a_n}{i+j} \leq \frac{\sum_{m=i}^{i+j+k} a_m}{j+k}$, *and equality only if* $a_1 = a_{i+j+k}$.

Proof. By induction on $i + j + k$, if $i + j + k = 1$ then the lemma holds. Assume that $i + j + k = n + 1$. If $a_1 = 0$ then $\frac{\sum_{n=1}^{i+j} a_n}{i+j} \leq \frac{\sum_{n=2}^{i+j} a_n}{i+j-1}$ and we can use the induction. If $a_1 > 0$ we can define a new variable $a_k' = a_k - a_1$ and we get that $a_1' = 0$ so we can use the previous argument.

In order to prove the correctness of the algorithm we need the following definition. Let the set of sensors $\Gamma_i = \{s : h(s, B_1) \leq i\}$ be the sensors that are in hop-distance $\leq i$ from B_1.

Lemma 3.2. $\mathcal{A}_j^t(\mathcal{X}) > 0$ *if and only if* $s \in \Gamma_t$.

Proof. The proof is by induction on the number t. For $t = 1$, the only sensors that don't change their estimation of their location are the ones that receive the transmission from B_1. This is exactly Γ_1. The rest of the sensors still estimate their location to be 0.

Now assume the lemma holds for $t = k$. From lemma 3.1 we get that for all $s_j \in \Gamma_k$, $\mathcal{A}_j^{k+1}(\mathcal{X}) > 0$. Now let $s \in S \setminus \Gamma_{k+1}$ since $N(s) \cap \Gamma_k = \emptyset$ and using the induction hypothesis we get that $\mathcal{A}_j^t(\mathcal{X}) = 0$. It remains to prove that for all $s_j \in \Gamma_{k+1} \setminus \Gamma_k$, $\mathcal{A}_j^{k+1}(\mathcal{X}) > 0$. In this case the lemma follows from the fact that $N(s) \cap \Gamma_k \neq \emptyset$ and lemma 3.1.

Using the previous lemma we get a bound on the time $stop_0$ signal is sent.

Initialize s_i,
  $\mathcal{A}_i^0(\mathcal{X}) \leftarrow 0$
  $rank(s_i) = \emptyset$
  $stop \leftarrow 0$
While $(stop = 0)$
  $\mathcal{A}_i^{t+1}(\mathcal{X}) \leftarrow \frac{\sum_{x_j \in N(x_i)} \mathcal{A}_j^t(\mathcal{X})}{|N(x_i)|}$
  if $(\exists s_j \in N(s_i) \wedge rank(s_j) \neq \emptyset \wedge rank(s_i) = \emptyset)$
    then $rank(s_i) \leftarrow rank(s_j) + |\{k \in N(s_i) : \mathcal{A}_j^t(\mathcal{X}) < \mathcal{A}_k^t(\mathcal{X}) < \mathcal{A}_i^t(\mathcal{X})\}|$
  if $(B_0 \in N(s_i) \wedge \forall j \in N(s_i), \mathcal{A}_i^t(\mathcal{X}) < \mathcal{A}_j^t(\mathcal{X}))$
    then $rank(s_i) \leftarrow 0$
  transmit $\mathcal{A}_i^{t+1}(\mathcal{X})$
  transmit $rank(s_i)$
  If(receive $stop_0 \wedge$ receive $stop_1$)
    then $stop \leftarrow 1$
  $t \leftarrow t + 1$
if(s_i receive $stop_k$ signal)
  then transmit $stop_k$ signal

Fig. 1. Pseudo code for algorithm for sensor s_i

Transmit the location i
If($\forall j \in N(B_i), \mathcal{A}_j^t(\mathcal{X}) > 0$)
  then transmit $stop_i$ signal.

Fig. 2. Pseudo code for algorithm for base station B_i

Corollary 3.3. B_0 transmit the $stop_0$ at time $h(B_0, B_1) + 1$

The next lemma shows that after n iterations Γ_n is sorted.

Lemma 3.4. After $t = h(s_1, B_1)$ iterations, if $x_i < x_j$ then $0 < \mathcal{A}^t(x_i) < \mathcal{A}^t(x_j)$.

Proof. The proof is by induction on the number of hops from the base station. We claim that after n iterations Γ_n is sorted. First we see that Γ_1 is sorted after the first iteration. Since all sensors in Γ_1 see B_1 and all the other values are 0, and so from lemma 3.1 the sensors in Γ_1 are sorted after applying the first iteration. Now assume that Γ_n is sorted at time $t = n$. For $s_i \in \Gamma_{n+1} \setminus \Gamma_n$, since $\mathcal{A}_i^n(\mathcal{X}) = 0$. From the induction hypothesis we get that for all $j \in \Gamma_n$, $\mathcal{A}_i^n(\mathcal{X}) > 0$. From the fact that the graph G_N is connected we get that $N(s_i) \bigcap \Gamma_n \neq \emptyset$. Now using lemma 3.1, we get that all sensors in $\Gamma_{n+1} - \Gamma_n$ will be sorted at time $n + 1$. Sensors in Γ_n will still be sorted as long as they compute their average at the same time step.

The next lemma shows that the algorithm stops after $2h(B_0, B_1) + 1$ iterations.

Lemma 3.5. After $t = 2h(B_0, B_1) + 1$ iterations, the algorithm stops and if $s_1 < s_1'$ then $rank(s) < rank(s')$

Proof. By corollary 3.3 B_0 sends the $stop_0$ signal at time $h(s_1, B_1) + 1$ and B_1 sends the $stop_1$ signal at time 2. In order for the sensors to stop computing their new location they have to receive both these signals. It follows that none of the sensors sets its $stop$ to 1 before time $h(s_1, B_1) + 1$, and the last of the $stop$ signals is set at time $2h(B_0, B_1) + 1$, since this is the time that it takes for the signal $stop_0$ to reach the last sensors. Note that all the sensors set their ranks before time $h(B_0, B_1) + 1$(before signal $stop_0$ was sent). From lemma 3.4 we get that the ranks are consistent i.e., if $s_1 < s_1'$ then $rank(s) < rank(s)$. Now the $stop_1$ signal and the ranks spread at the same speed and arrive at the last sensors at time $h(B_0, B_1)$. So the signal that makes the sensors set their $stop$ signals is actually $stop_0$.

Theorem 3.5.1. In one dimension the ranking problem can be solved in linear time.

3.2   Uniqueness of Localization

We remark that knowing only the graph G we can get a 2 approximation of the localization, indeed if $h(s, B_0) = k$ then $s(1)$ is at least $\frac{k-1}{2} \cdot r$ and at most $k \cdot r$. We show now that even if it is easy to find a localization of the sensors that is consistent with the observations, there exist instances for which one can find two consistent localizations in which the coordinates may differ by more than $1/6$ (and since the sensors are in $[0, 1]$ this is about 16% of the network width).

Note that sensor localization is not "hard"algorithmically, it's simply impossible since we miss information.

Theorem 3.6. *For any algorithm $\mathcal{A}$ that solves the localization problem in one dimension there exists a location of sensors $x_1, x_2, ..., x_n$ such that the error of the estimation $|x_i - \mathcal{A}(x_i)|$ for some i is bigger than $1/6$.*

Proof. Let ρ be the radius of the sensors. We put $n = \frac{1}{\rho + \frac{\rho + \epsilon}{2}}$ points in the interval $[0, 1]$. This means that $n\rho + \frac{n}{2}(\rho + \epsilon) = 1$. We consider two different scenarios, G and G'. The location in G is:

$$x_i = \begin{cases} i \cdot \rho & i \le \frac{n}{2} \\ \frac{n\rho}{2} + k(\rho + \epsilon) + \rho & i = \frac{n}{2} + (2k + 1) \\ \frac{n\rho}{2} + k(\rho + \epsilon) & i = \frac{n}{2} + 2k. \end{cases}$$

Similarly, the location of G' is:

$$x_i = \begin{cases} k(\rho + \epsilon) + \rho & i = (2k + 1) \le \frac{n}{2} \\ k(\rho + \epsilon) & i = 2k \le \frac{n}{2}. \\ \frac{n}{4}(\rho + \epsilon) + (i - \frac{n}{2}) \cdot \rho & i \ge \frac{n}{2} \end{cases}$$

If we let ϵ small enough, both locations give the same neighborhood graph and we can't differentiate them. We see that forcing $\epsilon \to 0$, for $i = n/2$, in G, x_i is placed on $2/3$ and in G', x_i is placed in $1/3$. Since the distance between them is $1/3$, the best estimation we can do will give an error of $1/6$. This proves the theorem.

3.3 Approximated Localization Under Distribution Assumption

We now use the rank in order to get a location estimator under the assumption that the sensors distribution is known. We illustrate the idea with an example that uses the uniform distribution.

Assume that the sensors are $i.i.d$ distributed, we can use the ranking algorithm combined with the distribution of the order statistics to solve the localization problem. Naturally the localization of the sensor with rank i will be $E[x^{(i)}]$ which is the expectation of the i-order statistics. We remark that the variance of the i-order statistics has been well studied (for a background on this subject we refer the reader to [1]. For example assume the sensors are placed by a uniform distribution in the $[0, 1]$. In this case it is well known that the ith order statistics has a $Beta(i, n - i + 1)$ distribution. Therefore the algorithm will compute the location of the sensor with rank i as being $\frac{i}{n+1}$, and the variance of this location is $\frac{i(n+1-i)}{(n+1)^2(n+2)}$. Simple calculation shows that the middle sensor with rank $\frac{n+1}{2}$ has the maximum variance, which is equal to $\frac{1}{4(n+2)}$, this variance is extremely small compared to the $1/6$ error that one can made in the deterministic case.

For a general distribution, one needs simply to compute an estimation of $E[x^{(i)}]$ and to output it as coordinate. The quality of this rank based estimator will depend on the variance of the i-order statistic.

3.4 Best Localization

According to section 3.2 we have many solutions that are consistent with the observation. Therefore we can not hope to find the sensor coordinates, but we may find a "best solution". Many different quality measures may be chosen. The most natural measure would be to choose the mean of all the solutions. We start with some definitions. A solution of the localization problem can be identified with a vector of $\mathbb{R}^n$. Let s_i denote the sensor with rank i, we associate to a solution A the vector $v(A)$ of $\mathbb{R}^n$ defined by $v(A)_i = s_i(1)$ $(v(S)_i$ is the coordinate of s_i in the solution). The solution set happens to be a convex polytope P, indeed a vector v represents a solution if and only if :

- $\forall i, j, |v_i - v_j| \geq r$ if s_i cannot transmit to s_j;
- $\forall i, j, |v_i - v_j| \leq r$ if s_i can transmit to s_j
- $v_1 = 0, v_n = 1, \forall i, v_i \geq 0, v_i \leq 1$.

This system is non linear but if one assumes that the sensors are ranked the sign of $v_i - v_j$ is known and the system becomes linear. The mean $\overline{v}$ of all solutions minimizes the expectancy of $\sum_{v \in P} |v - \overline{v}|^2$, assuming that all the solutions $v \in P$ are equally likely.

Finding the average solution reduces to computing the center of mass of polytope P. This problem is believed to be $\mathcal{NP}$ hard, but one can get a good approximation of the center of mass by sampling the points of the polytope and taking the sample average. In order to get in polynomial time a good sample of the polytope one can simply start from a point inside the polytope and walk randomly using a *hit and run walk* [12].

In our case we can get a better result, since the solution polytope is very particular. Indeed one can show that its vertices are integral and use Fourrier Motzkin [15] method to compute the integral $\int_{x \in P} x$ exactly in polynomial time.

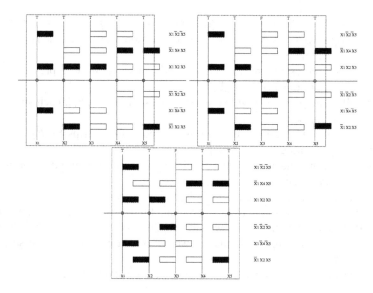

Fig. 3. Exempla of a Flip Game for the formula $(x_1 \vee x_2 \vee x_3) \wedge (\overline{x_1} \vee x_4 \vee x_5) \wedge (x_1 \vee \overline{x_2} \vee \overline{x_5})$

Theorem 3.6.1. In the one dimensional case one can compute the average solution to the localization problem in time $O(n^4)$.

4 Two Dimensional Case

In this section we sketch the proof that it is $\mathcal{NP}$-hard to solve the ranking and the localization problem in dimension 2. We also show that there exists no polynomial algorithm computing a localization that is approximately consistent with the observations unless $\mathcal{P} = \mathcal{NP}$, this implies that finding an approximated drawing of unit disk graphs is $\mathcal{NP}$ hard. This problem was mentioned as open in [14]; indeed the exact drawing problem was known to be $\mathcal{NP}$ hard [3], but the proof was much more complex than the one we give and could not be extended to the approximated case.

Note that we just give the main ideas of our proofs, a detailed version will be given in the final version.

We first give a sketch of the hardness proof. As in [2], we use a reduction to a specific form, a $3Sat$, the Jigsaw that we use is also directly inspired from this work.

The *Symmetric 3Sat problem* in which one wishes to satisfy not only any clause C of a formula F but also all the symmetric clauses $\overline{C}$ (eg. at least one literal is false in any clause) is $\mathcal{NP}$ Hard.

A Jigsaw Game to Emulate Symmetric 3Sat

The next Jigsaw that we will call *Flip game* is a game which emulates a symmetric $3Sat$ instance with a formula F. For an example of the game see figure 3.

- The Jigsaw is made with one horizontal axis on the $y = 0$ axis; on this axis n vertical rigid *bars* $B_1, B_2, \ldots, B_n$ can rotate (more exactly for each bar we can use the $y = 0$ symmetry).
- The bar B_i represents the variable x_i, and can be set up to represent the setting $x_i = TRUE$ or down $x_i = FALSE$.
- The half plane $y > 0$ is divided in to m strips $S_1, S_2, \ldots, S_m$, each strip S_i representing the clause C_i.
- symmetrically, the half plane $y < 0$ is divided in to m strips $\overline{S_1}, \overline{S_2}, \ldots, \overline{S_m}$, the strip $\overline{S_i}$ representing the clause $\overline{C_i}$.
- On each bar are attached *flippers* , a flipper is a rigid structure that can be flapped by symmetry around the bar axis.
- Flippers are attached to the bar with the following rules :
    If x_i is non negated in clause j we put a flipper on bar B_i in the up-strip j
    If x_i is negated in clause j we put a flipper on bar B_i in the down-strip j
    If x_i does not appear in clause j we put one flipper in the down-strip and the up-strip j.
    The space between two bars can contain at most one flipper.

The Flip game consists in rotating the bars and flipping the flippers so that flippers do not overlap.

Lemma 4.1. *The Flip game is $\mathcal{NP}$-Hard.*

Proof. Consider a winning position, we associate to it a Truth setting (x_i is true is the bar B_i is up). Since any strip can hold at most $n - 1$ non overlapping flippers, we conclude the bars are set so that in each strip we find at most $n - 1$ flippers. The number of flippers in any up (resp. down) strip i is respectively n minus the number of true literals (minus the number of false literals) in clause C_i. Hence flippers are non overlapping only if the associated truth setting satisfies F and $\overline{F}$.

Conversely, given any truth setting that satisfies the formula F we can set the bars according to it, them we have enough room to set the flippers in non overlapping position by flipping then left or right.

Emulating the Flip Game
Overall Argument

Consider a flip game, and any winning position drawn in the plane. We will associate to this drawing a sensor network with uniform transmission range 1, the key point is that the associated graph G does not depends on the winning position of the flip game.

Now, assume that we are given the graph G, due to rigidity constraints we can easily recover the flip game structure and moreover if we want to draw G and respect the sensor observations we have to draw it as a flip game in a winning position. Indeed the only freedom that we have when we try to draw G corresponds exactly to the possible moves of the flip game, and due to the sensors observations we cannot drawn overlapping flippers. Hence if we manage to draw G we have been able to find a winning position of the flip game.

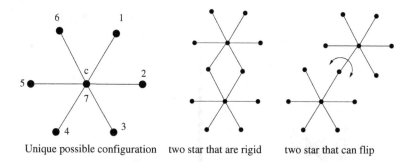

Unique possible configuration   two star that are rigid   two star that can flip

Fig. 4. Example of the three simple gadgets, the first is the rigid 7-star; this star has 7 leaves, note that in our drawing 2 points are placed at the center (the center of the star and one leave). The second gadget show how to connect two rigid 6-star and get a rigid shape. The last gadgets is the flipper gadget.

To emulate the flip game with sensors we will first force the sensors to be on the hexagon lattice using rigid graph structures; second we will maintains the jigsaw game degree of freedom. For this we use two gadgets; we call the first the rigid gadget, since it create a rigid bounding box which forces all the sensor to be set on the hexagon lattice. We call the second gadget flipper.

In what follows, we consider a drawing of a graph G modulo $\mathbb{R}^2$ isometries (for l_2 that is rotations translation), eg isometric drawings are considered as identical. When $d(u, v) = 1$ we may choose to have $[u, v]$ in $E(G)$ or not. We call *rigid* any graph which admits up to isometries one unique drawing.

Standard ball packing argument shows that the 7-star is rigid see figure 4, so if the graph G contains a 7 star it must drawn like on figure 4 up to translation or rotation.

Using those as basic blocks we can create rigid graphs of any shape on the hexagon lattice. We Remark that whenever we fix two vertices a and b of a rigid subgraph there are at most two ways to draw it, that are symmetric along the line (a, b).

Our graph is made from rigid components, components that are attached to the remaining of the structure by either 2 points (u, v) (so they can rotate along the (a, b) axis) or one point u (the can rotate around u) see figure 5.

We describe here how to simulate a Flip game (see figure 5, and figure 6 for an examples).

- We start defining our graph by making a large rigid bounding box. Inside we add a line of hexagons. At two opposite points of the box, this line is flexible but since it contains exactly enough hexagons to connect the two points, it must be drawn straight. We call this line the horizontal bar (see fig...). Note that each hexagon of the line still has the freedom to rotate around the y axis. Bars will be attached to those hexagons allowing them to rotate.
- Each Bar is built from a vertical line of hexagons, this line is associated to a rigid structure that force it to be drawn straight. On each bar each hexagon can be rotated around the bar axis.

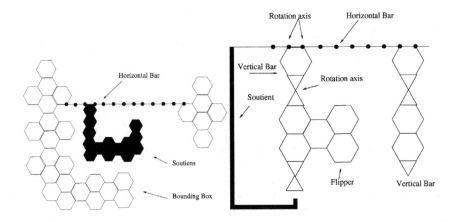

Fig. 5. Example of a Flip Game emulation

- Flippers are simply rigid structures that are attached to the bars, note that they can be flipped.

4.1 Range-Free Localization in Two Dimension Is APX

From our proof, it follows that we cannot approximate the ranking (that is find approximated ranks that correspond to localizations that are consistent with the data). Indeed, in our reduction ranking and localization are the same problem since the sensors are forced to be on the hexagonal lattice.

We now extend our result and sketch the proof that finding a localization that is *almost* consistent with the observations is $\mathcal{NP}$ hard.

Definition 1. *A localization is an ϵ approximated solution if s can transmit to s' when $d(s, s') \leq r$, s cannot transmit to s' if $d(s, s') \geq (1 - \epsilon)r$.*

The above definition simply means that the location of the sensors is such that the transmission range r is more or less respected.

Theorem 4.1.1.

- *In dimension 2, there exists an $\epsilon > 0$ such that finding an ϵ approximated localization is $\mathcal{NP}$ hard.*
- *finding an ϵ approximated realization of a unit disk graph is $\mathcal{NP}$ hard.*

Assume that we allow non edges to have length $1 - \varepsilon$ then the structures that we use are not perfectly rigid. But one can show that if ε is fixed and small enough the drawing of the rigid structures remains unique (up to isomorphism) and to some move of ε' that are sufficiently small so that disjoint structures cannot overlap. To avoid the small moves to add up and to bend the lines we use the next gadgets:

- We want to force the flippers of a given strip to be drawn in the same horizontal area, in order to bound the non rigidity drift we add on each bar and each strip a device that force the chunk i of two adjacent bars to be drawn in the same area (cf figure 7).

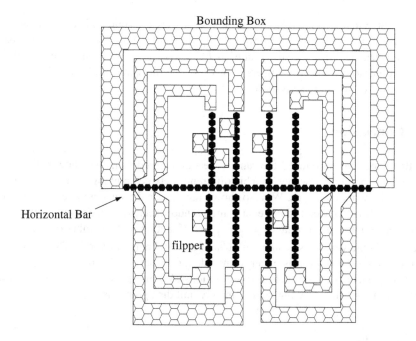

Fig. 6. The big picture

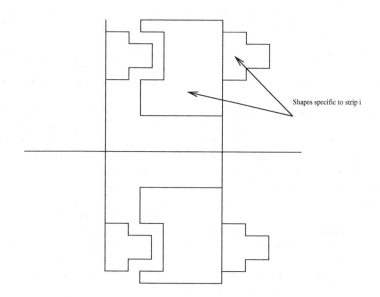

Fig. 7. Synchronizing bars with specific shapes

- We choose the flippers of adjacent bars large enough on the horizontal side so that they would overlap if they are flapped on the same size.

Assume that L is the distance of bars when they are rigid, L is also the horizontal size of a flipper. Since we have m strips, when the bars are not rigid the distance between them in some strip is at most $L(1 + \varepsilon') + \varepsilon' m$. We need it to be less than $2L$ in order to prevent two flippers to be drawn on the same side. If we choose $L = m$ both conditions are satisfied for $\varepsilon' \leq 1/2$.

5   Conclusion

In this paper we introduced the range free ranking problem. We showed that from a complexity point of view there is a big difference between ranking in one dimension and ranking in two dimensions. We also shown that in one dimension there is an important difference between worst case scenario and a random scenario, for example in the random uniform case the error (variance $< \frac{1}{4(n+2)}$) is a function of the number of sensors while the error in the worst case can be of 16% of the total network width. This result shows that in the random case it is better to include the sensors in the averaging process, and not to depend only on the anchors (base stations). We also proved some hardness results, and one implies that drawing approximately unit disk graphs is $\mathcal{NP}$ hard.

References

[1] Barry C. Arnold, N. Balakrishnan, H. N. Nagaraja July, A First Course in Order Statistics, Wiley Series in Probability and Mathematical Statistics: Probability and mathematical Statistics, 1992.

[2] N. Bhatt , Stavros S. Cosmadakis, The complexity of minimizing wire lengths in VLSI layouts, Information Processing Letters, v.25 n.4, pp. 263-267, June 17, 1987

[3] Heinz Breu and David G. Kirkpatrick. Unit Disk Graph Recognition is NP-Hard. Computational Geometry, 9:3-24, January 1998.

[4] N. Bulusu, J. Heidemann and D. Estrin, GPS-less Low CostOutdoor Localization for Very Small Devices, IEEE Personal Communications Magazine, 7(5):28-34, October 2000.

[5] N. Bulusu, J. Heidemann and D. Estrin, Density Adaptive Algorithms for Beacon Placement in Wireless Sensor Networks, In IEEE ICDCS '01, Phoenix, AZ, April 2001.

[6] T. Eren, D. Goldenberg, W. Whiteley, Y.R. Yang, A.S. Morse, B.D.O. Anderson, P. N. Belhumeur.Rigidity, Computation and Randomization in Network Localization, IEEE INFOCOM 2004.

[7] Tian He, Chengdu Huang, Brian M. Blum, John A. Stankovic, Tarek Abdelzaher, Range-free localization schemes for large scale sensor networks, MobiCom 2003, pp 81-95.

[8] J. Hightower and G. Boriello. Location systems for ubiquitous computing. IEEE Computer, 34(8) Aug. 2001, pp 57-66.

[9] N. Korte and R. H. Mohring. An incremental linear time algorithm for recognizing interval graphs. SIAM J. Comput., 18:68–81, 1989

[10] D. Niculescu and B. Nath, DV Based Positioning in Ad hoc Networks, In Journal of Telecommunication Systems, 2003.

[11] L. oherty, K. S. J. Pister, and L. E. Ghaoui. Convex position estimation in wireless sensor networks. In Proceedings of the IEEE Infocom, pages 1655-1663, Alaska, April 2001.

[12] L. Lovasz, Hit-and-run mixes fast, Mathematical Programming, Issue: Volume 86, Number 3, December 1999 pages 443-461.

[13] Teemu Roos, Petri Myllymaki, Henry Tirri, A Statistical Modeling Approach to Location Estimation, IEEE Transactions on Mobile Computing – January 01, 2002 Issue 1, pages 59-69.

[14] Thomas Rusterholz, Report On the Approximation of Unit Disk Graph Coordinates, http://dcg.ethz.ch/theses/ss03/virtualCoordinates_report.pdf.

[15] M. Schechter. Integration over Polyhedra, an Application of the Fourier-Motzkin Method", Am. Math. Monthly, 105(1997), pages 246-251.

[16] Yi Shang, Wheeler Ruml, Ying Zhang, Markus P. J. Fromherz, Localization from mere connectivity, MobiHoc 2003. pages 201-212.

[17] (Editor) Ivan Stojmenovi, Handbook of Wireless Networks and Mobile Computing. John Wiley and Sons. February 2002.

An Analytic Model Predicting the Optimal Range for Maximizing 1-Hop Broadcast Coverage in Dense Wireless Networks

Xiaoyan Li, Thu D. Nguyen, and Richard P. Martin

Department of Computer Science, Rutgers University
110 Frelinghuysen Rd, Piscataway, NJ 08854, USA
{xili,tdnguyen,rmartin}@cs.rutgers.edu

Abstract. We present an analytic model to predict the optimal range for maximizing 1-hop broadcast coverage in dense ad-hoc wireless networks, using information like network density and node sending rate. We first derive a geometric-based, probabilistic model that describes the expected coverage as a function of range, sending rate and density. Because we can only solve the resulting equations numerically, we next develop extrapolations that find the optimum range for any rate and density given a single precomputed optimum. Finally using simulation, we show that in spite of many simplifications in the model, our extrapolation is able to predict the optimal range to within 16%.

1   Introduction

We consider the problem of how networks of extremely dense embedded devices connected by a wireless network can set their output power to maximize the number of 1-hop receivers of a broadcast message. Many recent protocols proposed for use in ad-hoc wireless networks, such as directed diffusion [2] and ad-hoc positioning [8], rely on periodic broadcasts. In addition, many protocols, such as code propagation [6] and dynamic source routing (DSR) [3] rely on aperiodic broadcasts. Indeed, the fundamental broadcast nature of wireless networks makes broadcast an ideal building block for the discovery, routing and localization functions, which will be critical for future ad-hoc wireless systems. At the same time, as technology trends continue to reduce the size, power consumption and cost of embedded wireless network devices, the scale of the networks built with them will continue to grow. For example, sensors are rapidly approaching the small size and low cost necessary for novel usage such as *active tagging*, i.e., representing the state of common everyday objects such as furniture, pens, and books. Our initial measurements of a common indoor environment for the active tagging application show that the average degree of such a network could easily range up to several hundred.

Given the high densities of these future sensor networks, a resulting challenge for applications using broadcast will be how to manage channel capacity to ensure good performance in terms of throughput, fairness and broadcast coverage. This challenge arises because if all nodes act greedily, using the maximum range, the channel will collapse; that is, the likelihood of any neighbors receiving the message correctly quickly

I. Nikolaidis et al. (Eds.): ADHOC-NOW 2004, LNCS 3158, pp. 172–182, 2004.

approaches zero in dense networks due to collisions. As shown in [6], many protocols perform poorly as density increases because of this issue.

We are thus motivated to consider how devices in these networks can maximize the number of 1-hop receivers of a broadcast message. We refer to *broadcast* as the 1-hop message, as opposed to *flooding* which intends to cover all the nodes via multi-hop forwarding. We choose to examine broadcast coverage because it is an important metric for protocols using broadcast as a building block. In addition, a network without good broadcast coverage is unlikely to have good throughput or latency properties — if many nodes fail to receive a broadcast message they would most likely have trouble receiving a normal unicast message due to the fundamental channel properties (i.e., a single sender impacts many receivers) of wireless transmission.

Our approach centers on the *spatial reuse* of wireless resources. Specifically, given the surrounding sending rate, node density, and a simple geometric model of wireless communication to compute the radio range, each node can just set its range to the optimal value, which probabilistically maximizes the 1-hop *coverage* for a broadcast packet. We use radio range as a parameter because it is more tractable to analyze than output power directly; adjusting output power to achieve the computed optimal range is future work.

To derive our analytic approach, we first build a geometric model that predicts the likelihood of a collision given all nodes in a uniform density network broadcasting at the same rate using the same range. Using this model we can numerically solve for the optimal range for any combination of density and broadcast rate. Unfortunately, we cannot use the above model analytically since the optimal range is not easily computable using a closed-form formula. Thus, we next introduce two relations from which we can build an extrapolation that can be used in real-time to compute a range setting that is close to optimal based on information about the local density and sending rate. We show that simplifying assumptions made in deriving the model and the inexactness of one of the extrapolation relations leads to at most 16% error in estimating the optimal range.

We argue that our analytic approach is quite applicable to small wireless devices because of its simplicity; the complexity is buried offline in the numeric solver leaving the nodes to only perform the lighter weight extrapolation. In addition, we believe that estimations of the local density and sending rate are possible via local observations on the transmission channel. For example, distances to neighbor nodes can be computed using signal strength [1] or signal timing-differential approaches [9]. We leave the exact nature of density estimation as future work.

The rest of this paper are organized as following: In Section 2, we construct the geometric model. Section 3 introduces the extrapolation method. In Section 4 we evaluate the approach via simulation. We compare and contrast our work to related work in Section 5. Finally, in Section 6 we conclude.

2   Model Construction

In this section, we develop an analytic model to derive optimal transmission range based on the sending rate and node density. For simplicity, we develop only a two-dimensional model; extending the model to three-dimensions is straightforward. (All our simulations run in the full three dimensions.)

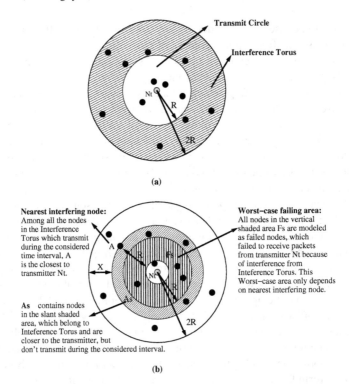

Fig. 1. Definitions and assumptions behind the derivation of our model in 2 dimensions. (a) shows the basic setup of our model, where N_t is the transmitting node, the white circle encapsulate nodes that should receive a transmission from N_t, called the transmit circle, and the slant-shaded area contains nodes that can interfere with the reception of N_t's packet. (b) shows what happens when a node A interferes with N_t's transmission causing some nodes in the transmit circle to fail. Here we assume that A is the interfering node closest to N_t, with distance X from the outer circle (2R) of the interference torus.

We use a set of assumptions to make an analytic solution tractable: (1) nodes are spatially distributed according to a Poisson distribution with an average density of λ_s; (2) applications running on each node generate packets to send according to a Poisson distribution with average rate λ_p; (3) all packets are of the same length and take time T to transmit; and, (4) all nodes use a CSMA protocol. We also use a fairly simple wireless communication model similar to the one in [11]: all nodes have the same radio range, where nodes within R distance from a transmitter will detect the packet transmission while those further away will not. More than one packet transmission within distance R to a receiver will cause collision and all overlapped packets at the receiver are corrupted.

Figure 1 shows the geometric reasoning underlying our model, which derives the expected coverage for a particular transmitter given the assumptions listed above. We call the area within distance R of the transmitter the *transmit circle*. During a packet transmission, nodes inside the transmit circle will detect the transmission and thus be blocked from transmitting via CSMA. Transmissions from within the *interference torus* (the slant shaded area bounded by circles with radius R and $2R$ from the transmitter, shown in Figure 1 (a)) will have a similar effect as hidden terminals, interfering with the receipt of the packet for some of the nodes in the transmit circle. We assume that the sending rates of nodes in the interference torus are not affected by CSMA; that is, transmission for each of those nodes is still independently defined by a Poisson distribution. This assumption means that we do not have to consider nodes that are further than $2R$ away from the transmitter, since their only effect on the nodes under consideration is through CSMA. This assumption is conservative since CSMA effectively slows down the sending rate, implying that the model will overestimate the number of interfering packets.

We call a node within the transmit circle that does not correctly receive the transmitted packet a *failed* node. Thus, we model packet coverage as the number of nodes that can detect a particular packet transmission minus failed nodes, which gives the following equation:

$$E(C) = \lambda_s(\pi R^2) - E(\text{number of failed nodes})$$

where $\lambda_s(\pi R^2)$ is just the expected number of nodes in the transmit circle. As can be seen, the task of modeling coverage really becomes a task of deriving the number of failed nodes. An exact derivation, however, would be quite challenging because in the general case, we would have to account for multiple overlapping circular interference regions caused by colliding transmissions from the interference torus. One method to make the analysis tractable is to assume that if a node with distance d to the transmitter fails, then all nodes i with distance $d_i \geq d$ also fail. This assumption effectively means that the modeled total failing area (shown as the vertically shaded *worst-case failing area* in Figure 1 (b)) will be decided by the distance from the transmitter to the *nearest interfering node*. Clearly this simplification is conservative; the modeled failing area will always be larger than or equal to the true failing area. However, the worst-case failing area computation is now easily tractable.

Let X be a random variable that represents the distance from the nearest interfering node (A in Figure 1(b)) to the outer ring of the torus. Let A_s (the slant shaded area in Figure 1(b)) be the area of the Interference torus containing nodes which are closer to the transmitter than A is to the transmitter, such that

$$A_s = \pi((2R - x)^2 - R^2)$$

Then, the CDF for X is:

$$P(X \leq x) = \sum_{k=0}^{\infty} \left[P(k \text{ nodes in } A_s)(P(a \text{ node does not transmit for } 2T))^k \right]$$

It gives the probability that the closest interfering node is at least $(2R - x)$ away from the transmitter; i.e., the probability that no nodes within A_s transmit during the current transmission. (We use $2T$ as the bounding time range since unslotted CSMA needs both the current transmission period T and the previous T period to be quiet for a successful reception.) Substituting in the appropriate parameters, this equation becomes

$$P\left(X \leq x\right) = \sum_{k=0}^{\infty} \left[\left(\frac{(\lambda_s A_s)^k e^{-\lambda_s A_s}}{k!} \right) \left(e^{-\lambda_p 2T} \right)^k \right]$$

$$= e^{-\lambda_s A_s} \sum_{k=0}^{\infty} \left[\frac{(\lambda_s A_s e^{-\lambda_p 2T})^k}{k!} \right]$$

$$= e^{-\lambda_s A_s} e^{\lambda_s A_s e^{-\lambda_p 2T}}$$

$$= e^{-\lambda_s A_s \left(1 - e^{-\lambda_p 2T}\right)} \tag{1}$$

Thus, the expected number of failed nodes in the worst case can be readily computed as:

$$NF_w = \int_0^R \left[\lambda_s \pi (R^2 - (R - x)^2) P(x \leq X \leq x + dx) \right] \tag{2}$$

where $\pi(R^2 - (R - x)^2)$ represents the *worst-case failing area* (shown as the vertical shaded area in Figure 1(b)).

As mentioned above, approximating the failing area with *worst-case failing area* is conservative. Particularly, when there are very few interfering transmitters, this simplification is too conservative, and thus unable to provide a good approximation. To account for the inaccuracy, we set a threshold for the expected number of interfering transmitters. Only when the expected number reaches this threshold, we use the worst-case failing area as the modeled failing area; otherwise we use portions of the worst-case failing area.

To compute the expected number of interfering transmitters, we need to derive an expression for the *aggregate sending rate* given an average node density λ_s and average individual sending rate λ_p. Recall that we model the transmission at each node as an independent Poisson process with intensity λ_p, which will transmit with probability $1 - e^{-\lambda_p t}$. Given a density, the aggregate sending rate observed per unit area then is simply $\lambda_s(1 - e^{-\lambda_p t})$. Thus the expected number of interfering transmitters is the aggregate transmission from the interference torus area:

$$E(Tr) = \lambda_s(1 - e^{-\lambda_p 2T})\pi((2R)^2 - R^2) \tag{3}$$

We derive the interfering transmitters threshold by assuming all the interfering transmitters sit in the middle of the interference torus radius [1] (i.e., $\frac{3}{2}R$ to the center of the transmit circle). We can then compute the worst-case failing area and the interfered area from a single interfering transmitter respectively. The ceiling for the quotient of these two areas is used as the threshold. The threshold for 2 dimensions is 6 [2], and for 3 dimensions is 11.

---

[1] Through similar simulation validations as in Section 4 we find that assuming the interfering transmitters are in the middle of the interference torus area/volume would give a better approximation, but that underestimates the collisions in a few cases, which is not conservative with respect to R.

[2] For 2 dimensions, the worst-case failing area in this case is $\frac{3}{4}\pi R^2$. For 2 circles with radius R and the centers d apart $(d < 2R)$, the intersection area can be computed as $(2R^2 \cos^{-1}(\frac{d}{2R}) - \frac{d}{2}\sqrt{4R^2 - d^2})$. Thus, the interfered area from a single interfering transmitter is $(R^2(\cos^{-1}(\frac{3}{4}) - \frac{3\sqrt{7}}{8}))$. So the threshold is $\lceil 5.19 \rceil = 6$.

Finally, we complete the model as the following piecewise function:

$$E(C) = \begin{cases} \lambda_s(\pi R^2) - \frac{E(Tr)}{6} NF_w & E(Tr) < 6 \\ \lambda_s(\pi R^2) - NF_w & E(Tr) \geq 6 \end{cases} \tag{4}$$

3 Practical Usage of the Model

The model we have just derived gives us a mathematical means to estimate the coverage given λ_s, λ_p and T. Unfortunately, the final result is not a closed form formula and so requires numeric methods to find the R value (R_o) that maximizes the coverage. This makes it impractical for the embedded devices to use the model directly at run-time. Thus, in this section, we derive two relations that will allow us to find any R_o given a single pre-computed R_o for a particular tuple of (λ_s, λ_p, T). These extrapolations thus make it tractable to find R_o in the context of a running protocol stack on resource-constrained devices.

3.1 Relating Density to Coverage

We begin by deriving a relationship between the expected coverage of two scenarios with the same sending rate (λ_p) but different densities(λ_s). In particular, given two scenarios ($\lambda_{p1}, \lambda_{s1}, R_1$) and ($\lambda_{p2}, \lambda_{s2}, R_2$) with the same sending rate, the expected coverage between the two scenarios is the same when the ranges encircle the same expected number of nodes. Specifically:

$$\lambda_{p1} = \lambda_{p2} , \ \lambda_{s1} R_1^2 = \lambda_{s2} R_2^2 \ \Rightarrow \ E(C_1) = E(C_2) \tag{5}$$

To show that the above relationship holds, observe that since $\lambda_{s1} R_1^2 = \lambda_{s2} R_2^2$, the two scenarios are geometrically identical given our model represented by Figure 1. That is, the expected number of nodes in the transmit circle and in the interference torus are exactly the same across the two scenarios. Further, since all nodes are spatially distributed according to a Poisson distribution and have the same packet sending rate, λ_p, we can conclude that each node's transmission will be equally affected in the two cases, which means that the expected coverage in the two cases will be the same.

3.2 Relating the Sending Rate to Coverage

Having derived a relationship between density and coverage, we now consider the other dimension, the sending rate. The intuition is to derive a relationship describing the expected coverage change as we vary the number of interfering transmissions while keeping the range constant.

We have shown in Section 2 that given an average node density λ_s and average individual sending rate λ_p, the aggregate sending rate observed per unit area is simply $\lambda_s(1 - e^{-\lambda_p t})$. Now, given two scenarios ($\lambda_{p1}, \lambda_{s1}, R_1$) and ($\lambda_{p2}, \lambda_{s2}, R_2$), we wish to show that

$$\lambda_{s1} \left(1 - e^{-\lambda_{p1} 2T}\right) = \lambda_{s2} \left(1 - e^{-\lambda_{p2} 2T}\right) , \ R_1 = R_2 \ \Rightarrow \ \frac{E(C_1)}{E(C_2)} = \frac{\lambda_{s1}}{\lambda_{s2}} \tag{6}$$

To show that this relationship holds, observe that $E(C)$ as defined in equation 4 can be rewritten as:

$$E(C) = \lambda_s \pi \Psi(R, \lambda_s, \lambda_p), \; with \; \Psi(R, \lambda_s, \lambda_p) = \begin{cases} R^2 - \frac{E(Tr)}{6} \frac{NF_w}{\lambda_s \pi} & E(Tr) < 6 \\ R^2 - \frac{NF_w}{\lambda_s \pi} & E(Tr) \geq 6 \end{cases}$$

Since $\lambda_{s1}(1 - e^{-\lambda_{p1} 2T}) = \lambda_{s2}(1 - e^{-\lambda_{p2} 2T})$ and $R_1 = R_2$, combining the above equation with equations 1, 2 and 3 shows that the $\Psi(R, \lambda_s, \lambda_p)$ portion of the above equation is the same for the two scenarios. Thus, we conclude that $\frac{E(C_1)}{E(C_2)} = \frac{\lambda_{s1}}{\lambda_{s2}}$.

Note that since the effect of CSMA on the observed aggregate sending rate changes with the density, our assumption of independent transmission rate in the interference torus translates to different levels of inaccuracy for different densities. Thus, the relationship just derived is not exact, whereas the first relationship is exact.

3.3 Extrapolation Using the Derived Relations

We now show how the above two relations can be used to find the optimal transmission range R_o for an arbitrary network defined by the tuple $(\lambda_s, \lambda_p, T)$.

As described in Section 1 and verified in our experiments, the curve plotting expected coverage against the transmission range in a specific setting would follow a bell shape. The range corresponding to the maximum expected coverage is the R_o for this specific setting.

Using equation 5, if we keep λ_p fixed while changing λ_s to λ_s', then

$$E(C(\lambda_s', \lambda_p, r')) = E(C(\lambda_s, \lambda_p, r)), \; where \; \lambda_s' r'^2 = \lambda_s r^2$$

In essence, when we increase λ_s, the bell curve moves left and becomes sharper; when we decrease λ_s, the curve moves right and becomes flatter. However, the curve will remain bell shaped such that the new R_o' is given by

$$R_o' = \sqrt{\frac{\lambda_s R_o^2}{\lambda_s'}} \tag{7}$$

Using equation 6, if we change both λ_p and λ_s but keep $\lambda_s(1 - e^{-\lambda_p 2T})$ as well as R constant, then the expected coverage for each transmission range value will scale according to

$$E(C(\lambda_s', \lambda_p', r)) = E(C(\lambda_s, \lambda_p, r))\frac{\lambda_s'}{\lambda_s}, \; where \; \lambda_s\left(1 - e^{-\lambda_p 2T}\right) = \lambda_s'\left(1 - e^{-\lambda_p' 2T}\right)$$

The above relation means that the new curve for expected coverage will change along the y-axis but not the x-axis: it will shrink when we decrease λ_s and grow when we increase λ_s. So, the original R_o will still correspond to the maximum coverage, i.e.:

$$\lambda_s(1 - e^{-\lambda_p 2T}) = \lambda_s'(1 - e^{-\lambda_p' 2T}) \quad \Rightarrow \quad R_o' = R_o \tag{8}$$

Putting the above reasoning together, if we know R_{o1} for a particular tuple $(\lambda_{s1}, \lambda_{p1})$, then we can compute R_{o2} for any other tuple $(\lambda_{s2}, \lambda_{p2})$, assuming that packets are of the same size and so T stays constant. This computation, which is just an extrapolation along the dual dimensions of λ_s and λ_p, is done as follows. Let

$$\lambda_{p3} = \lambda_{p1} \quad and \quad \lambda_{s3}(1 - e^{-\lambda_{p3}2T}) = \lambda_{s2}(1 - e^{-\lambda_{p2}2T})$$

Then, using equations 7 and 8, we can compute that

$$R_{o3} = \sqrt{\frac{\lambda_{s1} R_{o1}^2}{\lambda_{s3}}} \qquad and \qquad R_{o2} = R_{o3}$$

Thus,

$$R_{o2} = \sqrt{\frac{\lambda_{s1} R_{o1}^2}{\lambda_{s3}}} = \sqrt{\frac{\lambda_{s1}(1 - e^{-\lambda_{p1}2T}) R_{o1}^2}{\lambda_{s2}(1 - e^{-\lambda_{p2}2T})}}$$

shows that we can derive the optimal transmission range for any tuple (λ_s, λ_p) given a known optimal range for one such tuple. In the remainder of the paper, we will use C_o to denote the quantity $\lambda_{s1}(1 - e^{-\lambda_{p1}2T}) R_{o1}^2$.

4   Model Validation

We now proceed to quantify the error introduced by the various simplifying assumptions that make the derivation and analysis of the model tractable; specifically: (1) the assumption that sending rates of nodes in the interference torus are not affected by CSMA, and (2) approximating the failing area with portions of the whole worst-case failing area. We use a simulation-based approach, where we simulate a large number of scenarios and then compare the observed optimal range with that predicted by the model. In particular, our simulator captures the rate-limiting effect of CSMA and there is no approximation involved in counting successful or failed receptions. We shall see that under a wide range of conditions, the modeling error never results in a range prediction error of more than 16%.

Our validation process is as follows: We first numerically solve C_o for a number of different (λ_s, λ_p) tuples with a constant T (0.04 sec.). Although our extrapolations are not exact, we found that all the numerically computed C_o values agreed to within 4 digits, and thus we always use $C_o = 0.188$. Next, we simulated various network scenarios to empirically find the optimal R_o for different settings of (λ_s, λ_p) and compare them against the ones computed using extrapolation from C_o. To find the empirical R_o, we experimentally plot the coverage vs. range curve while varying R to find the one that maximizes coverage. (In both experiment sets, the explored granularity for transmission ranges is 0.1m.)

Our simulator models 3 dimensional spaces. The network configuration and wireless communication follow the same basic model as described in Section 2. To account for boundary effects, we implemented spatial wrap-around, where transmissions propagate at the edge of the space as if space is a torus. In all of the experiments, we make sure

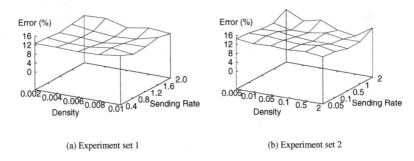

(a) Experiment set 1   (b) Experiment set 2

Fig. 2. Validation results. The Figure shows the percentage difference in the optimal ranges, obtained via simulation and analytic model, vs. density and sending rate.

that the tested range is fairly small compared to the modeled space so that the spatial correlations resulted from wrap-around do not significantly affect the results.

Figure 2 presents our validation results, showing the percentage difference in the optimal range obtained via simulation and the analytic model; here the analytic model always under-predicts the range, which we believe to be good since a small under utilization of the channel is preferable to an over-aggressive range setting that could lead to channel collapse.

Experiment set 1 corresponds to a small fraction of the parameter space, which allows us to extensively test each point. All the experiments in set 1 ran in a $100 \times 100 \times 100m^3$ volume. Each parameter setting is tested with multiple spatial configurations. Specifically, 10 randomized configurations if the number of nodes is less than 5000; otherwise 5 configurations. The experiments ran until the time when each node is expected to have sent 50 messages. The final coverage for a particular setting is averaged over all nodes.

Experiment set 2 corresponds to broader settings along both λ_s and λ_p dimensions, covering most practical scenarios (Notice the axes in Figure 2(b) are not linear). Compared to experiment set 1, this set extends to much higher densities and slower sending rates. However, because some of the settings require simulating tens of thousands nodes we have not been able to test each point as extensively as we did for set 1. For example, the simulations are run on only one spatial configuration. We believe the average over the large number of nodes involved in each simulation will average out the effects of these limitations.

In all, our validation shows that the extrapolation can predict R_o to within 16%. We believe that this is reasonably accurate for applying our model to real practical settings.

5   Related Work

We categorize related work into three general areas: topology control, MAC layer improvements, and analytic modeling.

Topology Control. While all the range adjustment to control topology work below improves spatial reuse, their more prominent goal has been energy conservation while maintaining connectivity. Our work does not consider energy use; we instead maximize the 1-hop broadcast coverage at the expense of energy. Our work also does not provide connectivity guarantees.

The work in [10] formulated topology control as an optimization problem and provided a centralized algorithm to find the minimum power that should be used by each node to achieve the objective. [12] used an orientation approach, where each node keeps increasing its range until the collected orientation information guarantees connectivity. The algorithm in [5] works closely with the routing layer rather than at the MAC layer. By locally comparing the multiple routing tables, a node could choose the minimum power level needed to forward a packet. A variety of centralized or distributed scheduling methods [13] have been used to control the topology for wireless networks. However, we consider centralized approach prohibitive in dense wireless sensor networks. In addition, the above distributed reservation approach relies on feedback, which is impossible to manage for broadcast messages that we considered here.

MAC layer improvements. Typical 802.11 MAC layers use RTS and CTS control packets set at maximal power to eliminate the hidden terminal problem. RTS/CTS schemes lower spatial reuse by blocking possible concurrent transmissions. One scheme [4] proposed to transmit data packets at the minimum required power level thus reducing energy use, however, there was no spatial reuse improvement. Another work [7] proposed a separate busy tone channel, which was used by each node to advertise the additional noise level that could be accommodated and thus could support more concurrent transmissions, i.e., better spatial reuse. Instead of modifying the MAC layer directly, our range control scheme could work between the MAC and routing layers.

Analytic Modeling. The analytic modeling work most related to ours is [11], which found the optimum degree where packet forwarding would have the best expected progress. Our analytic modeling methods are similar in that we reason about the best expected likelihood of packet reception.

6 Conclusion

In this work we presented an analytic model for ad-hoc wireless networks to predict the optimal range to maximize the 1 hop broadcast coverage. Our approach is in effect performing topology control on the broadcast packets. Unlike most topology control approaches, however, we took a geometric rather than a graph theoretic approach to reason about the impact of the effects of range on coverage.

We developed a geometric-based analytic model that describes the relationship between sending rate, density, range and coverage. Although our model is not a closed-form solution, we presented a simple method to extrapolate the optimal range given a node's local sending rate and density observations. Through simulations we demonstrated that despite its numerous simplifications, the model predicts the optimal range setting to within 16% across an order of magnitude set of rates and densities which could be realistically found in these networks.

References

1. J. Hightower, C. Vakili, G. Borriello, and R. Want. Design and Calibration of the SpotON Ad-Hoc Location Sensing System, unpublished., 2001.
2. C. Intanagonwiwat, R. Govindan, and D. Estrin. Directed Diffusion: A Scalable and Robust Communication Paradigm for Sensor Networks. In *In Proceedings of the Sixth Annual International Conference on Mobile Computing and Networking*, Boston, MA, Aug. 2000.
3. D. Johnson and D. Maltz. Dynamic Source Routing in Ad Hoc Wireless Network. In *Proceedings of the 4th Annual ACM/IEEE International Conference on Mobile Computing and Networking (Mobicom)*, Rye, New York, Nov. 1996.
4. E.-S. Jung and N. Vaidya. A Power Control MAC Protocol for Ad Hoc Networks. In *ACM International Conference on Mobile Computing and Networking*, 2002.
5. V. Kawadia and P. R. Kumar. Power Control and Clustering in Ad Hoc Networks. In *INFOCOM*, 2003.
6. P. Levis, N. Patel, D. Culler, and S. Shenker. Trickle: A Self-Regulating Algorithm for Code Propagation and Maintenance in Wireless Sensor Networks. In *First Symposium on Network Systems Design and Implementation (NSDI)*, Mar. 2004.
7. J. Monks, V. Bharghavan, and W. mei W. Hwu. A Power Controlled Multiple Access Protocol for Wireless Packet Networks. In *INFOCOM*, pages 219–228, 2001.
8. D. Niculescu and B. Nath. Ad hoc positioning system (APS). In *GLOBECOM (1)*, pages 2926–2931, 2001.
9. N. Priyantha, A. Chakraborty, and H. Balakrishnan. The Cricket Location-Support system. In *ACM International Conference on Mobile Computing and Networking (MobiCom)*, Boston, MA, Aug. 2000.
10. R. Ramanathan and R. Hain. Topology Control of Multihop Wireless Networks Using Transmit Power Adjustment. In *INFOCOM (2)*, pages 404–413, 2000.
11. H. Takagi and L. Kleinrock. Optimal Transmission Ranges for Randomly Distributed Packet Radio Terminals. *IEEE Transactions on Communications*, 32(3):246–257, March 1984.
12. R. Wattenhofer, L. Li, P. Bahl, and Y.-M. Wang. Distributed Topology Control for Wireless Multihop Ad-hoc Networks. In *INFOCOM*, pages 1388–1397, 2001.
13. C. Zhu and M. S. Corson. A Five-phase Reservation Protocol (FPRP) for Mobile Ad Hoc Networks. In *Wireless Networks*, July 2001.

Experimental Comparison of Algorithms for Energy-Efficient Multicasting in Ad Hoc Networks*

Stavros Athanassopoulos, Ioannis Caragiannis, Christos Kaklamanis, and
Panagiotis Kanellopoulos

Research Academic Computer Technology Institute and
Dept. of Computer Engineering and Informatics
University of Patras, 26500 Rio, Greece

Abstract. Energy is a scarce resource in ad hoc wireless networks and it
is of paramount importance to use it efficiently when establishing communication patterns. In this work we study algorithms for computing
energy-efficient multicast trees in ad hoc wireless networks. Such algorithms either start with an empty solution which is gradually augmented
to a multicast tree (*augmentation algorithms*) or take as input an initial
multicast tree and 'walk' on different multicast trees for a finite number
of steps until some acceptable decrease in energy consumption is achieved
(*local search algorithms*).
We mainly focus on augmentation algorithms and in particular we have
implemented a long list of existing such algorithms in the literature and
new ones. We experimentally compare all these algorithms on random
geometric instances of the problem and obtain results in terms of the
energy efficiency of the solutions obtained. Additional results concerning
the running time of our implementations are also presented. We also explore how much the solutions obtained by augmentation algorithms can
be improved by local search algorithms. Our results show that one of our
new algorithms and its variations achieve the most energy-efficient solutions while being very fast. Our investigations shed some light to those
properties of geometric instances of the problem which make augmentation algorithms perform well.

1 Introduction

Wireless networks have received significant attention during the recent years.
Especially, ad hoc wireless networks emerged due to their potential applications
in battlefield, emergency disaster relief, etc. [15]. Unlike traditional wired networks or cellular wireless networks, no wired backbone infrastructure is installed
for ad hoc wireless networks.

A node (or station) in these networks is equipped with an omnidirectional
antenna which is responsible for sending and receiving signals. Communication

---

* This work was partially supported by the European Union under IST FET Project
CRESCCO and RTN Project ARACNE.

I. Nikolaidis et al. (Eds.): ADHOC-NOW 2004, LNCS 3158, pp. 183–196, 2004.

is established by assigning to each station a transmitting power. In the most common power attenuation model, the signal power falls as $1/r^{\alpha}$, where r is the distance from the transmitter and α is a constant which depends on the wireless environment (typical values of α are between 1 and 6). So, a transmitter s can successfully send a signal to a receiver t if $P_s \geq \gamma \cdot d(s,t)^{\alpha}$, where P_s is the power of the signal transmitted, $d(s,t)$ is the Euclidean distance between the transmitter and the receiver, and γ is the receiver's power threshold for signal detection which is usually normalized to 1. In this case, we say that node s establishes a direct link to node t. So, communication from a node s to another node t may be established either directly if the two nodes are close enough and s uses adequate transmitting power, or by using intermediate nodes. Observe that due to the nonlinear power attenuation, relaying the signal between intermediate nodes may result in energy conservation (we use the terms energy and power interchangeably).

A crucial issue in ad hoc wireless networks is to support communication patterns that are typical in traditional networks. These may include broadcasting, multicasting, gossiping (all–to–all communication) and more. Since establishing a communication pattern strongly depends on the use of energy, the important engineering question to be solved is to guarantee a desired communication pattern minimizing the total energy consumption.

An ad hoc wireless network is usually modelled as a complete directed graph $G = (V, E)$, with a non-negative edge cost function $c : E \to R^{+}$. Intuitively, V is the set of stations or nodes, the edges in E correspond to potential direct links, and the function c denotes the minimum energy required for establishing a direct link between any possible transmitter-receiver pair. Usually, the edge cost function is symmetric (i.e., $c(u,v) = c(v,u)$). An important special case, which usually reflects the real-world situation, henceforth called geometric case, is when nodes of G are points in a Euclidean space and the cost of an edge (u, v) is defined as the Euclidean distance between u and v raised to a fixed power α, i.e. $c(u,v) = d(u,v)^{\alpha}$. Asymmetric edge cost functions can be used to model medium abnormalities or batteries with different energy levels [12].

Consider a guest network denoted by a graph $H = (U, A)$, with $U \subseteq V$ and $A \subseteq E$. In order to establish the guest network H on the ad hoc network G, we must set the energy level of node u to $\max_{v \in U : (u,v) \in A} c(u, v)$. In other words, the energy level at which node u operates should be such that it can establish as direct links all edges of A directed out of u. The total energy needed for establishing H on G is the sum of the energy levels of all nodes.

The optimization problem we study in this paper can be stated as follows. Given an ad hoc network represented by a graph $G = (V, E)$, with a non-negative edge cost function $c : E \to R^{+}$, a special node $r \in V$ called *root*, and a set of *terminals* $D \subseteq V - \{r\}$, find a multicast tree, i.e., a tree rooted at r and spanning all the nodes in D, which can be established as a guest network in G with the minimum total energy. This problem is known as MINIMUM ENERGY MULTICAST TREE (MEMT). The special case of MEMT where $D = V - \{r\}$ is known as MINIMUM ENERGY BROADCAST TREE (MEBT).

In the case of symmetric edge cost functions, Caragiannis et al. [4] present logarithmic (in the number n of stations) approximation algorithms for MEMT and MEBT. These results are asymptotically optimal since MEBT in symmetric graphs has been proven to be inapproximable within a sublogarithmic factor [6]. In [4] it is also shown that, in the case of asymmetric edge cost functions, MEMT is hard to approximate within $O(\log^{2-\epsilon} n)$, while Liang [12] presents an $O(|D|^\epsilon)$ approximation algorithm. For MEBT in the case of asymmetric edge costs, logarithmic approximation algorithms have been presented in [2,4]. The algorithm in [4] for the symmetric case of MEMT and the algorithm in [2] for the asymmetric case of MEBT borrow ideas from algorithms for a natural combinatorial problem known as NODE WEIGHTED STEINER TREE (NWST) [9,11].

Although the above cases are very interesting from the theoretical point of view, the most important questions in practice concern the geometric case. This case was first considered in [10] in a slightly different context. Geometric cases of MEBT have received significant attention in the literature. When the nodes are points on a line, MEBT can be optimally solved in polynomial time [3,7]. The case where the nodes are points in the Euclidean plane has been much studied. In this case, MEBT was proved to be NP-hard in [6]. The first algorithms were proposed in the seminal work of Wieselthier et al. [15]. These algorithms were based on the construction of minimum spanning trees (MST) and shortest path trees (SPT) on the graph representing the ad hoc network. The approach followed in [15] for computing solutions of MEMT was to prune the trees obtained in solutions of MEBT. Experimental results showed that the algorithm *Broadcast Incremental Power* (BIP) outperforms algorithms MST and SPT. In subsequent work Wan et al. [14] study the algorithms presented in [15] in terms of efficiency in approximating the optimal solution. Their main result is an upper bound of 12 on the approximation ratio of algorithm MST. This result implies a constant approximation algorithm for MEMT as well. Slightly weaker approximation bounds for MST have been presented in [6]. In [14], it is also proved that the approximation ratio of BIP is not worse than that of MST, and that other intuitive algorithms have very poor approximation ratio.

However, several intuitive algorithms have been experimentally proved to work very well on random instances of the problem. In [13,16], algorithms based on shortest paths are enhanced with the *potential power saving* idea and are experimentally shown to outperform most of the known algorithms. In [1], Cagalj et al. introduced a heuristic called *Embedded Wireless Multicast Advantage* (EWMA) for computing efficient solutions to MEBT instances. This algorithm takes as input a spanning tree and transforms it to an energy-efficient broadcast tree by performing local improvements. As we discuss in Section 2, EWMA can be easily converted to work for MEMT as well. Another heuristic called Sweep was proposed in [15]; this also takes as input a tree and transforms it to an energy-efficient tree by performing local improvements. In contrast to these two algorithms, most of the algorithms discussed above are based on the idea of constructing a tree gradually. This means that, starting from an empty solution, a tree is augmented by repeatedly including new structures (i.e., new nodes and edges) until connectivity from the root to the terminals is established. Another issue of apparent importance is to design algorithms for MEMT that are amenable to implement in a distributed environment (see e.g., [1,5,16]).

In this work, we divide algorithms presented in the literature in two categories: *local search algorithms* and *augmentation algorithms*. We describe the general features of both categories and report on the implementation of many algorithms, both existing and new ones. Our implementations include algorithms designed for approximating MEMT in the more general symmetric case as well as algorithms which are more intuitive for the geometric case. Our purpose is to experimentally compare all these algorithms in terms of the energy efficiency of the solutions obtained on random instances of MEMT on the Euclidean plane. An evaluation of the running time of the algorithms is also presented. The rest of the paper is structured as follows. We devote Section 2 to local search algorithms, while augmentation algorithms are discussed in Section 3. The experimental results are presented and commented in Section 4.

2   Local Search Algorithms

Local search algorithms perform a 'walk' on multicast trees. The walk starts from a multicast tree given as input. In each step, a local search algorithm moves to a new multicast tree obtained by removing some of the edges of the previous one and adding new edges, so that the necessary connectivity properties are maintained. The rule used in each move for selecting the next multicast tree is related to energy. Since local search algorithms require a multicast tree to start walking on, they are usually called after an augmentation algorithm. Typical representatives of this category are the algorithms Prune, EWMA and Sweep.

Algorithm Prune has been extensively used (see e.g., [13,15]) for obtaining a multicast tree from a broadcast tree. In each step the algorithm performs the following operation. For each leaf which is not a terminal, it removes it from the tree together with its incoming edge. The algorithm terminates when all leaves are terminals. Prune can be easily implemented to run in linear time.

EWMA was proposed in [1] for solving MEBT, where it was assumed that the tree to start with is a minimum spanning tree. However, as it is clear in the following description of the algorithm, it can be used for MEMT and can start with any multicast tree. Starting with a multicast tree, EWMA walks on multicast trees by performing the following two types of changes in each step: (1) outgoing edges are added to a single node v; this node is said to be *extended* and (2) all outgoing edges are removed from some descendants of v; in this case we say that the particular descendants of v are *excluded*. Throughout its execution, EWMA uses three sets C, F and E. Intuitively, C is the set of nodes which have been considered by the algorithm, F is the set of nodes that were extended at least once in some previous step and were never excluded, and E is the set of nodes that were excluded in some previous step. Initially, the algorithm sets $C = \{r\}$ and $F = E = \emptyset$. In each step, EWMA takes as input the multicast tree produced in the previous step together with the sets C, F and E. Define the *gain* of a node v as the decrease in the energy of the multicast tree obtained by excluding some of the nodes of the tree, in exchange for the increase in node v's energy in order to establish edges to all excluded nodes and their children. If no node in $C - F - E$ has positive gain, the node with the minimum energy is included in F and its children are included in C. Otherwise, the node with maximum gain is included in the set F, the excluded nodes are included in the set E, while both the excluded nodes and their children are included in the set C. The edges of the excluded nodes to their children are removed (changes of type (2)) and outgoing edges from v to all excluded nodes and their children are established (change of type (1)). The new multicast tree together with the updated sets C, F and E are passed as input to the next step. The algorithm terminates when the root and all terminals are contained in C. Our implementation of EWMA has running time $O(n^3)$ in the worst case.

Sweep was proposed in [15] as a simple heuristic for improving solutions of MEBT obtained by spanning tree and shortest path algorithms. Clearly, it can be used on any multicast tree as well. Sweep works as follows. It first assigns distinct IDs (consecutive integers $0, 1, ...$) to all nodes. Starting from the multicast tree, it proceeds in steps. In the i-th step, it examines the node v_i having ID equal to i. If, for some nodes $v_{i_1}, v_{i_2}, ...$ which are not ancestors of v_i, v_i's energy in the multicast tree is not smaller than the cost of all the edges from v_i to $v_{i_1}, v_{i_2}, ...$, Sweep removes the incoming edges of $v_{i_1}, v_{i_2}, ...$ and adds edges from v_i to $v_{i_1}, v_{i_2}, ...$ in the multicast tree. The algorithm terminates when all nodes have been examined. Although several variations of Sweep seem natural (this is also mentioned in [15]), somewhat surprisingly, none of several variations we implemented is better than Sweep in terms of the energy efficiency of the solution obtained. A simple implementation of Sweep runs in time $O(n^2)$.

3   Augmentation Algorithms

Augmentation algorithms build a multicast tree by starting from an empty solution which is gradually augmented until a guest network having directed paths from the root to the terminals is established. Clearly, once such a guest network is available, it can be easily converted to a multicast tree. The solution is augmented in phases. In each phase, an augmentation algorithm adds to the solution a structure. This may be an edge to a node (e.g., in BIP and well known implementations of MST [8]), a set of edges directed out of the same node (e.g., in a variation of BIP called BAIP [15]), a path (e.g., in algorithms presented in [13,16]), a spider (a special directed graph used in [2] and implicitly in [4]), etc. The structure is selected among all candidate structures so that a local objective is minimized. The local objective is usually related to the energy needed in order to establish the edges of the structure. We devote the rest of this section to the detailed description of several augmentation algorithms. These include existing algorithms in the literature as well as new ones with several variations.

Basic algorithms. Starting from a multicast tree containing only the root, the algorithms *Shortest Path First* (SPF), *Multicast Incremental Power* (MIP), *Dens-*

est Shortest Path First (DSPF) and *Densest Two Shortest Paths First* (D2SPF) augment the multicast tree in phases, until all terminals are included in the tree. In each phase, algorithm SPF adds to the multicast tree the shortest directed path (i.e., the path whose establishment requires the minimum amount of energy) that connects some node of the tree to a terminal out of the tree. Algorithm MIP adds the path requiring the minimum additional energy that connects the tree to a terminal out of the tree. This means that after each phase the cost of the edges directed out of each node v in the path selected in the phase is decreased by the cost of the edge outgoing from v in the path. Algorithm DSPF selects among the minimum additional energy paths from the tree to the terminals out of the tree, the one minimizing the ratio of additional energy over the number of new terminals that are included in the tree. Algorithm D2SPF is similar to DSPF; the difference being that, for any node v and any terminal t both not in the multicast tree, D2SPF considers paths from the tree to t passing through v.

In our implementations, we make extensive use of algorithms for computing shortest paths. Computing shortest paths from a node to all other nodes in a complete directed graph can be done in time $O(n^2)$ using a simple implementation of Dijkstra's algorithm using binary heaps as priority queues [8]. More complex heaps which have been proved to yield faster implementations of Dijkstra's algorithm (e.g., Fibonacci heaps) do not decrease the running time substantially in our case since the graph representing the ad hoc network is complete. SPF, MIP and DSPF run in time $O(mn^2)$, where m is the number of terminals. In each of the at most m phases, the algorithms perform a shortest path computation. The processing of the graph (i.e., the decrease to the edge costs) required in MIP and DSPF does not affect their asymptotic running time compared to SPF. Each phase of D2SPF requires the computation of all-pairs shortest paths which needs time $O(n^3)$ leading to overall running time of $O(mn^3)$.

The potential power saving idea. The basic algorithms do not examine whether establishing a new path could also include nodes which had been included in the multicast tree in previous phases, and could now be connected to the multicast tree as children of some node in the path. In this way, the energy of their previous parent could be decreased. The total decrease is denoted as *potential power saving* ([13,16]). The algorithms described here are variations of the basic algorithms; their difference being that, when computing the additional energy of a candidate path, they subtract the potential power saving. When a new path is established, the multicast tree is modified accordingly, i.e., nodes previously included in the multicast tree might now be connected as children of nodes in the new path.

Algorithms SP3SF, DSP3SF and D2SP3SF are variations of MIP, DSPF and D2SPF with potential power saving, respectively. Algorithm SP3SF was originally proposed in [13] and, according to our knowledge, computes the most energy-efficient solutions in the setting studied here. Algorithm 2SP3SF is a variation of D2SP3SF where the local objective in each phase is to minimize the additional energy minus the potential power saving. The necessary computations for computing the potential power saving increase the running time of algorithms

SP3SF and DSP3SF to $O(m^2n^2)$, while algorithms 2SP3SF and D2SP3SF run in time $O(mn^3)$. However, in practice the running time of SP3SF and DSP3SF is only slightly worse than that of MIP and DSPF, respectively.

NWST-based algorithms. We now present six algorithms which borrow ideas from approximation algorithms for NWST presented in [9,11]. In fact, the algorithm of [4] for the symmetric case of MEMT reduces instances of MEMT to instances of NWST. Although, the size of the resulting instance of NWST is polynomial in theory, even for small instances of MEMT the corresponding instance of NWST is intractable to be solved with the algorithms of [9,11] in practice. The algorithms presented below apply ideas from [9,11] to the MEMT instance directly. A similar algorithm has been designed for MEBT in [2]. All these algorithms use the idea of gradually augmenting a guest network by repeatedly adding spiders or forks of small density.

A *spider* is a directed graph consisted of a node called *head* and a set of directed paths called *legs*, each of them from the head to the nodes called *feet* of the spider. The definition allows legs to share nodes and edges. The weight of a spider is the maximum cost of the edges leaving the head plus the sum of costs of the legs, where the cost of a leg is the sum of the cost of its edges without the edge leaving the head. A *fork* is a spider having a *center* node that is reached through a shortest directed path from the head so that the subgraph of the fork induced by removing the edges of this path and all nodes in the path but the center is a spider having the center as its head. We call this spider the *subspider* of the fork. The weight of the fork is the cost of the edges on the directed path from the head to the center plus the weight of its subspider. We say that a node u can be connected *for free* to a path, if the cost $c(v_i, u)$ of the edge from a node v_i in the path to u is smaller than the cost of the edge directed out of v_i in the path.

Algorithm *Densest Spider First* (DSF) establishes a guest network H in which the root and the terminals are contained in the same loosely connected component. Loose connectivity of a directed graph means that its undirected counterpart (i.e., the graph obtained by substituting directed edges by undirected ones) is connected. Then, for establishing a multicast tree in G, the algorithm computes a tree directed from the root to the terminals in the supergraph of H containing all the edges in H and their opposite-directed edges. The algorithm assigns indices to nodes to keep track of which nodes belong to the same loosely connected component in H. Each index is a non-negative integer; nodes having the same finite index are loosely connected and nodes having different or infinite indices are not connected at all. Initially, the index of the root is 0, terminals have finite, positive and distinct integer indices, while all other nodes have infinite index. The algorithm proceeds in steps until the root and all terminals have index 0. In each step, the algorithm finds a node v and a spider having v as its head and nodes having pairwise distinct, different than v and not infinite indices as its feet, so that the ratio of the weight of the spider over the number of its feet (called the density of the spider) is minimized. Let $i_1, i_2, ..., i_k$ be the indices of the nodes in the spider in non-decreasing order. For each node in the graph with index i_1 or i_2, ..., or i_k the algorithm sets its index to i_1 and adds the edges of the spider to H. Algorithms DSF-2 and DSF-3 are slight variations of DSF; the only difference being that the node selected as the head of the spider in each step is constrained to have finite index in DSF-2 and index 0 in DSF-3.

In our implementations, these algorithms first perform a preprocessing to compute, for each node and each possible energy level of this node, the shortest paths from this node to all other nodes. This requires time $O(n^4)$ and dominates the asymptotic running time of the algorithms. Once the length of all shortest paths has been computed, then computing the minimum density spider having as head a particular node with a particular energy level can be done in time $O(m)$. Hence, the minimum density spider in each of the at most m phases is computed in time $O(mn^2)$.

Algorithm *Densest Incremental Spider First* (DISF) is a variation of DSF-3. In each step, after the minimum density spider has been selected, for each node u in the spider, the cost of the edges directed out of u in G is decreased by the maximum cost of the edges directed out of u in the spider. Intuitively, the weight of the spider in algorithms DSF, DSF-2 and DSF-3 is an upper bound on the energy required to establish the edges of the spider. In DISF, the weight of the spider computed in each step is an upper bound on the additional energy required to establish the edges of the spider, given that edges that were included in H in previous steps have already been established. Due to the update in the edge cost function required in each phase of algorithm DISF, a similar preprocessing to that used in DSF is required in each phase. This needs time $O(mn^4)$ and dominates the asymptotic running time of the algorithm.

Algorithm *Densest Fork First* (DFF) is a variation of DISF. In each step a fork having as head a node of index 0 minimizing the ratio of the weight of the fork over the number of feet is added to the guest network H. Finally, algorithm DFF-2 is similar to DFF. The main difference is that the number of feet plus the number of non-zero finite indices of nodes which can be connected for free to the path from the head to the center is used in the denominator of the local objective. The running time of algorithms DFF and DFF-2 is asymptotically the same with that of DISF. In each phase the time required for preprocessing asymptotically dominates the time required for computing the densest fork.

Constrained and iterative versions. In many of the above algorithms, more than one terminals can be added to the multicast tree in each phase. It may be the case that adding a structure with many new terminals in a phase worsens the final solution. We have implemented constrained versions of the algorithms described above. which take as input an *augmentation constraint parameter* denoting the maximum number of new terminals allowed to be included in each phase and constrain appropriately the space of candidate structures. Iterative versions of algorithms DSPF, DSP3SF, SP3SF, D2SPF, D2SP3SF, 2SP3SF, DISF, DFF and DFF-2 repeatedly run their constrained versions for all possible values of the augmentation constraint parameter and output the best solution. These algorithms are called iDSPF, iDSP3SF, iSP3SF, iD2SPF, iD2SP3SF, i2SP3SF, iD-ISF, iDFF and iDFF-2, respectively. Clearly, an iterative version of an algorithm is superior to its unconstrained version in terms of energy efficiency; however, it may require running time proportional to the number of terminals times the running time of the unconstrained version.

4   Experimental Results

In this section we discuss the outcome of our experimentation with the algorithms presented in the previous sections. Results of our experiments are depicted in Tables 1-5. We have also implemented a few more broadcasting algorithms (e.g., variations of algorithms BIP and BAIP) and performed experiments with them by running algorithm Prune to their solutions in order to obtain multicast trees. These results are certainly inferior to those presented below and will not be further discussed. In addition, we have considered variations of the algorithms which are appropriate to implement in a distributed setting. The corresponding experimental results are usually worse than those of the centralized algorithms. We will not discuss distributed implementation issues here; we prefer to focus on centralized implementations since they give lower bounds on the energy efficiency of the solutions and the overall work required in their distributed implementations.

The algorithms were executed on geometric instances of the problem of different size. Input instances consist of nodes corresponding to points with uniformly random coordinates in $[0, 5)$. A node is randomly selected to be the root and terminals are selected uniformly at random without replacement among all other nodes. For each instance, we use the term *group* to denote the set of terminals together with the root.

Basic algorithms and algorithms with potential power saving were executed on instances of size 100 for groups of size $10, 20, ..., 100$ and values 2 and 4 for α. Table 1 shows the performance of the basic algorithms and algorithms with potential power saving with respect to the energy efficiency. For each group size, 100 random instances with the particular group size are constructed and all algorithms are executed on these random instances. The energy values shown in Table 1 for each algorithm in its executions on a particular group size, is the average energy of the multicast trees computed by the algorithm in its execution on all random instances of the particular group size.

All potential power saving algorithms perform better than basic algorithms (this had been experimentally observed for SP3SF and MIP in [13,16]). This difference is larger in the case $\alpha = 2$. Interestingly, D2SPF is at most 1% worse than SP3SF in this case. D2SP3SF outperforms all potential power saving algorithms in the case $\alpha = 2$; it is significantly better than SP3SF and 2SP3SF and slightly better than DSP3SF. Also, 2SP3SF is marginally better than SP3SF. In the case $\alpha = 4$, SP3SF and 2SP3SF seem to produce the most energy-efficient solutions (with DSP3SF being marginally worse). In our experiments, we observed that algorithms using two concatenated shortest paths as structures are always much slower than algorithms using as structures single shortest paths (see e.g., Table 5). This is interesting (and somewhat surprising) since the asymptotic running time of all algorithms using the potential power saving idea is essentially the same for large group sizes.

Table 1. Comparison of basic algorithms, algorithms using the potential power saving idea and their iterative versions on random instances with 100 nodes, $\alpha = 2$ and $\alpha = 4$ and group sizes $10, 20, ..., 100$.

$\alpha = 2$	10	20	30	40	50	60	70	80	90	100
SPF	5.27	7.19	8.49	9.51	10.07	10.56	11.26	11.55	12.09	12.45
MIP	4.98	6.78	7.96	8.85	9.47	9.89	10.52	10.84	11.26	11.61
DSPF	4.93	6.64	7.83	8.66	9.22	9.61	10.17	10.40	10.73	11.07
D2SPF	5.01	6.59	7.80	8.56	9.06	9.46	10.00	10.22	10.53	10.86
SP3SF	4.92	6.59	7.66	8.50	9.03	9.38	9.90	10.10	10.49	10.79
DSP3SF	4.87	6.51	7.59	8.43	8.88	9.25	9.76	10.01	10.26	10.63
2SP3SF	4.90	6.56	7.62	8.45	8.99	9.32	9.89	10.07	10.39	10.74
D2SP3SF	4.96	6.50	7.60	8.34	8.87	9.18	9.69	9.90	10.18	10.49
iDSPF	4.89	6.55	7.70	8.52	9.07	9.44	9.97	10.23	10.56	10.92
iD2SPF	4.87	6.49	7.61	8.41	8.89	9.30	9.78	10.05	10.39	10.71
iSP3SF	4.92	6.59	7.66	8.50	9.03	9.38	9.90	10.10	10.48	10.79
iDSP3SF	4.81	6.37	7.38	8.16	8.58	8.94	9.41	9.63	9.92	10.28
i2SP3SF	4.90	6.55	7.62	8.45	8.99	9.32	9.88	10.07	10.38	10.73
iD2SP3SF	4.81	6.38	7.40	8.16	8.64	8.98	9.47	9.68	9.96	10.25

$\alpha = 4$	10	20	30	40	50	60	70	80	90	100
SPF	1.51	1.97	2.46	2.65	2.89	2.99	3.19	3.25	3.40	3.61
MIP	1.48	1.93	2.41	2.59	2.82	2.93	3.10	3.18	3.32	3.52
DSPF	1.49	1.93	2.41	2.61	2.82	2.93	3.11	3.19	3.32	3.52
D2SPF	1.49	1.94	2.42	2.60	2.81	2.91	3.10	3.18	3.29	3.52
SP3SF	1.47	1.91	2.36	2.55	2.74	2.85	3.02	3.08	3.21	3.41
DSP3SF	1.46	1.92	2.38	2.55	2.75	2.87	3.04	3.12	3.25	3.43
2SP3SF	1.46	1.90	2.36	2.55	2.74	2.85	3.02	3.09	3.21	3.41
D2SP3SF	1.48	1.92	2.38	2.55	2.75	2.86	3.03	3.12	3.22	3.41
iDSPF	1.47	1.91	2.38	2.56	2.78	2.89	3.06	3.13	3.27	3.46
iD2SPF	1.47	1.90	2.37	2.55	2.77	2.87	3.05	3.12	3.24	3.44
iSP3SF	1.47	1.91	2.36	2.55	2.74	2.85	3.02	3.08	3.21	3.41
iDSP3SF	1.45	1.88	2.33	2.50	2.70	2.81	2.96	3.04	3.16	3.34
i2SP3SF	1.46	1.90	2.36	2.55	2.74	2.85	3.02	3.09	3.21	3.41
iD2SP3SF	1.46	1.89	2.34	2.51	2.71	2.82	2.99	3.06	3.17	3.35

In general, NWST-based algorithms are slow. This has been already justified in Section 3 where we discuss their running time (see also Table 5). This fact did not allow us to perform large experiments. Our experiments with instances of the problem with size 40 (see Table 2) show that NWST-based algorithms are inferior to most of the algorithms discussed above. DSF, DSF-2 and DSF-3 are not much slower than D2SP3SF and 2SP3SF but their solutions are much worse in terms of energy efficiency (in particular in the case $\alpha = 2$). DISF is slightly better in the case $\alpha = 2$ and rather worse in the case $\alpha = 4$. Its running time is huge. DFF and DFF-2 are even slower but seem to be the best among the NWST-based algorithms in terms of energy efficiency in their solutions. Overall, all NWST-based algorithms are worse than algorithms using the potential power saving idea. Algorithms DISF, DFF and DFF-2 include some of those properties which make basic algorithms with potential power saving perform well, i.e., they augment a multicast tree containing the root by including in it new terminals in each phase. Unfortunately, the local objective used in NWST-based algorithms contains the weight of spiders or forks and this may not always be proportional to the energy. In addition, we see no clear way of incorporating the potential power saving (or a similar) idea to NWST-based algorithms. Recall that NWST-based algorithms are variations of DSF which was designed to efficiently approximate optimal solutions of MEMT in the more general symmetric case. The performance of NWST-based algorithms indicates that the particular geometric version of the problem we consider here has certain properties which are better exploited by algorithms with simple and intuitive local objectives.

Table 2. Comparison of NWST-based algorithms and their iterative versions with algorithms MIP, SP3SF and DSP3SF on random instances with 40 nodes, $\alpha = 2$ (left) and $\alpha = 4$ (right) and group sizes $10, 20, 30$ and 40.

$\alpha = 2$	10	20	30	40	$\alpha = 4$	10	20	30	40
MIP	7.16	9.61	10.99	11.97	MIP	5.66	8.22	9.25	10.00
SP3SF	6.94	9.09	10.28	11.18	SP3SF	5.55	7.96	8.98	9.66
DSP3SF	6.83	8.87	9.97	11.04	DSP3SF	5.51	7.87	9.06	9.70
DSF	8.49	11.29	12.31	12.57	DSF	6.52	9.13	9.99	10.27
DSF-2	8.11	10.54	11.62	12.57	DSF-2	5.83	8.36	9.50	10.27
DSF-3	8.12	10.40	11.28	11.95	DSF-3	6.15	8.91	9.99	10.65
DISF	7.72	10.06	11.00	11.79	DISF	6.01	8.76	9.74	10.49
DFF	7.20	9.63	10.69	11.79	DFF	5.74	8.37	9.38	10.11
DFF-2	7.19	9.54	10.66	11.72	DFF-2	5.71	8.33	9.37	10.11
iDISF	6.94	9.15	10.18	10.95	iDISF	5.57	8.05	9.11	9.79
iDFF	7.06	9.47	10.50	11.55	iDFF	5.67	8.20	9.33	10.05
iDFF-2	7.04	9.39	10.44	11.53	iDFF-2	5.65	8.20	9.32	10.00

We also investigated iterative versions of our algorithms. By the definition of these algorithms, it is clear that they always perform better than their unconstrained counterparts in terms of energy efficiency at the expense of multiplying the running time with the group size. In practice, in most of our implementations, the running time of iterative algorithms is closer to the running time of their unconstrained counterparts (see Table 5). iDSP3SF computes the most efficient solutions with respect to energy efficiency. iD2SP3SF is slightly worse than iDSP3SF in terms of energy efficiency but its running time is enormous. The only iterative algorithm with running time comparable to that of iDSP3SF is iSP3SF which does not actually improve its unconstrained counterpart SP3SF. Overall, the solutions obtained by iDSP3SF are up to 5% (and about 2%) better than those of iSP3SF in the case of $\alpha = 2$ (and $\alpha = 4$, respectively). The corresponding results are depicted in Tables 1. The solutions obtained by iterative versions of NWST-based algorithms are usually much worse while their running time is huge. It is interesting, however, that the solutions obtained by iDISF significantly improve the results obtained by DISF. Unfortunately, this seems to be the slowest among all algorithms we implemented.

Our next investigation probably answers why iDSP3SF is superior to iSP3SF while DSP3SF and SP3SF compute solutions of comparable energy efficiency. In Table 3 we present the performance of constrained versions of algorithms DSP3SF and SP3SF (additional results for the constrained version of DSPF are also presented). It is clear that the constrained version of SP3SF computes solutions of almost the same energy regardless of the augmentation constraint parameter, while this is not the case for DSP3SF. This indicates that, given an instance of the problem, many different augmentation constraint parameter values are possible to give the best solution of iDSP3SF with respect to energy efficiency, while the solution obtained by iSP3SF is marginally better than the solution obtained by the constrained version of SP3SF with augmentation parameter constraint equal to 1.

Table 3. The energy efficiency of constrained versions of algorithms DSPF, SP3SF and DSP3SF on random instances with 100 nodes, $\alpha = 2$ and $\alpha = 4$, and group sizes 50 and 100 for different augmentation constraint parameter values. The last column contains the energy of the best solution for parameter values greater than 10.

$\alpha = 2$, group size: 50	1	2	3	4	5	6	7	8	9	10	> 10
DSPF	9.46	9.36	9.32	9.27	9.26	9.25	9.24	9.23	9.24	9.23	9.22
SP3SF	9.04	9.03	9.03	9.03	9.03	9.03	9.03	9.03	9.03	9.03	9.03
DSP3SF	9.04	9.02	8.91	8.88	8.85	8.85	8.86	8.86	8.88	8.88	8.87
$\alpha = 2$, group size: 100	1	2	3	4	5	6	7	8	9	10	> 10
DSPF	11.61	11.43	11.32	11.25	11.21	11.18	11.15	11.15	11.12	11.11	11.07
SP3SF	10.83	10.80	10.80	10.79	10.79	10.79	10.79	10.79	10.79	10.79	10.79
DSP3SF	10.83	10.85	10.74	10.71	10.70	10.64	10.68	10.64	10.64	10.67	10.62
$\alpha = 4$, group size: 50	1	2	3	4	5	6	7	8	9	10	> 10
DSPF	2.82	2.83	2.82	2.82	2.82	2.82	2.82	2.82	2.82	2.82	2.82
SP3SF	2.74	2.74	2.74	2.74	2.74	2.74	2.74	2.74	2.74	2.74	2.74
DSP3SF	2.74	2.75	2.74	2.75	2.75	2.75	2.76	2.76	2.76	2.75	2.75
$\alpha = 4$, group size: 100	1	2	3	4	5	6	7	8	9	10	> 10
DSPF	3.52	3.52	3.52	3.53	3.53	3.52	3.52	3.52	3.52	3.52	3.51
SP3SF	3.41	3.41	3.41	3.41	3.41	3.41	3.41	3.41	3.41	3.41	3.41
DSP3SF	3.41	3.43	3.43	3.42	3.43	3.42	3.43	3.43	3.42	3.43	3.42

The effect of local search algorithms on solutions obtained by the augmentation algorithms we have implemented is important in the case $\alpha = 2$ while it seems to be marginal in the case $\alpha = 4$ (see Table 4). Such algorithms do not add significant overhead to the overall running time and usually lead to much better solutions. EWMA seems to be appropriate for the case $\alpha = 2$ and in particular for broadcasting instances for which it was originally designed, while Sweep seems to be slightly better in the case $\alpha = 4$. The improvement in solutions of augmentation algorithms achieved after running EWMA and/or Sweep is larger for the augmentation algorithms which are worse in terms of the energy efficiency and smaller for those algorithms which output more efficient solutions. Usually, running repeatedly local search algorithms can improve a solution further. This improvement starts to be negligible after the first two or three executions. An interesting question is whether running EWMA and/or Sweep after the unconstrained version of an algorithm is better than its iterative version. We observed that this is the case for algorithms SP3SF and 2SP3SF (we have already seen that iterative versions of these algorithms do not significantly improve on the energy efficiency of the solutions), but this is not clear for algorithms DSPF, DSP3SF,

D2SPF and D2SP3SF, even if we compare these algorithms followed by several calls to EWMA and Sweep with their iterative versions. Also, running EWMA after e.g., iDSP3SF improves the solutions further. This discussion implies that local search algorithms cannot substitute iterative algorithms; however they can be used to slightly improve their performance with respect to energy efficiency of the solutions obtained.

In conclusion, iDSP3SF followed by EWMA seems to give the most energy-efficient solutions. Algorithms SP3SF and DSP3SF followed by local search algorithms provide a good compromise between energy efficiency and running time.

Table 4. Effects of local search algorithms on augmentation algorithms for random instances with 100 nodes, $\alpha = 2$ and $\alpha = 4$, and group sizes 10, 50 and 100. The four columns for each group size denote the energy of the algorithm, the algorithm followed by Sweep, the algorithm followed by EWMA, and the algorithm followed by executions of EWMA, Sweep and EWMA, respectively.

$\alpha = 2$	10				50				100			
SPF	5.27	5.13	5.15	5.04	10.07	9.55	9.40	9.06	12.45	11.76	11.00	10.77
MIP	4.98	4.91	4.92	4.87	9.47	9.09	9.06	8.82	11.61	10.93	10.72	10.46
DSPF	4.93	4.87	4.89	4.85	9.22	8.98	8.90	8.75	11.07	10.70	10.51	10.31
iDSPF	4.89	4.84	4.85	4.81	9.07	8.82	8.79	8.63	10.92	10.57	10.37	10.19
DSP3SF	4.87	4.87	4.84	4.83	8.88	8.88	8.74	8.71	10.63	10.63	10.30	10.23
iDSP3SF	4.81	4.81	4.79	4.78	8.58	8.58	8.48	8.46	10.28	10.28	10.05	10.02
SP3SF	4.92	4.92	4.89	4.89	9.03	9.02	8.85	8.79	10.79	10.78	10.39	10.32
iSP3SF	4.92	4.92	4.89	4.89	9.03	9.02	8.85	8.79	10.79	10.78	10.39	10.31

$\alpha = 4$	10				50				100			
SPF	1.51	1.49	1.50	1.48	2.89	2.81	2.85	2.79	3.61	3.49	3.53	3.45
MIP	1.48	1.47	1.48	1.47	2.82	2.75	2.79	2.74	3.52	3.41	3.49	3.40
DSPF	1.49	1.48	1.48	1.47	2.82	2.76	2.80	2.76	3.52	3.42	3.49	3.42
iDSPF	1.47	1.46	1.47	1.46	2.78	2.73	2.76	2.73	3.46	3.37	3.43	3.36
DSP3SF	1.47	1.47	1.46	1.46	2.75	2.75	2.75	2.74	3.43	3.43	3.43	3.42
iDSP3SF	1.45	1.45	1.45	1.45	2.70	2.69	2.69	2.69	3.34	3.34	3.34	3.33
SP3SF	1.47	1.47	1.46	1.46	2.74	2.74	2.74	2.73	3.41	3.41	3.40	3.39
iSP3SF	1.47	1.47	1.46	1.46	2.74	2.74	2.74	2.73	3.41	3.41	3.40	3.39

Table 5. Running time of the algorithms on random instances of 40 nodes and group size 40.

Algorithm	Time	Algorithm	Time	Algorithm	Time
SPF	12 msec	iDSPF	95 msec	DSF-3	1.2 sec
MIP	14 msec	iD2SPF	6 sec	DISF	11.6 sec
DSPF	6 msec	iSP3SF	250 msec	DFF	15.7 sec
D2SPF	265 msec	iDSP3SF	130 msec	DFF-2	18.1 sec
SP3SF	22 msec	i2SP3SF	10.7 sec	iDISF	349 sec
DSP3SF	6 msec	iD2SP3SF	6.7 sec	iDFF	137.6 sec
2SP3SF	1 sec	DSF	1.2 sec	iDFF-2	194.8 sec
D2SP3SF	280 msec	DSF-2	1.2 sec		

References

1. M. Cagalj, J.P. Hubaux and C. Enz. Minimum-Energy Broadcast in All-Wireless Networks: NP-completeness and Distribution Issues. In *Proc. of the 8th ACM International Conference on Mobile Networking and Computing (Mobicom '02)*, pp. 172–182, 2002.

2. G. Călinescu, S. Kapoor, A. Olshevsky and A. Zelikovsky. Network Lifetime and Power Assignment in Ad-Hoc Wireless Networks. In *Proc. of the 11th Annual European Symposium on Algorithms (ESA '03)*, LNCS 2832, Springer, pp. 114–126, 2003.

3. I. Caragiannis, C. Kaklamanis and P. Kanellopoulos. New Results for Energy-Efficient Broadcasting in Wireless Networks. In *Proc. of the 13th Annual International Symposium on Algorithms and Computation (ISAAC '02)*, LNCS 2518, Springer, pp. 332–343, 2002.

4. I. Caragiannis, C. Kaklamanis and P. Kanellopoulos. Energy-Efficient Wireless Network Design. *In Proc. of the 14th Annual International Symposium on Algorithms and Computation (ISAAC '03)*, LNCS 2906, Springer, pp. 585–594, 2003.

5. J. Cartigny, D. Simplot, I. Stojmenovic. Localized minimum energy broadcasting in ad hoc networks. In *Proc. of IEEE INFOCOM 2003*, 2003.

6. A.E.F. Clementi, P. Crescenzi, P. Penna, G. Rossi, and P. Vocca. On the Complexity of Computing Minimum Energy Consumption Broadcast Subgraphs. In *Proc. of the 18th Annual Symposium on Theoretical Aspects of Computer Science (STACS '01)*, LNCS 2010, Springer, pp. 121–131, 2001.

7. A.E.F. Clementi, M. Di Ianni, R. Silvestri. The Minimum Broadcast Range Assignment Problem on Linear Multi-Hop Wireless Networks. *Theoretical Computer Science*, 299 (1-3), pp. 751–761, 2003.

8. T.H. Cormen, C.E. Leiserson, R.L. Rivest and C. Stein. Introduction to Algorithms. *The MIT Press*, Second Edition, 2001.

9. S. Guha and S. Khuller. Improved Methods for Approximating Node Weighted Steiner Trees and Connected Dominating Sets. *Information and Computation*, 150(1), pp. 57–74, 1999.

10. L. M. Kirousis, E. Kranakis, D. Krizanc, and A. Pelc. Power Consumption in Packet Radio Networks. *Theoretical Computer Science*, 243(1-2), pp. 289–305, 2000.

11. P.N. Klein and R. Ravi. A Nearly Best Possible Approximation Algorithm for Node-Weighted Steiner Trees. *Journal of Algorithms*, 19(1), pp. 104–115, 1995.

12. W. Liang. Constructing Minimum-Energy Broadcast Trees in Wireless Ad Hoc Networks. In *Proc. of 3rd ACM International Symposium on Mobile Ad Hoc Networking and Computing (MOBIHOC '02)*, pp. 112–122, 2002.

13. P. Mavinkurve, H.Q. Ngo and H. Mensa. MIP3S: Algorithms for Power-conserving Multicasting in Wireless Ad Hoc Networks. In *Proc. of the 11th IEEE International Conference on Networks (ICON '03)*, 2003.

14. P.-J. Wan, G. Călinescu, X.-Y. Li, and O. Frieder. Minimum-Energy Broadcasting in Static Ad Hoc Wireless Networks. *Wireless Networks*, 8(6), pp. 607–617, 2002.

15. J. E. Wieselthier, G. D. Nguyen, and A. Ephremides. On the Construction of Energy-Efficient Broadcast and Multicast Trees in Wireless Networks. In *Proc. of IEEE INFOCOM 2000*, pp. 585–594, 2000.

16. V. Verma, A. Chandak and H.Q. Ngo. DIP3S: A Distributive Routing Algorithm for Power-Conserving Broadcasting in Wireless Ad Hoc Networks. In *Proc. of the Fifth IFIP-TC6 International Conference on Mobile and Wireless Communications Networks (MWCN '03)*, pp. 159–162, 2003.

Improving Distance Based Geographic Location Techniques in Sensor Networks

Michel Barbeau[1], Evangelos Kranakis[1], Danny Krizanc[2], and Pat Morin[1]

[1] School of Computer Science, Carleton University,
Ottawa, Ontario, Canada, K1S 5B6.
[2] Department of Mathematics, Wesleyan University, Middletown CT, 06459.

Abstract. Supporting nodes without Global Positioning System (GPS) capability, in wireless ad hoc and sensor networks, has numerous applications in guidance and surveying systems in use today. At issue is that a procedure be available so that the subset of nodes with GPS capability succeed in supporting the maximum possible number of nodes without GPS capability and as a result enable the highest connectivity of the underlying network infrastructure. In this paper, we identify incompleteness in the standard method for computing the position of a node based on three GPS enabled neighbors, in that it may fail to support the maximum possible subset of sensors of the wireless network. We give a new complementary algorithm (the three/two neighbor algorithm) that indeed succeeds in providing a higher fraction of nodes (than the 3-Neighbour algorithm) with their position. We prove its correctness and test its performance with simulations.

1 Introduction

In wireless ad hoc systems, location determination can be an important parameter in reducing information overhead, thus simplifying the distribution of information and limiting infrastructure reliance. Location awareness has proven to be an important component in designing communication algorithms in such systems and there have been many papers making use of this paradigm (e.g., Kranakis et al. [13] Bose et al. [4], Kuhn et al. [14], and Boone et al. [3], to mention a few) thus making possible the execution of *location based* routing using only local (i.e., information in the vicinity of the node) as opposed to global knowledge on the status of the nodes. In addition, guidance and surveying systems in use today have numerous military and civilian applications. The current Global Positioning System (GPS) is satellite based and determines the position of a GPS equipped device using the radiolocation method. However, there are instances where devices may not have GPS capability either because the signal is too weak (due to obstruction) or integration is impossible. Adding to these the fact that such devices are easy to jam and there have been calls to declare the GPS a critical infrastructure.

Typically, sensors so equipped can determine their position with their GPS devices. The remaining nodes have no option but to query neighbors for their

I. Nikolaidis et al. (Eds.): ADHOC-NOW 2004, LNCS 3158, pp. 197–210, 2004.
© Springer-Verlag Berlin Heidelberg 2004

location and thus determine their position using radiolocation. There are several radiolocation methods in use (see Bahl et al. [2]) including signal strength, angle of arrival, time of arrival, and time difference of arrival. However, despite the method used at issue is that a procedure be available so that the subset of nodes with GPS capability succeed in supporting the maximum possible number of nodes without GPS capability and as a result enable the highest connectivity of the underlying network infrastructure.

Consider a set S of n sensors in the plane. Further assume that the sensors have the same reachability radius r. This paper addresses the problem of supporting nodes that do not have GPS capability within a sensor network. We assume that all the sensors have identical reachability radius r. We are interested in algorithms that will position the maximum number of sensors in the set S. The paper is organized as follows. In Section 2, we give an overview of radiolocation.

In Section 3, we show first that the usual positioning algorithm that determines the position of a node based on the presence of its three GPS enabled neighbors may fail to determine the position of the maximum number possible of sensors. We give a new algorithm that is based on using the distance one neighborhood. It may outperform the traditional algorithm in the sense that it determines correctly the position of more sensors than the usually used three neighbor algorithm. Later we explore the distance k ($k \geq 1$) neighborhood of a node and derive an algorithm that captures the maximum number of nodes of a sensor system that can compute their geographic position. In Section 4, we explore the performance of these algorithms in a random setting whereby sensors are dropped randomly and independently with the uniform distribution in the interior of a unit square. We investigate what is the impact of the size of the reachability radius r of the sensors as a function of the number of sensors so that with high probability all the sensors in the unit square determine correctly their position. Our approach to improve distance based geographic location techniques is fairly general and can be applied over an existing algorithm like the one proposed by Capkun et al. [9].

2   Overview of Radiolocation

The position of wireless nodes can be determined using one of four basic techniques, namely time of arrival (TOA), time difference of arrival (TDOA), angle of arrival (AOA) or signal strength.

The TOA technique is pictured in Figure 3. Node A is unaware of its position. Three position aware nodes are involved, let us say B_1, B_2, and B_3. Each position aware node B_i sends a message to A and the trip time of the signal is measured. The trip time multiplied by the propagation speed of signals (i.e. the speed of light) yields a distance d_i. The distance $d(A, B_i)$ defines a circle around node $d(A, B_i)$. The position of A is on the circumference of this circle. In a two dimensional model, the position of A is unambiguously determined as the intersection of three such circles.

Trip time measurement from each node B_i to node A requires synchronized and accurate clocks at both locations. If round-trip time is measured instead (halved to obtain trip time), then this requirement is relaxed. No clock synchronization is required and accurate clocks are required only at the B_i's.

TDOA is pictured in Figure 1. Nodes B_1, B_2, and B_3 are aware of their position while node A is not. B_1 and B_2 simultaneously send a signal. Times of arrivals t_1 and t_2 of signals from B_1 and B_2, respectively, are measured by A. Node A has the capability to recover the identity of the sender of a signal and its position. The time difference of arrival is calculated, i.e. $\delta_t = t_2 - t_1$. The time difference δ_t multiplied by the speed of light is mapped to the distance difference δ_d. The position (x_1, y_1) of B_1, position (x_2, y_2) of B_2 and δ_d define a hyperbola h with equation:

$$\sqrt{(x - x_1)^2 + (y - y_1)^2} - \sqrt{(x - x_2)^2 + (y - y_2)^2} = \delta_d \qquad (1)$$

The positions of B_1 and B_2 are at the foci of the hyperbola. The position of node A is a solution of Equation 1. The geometrical properties of the hyperbolas are such that all points located on the line of h are of equal time difference δ_t and equal distance difference δ_d.

Two such hyperbolas can be defined by involving two different pairs of nodes (B_1, B_2) and (B_2, B_3) which produce two time differences of arrival δ_1 and δ_2. In a two dimensional model, the observer of the times of arrival δ_1 and δ_2, i.e. node A, is at the position corresponding to the intersection of the two hyperbolas. Note that there are cases in which the two hyperbolas intersect at two points. In these cases, a third independent measurement is required to resolve the ambiguity.

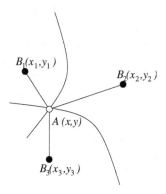

Fig. 1. The TDOA technique.

The AOA technique is pictured in Figure 2. Two position aware nodes, let us say B_1 and B_2, are required to determine the position of a node A. Nodes B_1 and B_2 have to be able to determine the direction from which a signal is coming. This can be achieved with an array antenna [10]. An imaginary line is drawn from B_1

to A and another imaginary line is drawn from B_2 to A. The angle of arrival is defined as the angle that each of these lines make with a line directed towards a common reference. The intersection of these two lines unambiguously determines the position of A. Note however, that if A, B_1, and B_2 all lie on the same straight line, another independent measurement is required to resolve the ambiguity. Accuracy of the AOA technique is largely dependent on the beamwidth of the antennas. According to Pahlavan and Krishnamurthy [15] the TOA technique is superior to the AOA technique. In CDMA cellular networks, Caffery and Stüber [8] come to similar conclusions.

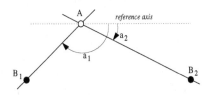

Fig. 2. The AOA technique.

The signal strength based technique exploits the fact that a signal loses its strength as a function of distance. Giving the power of a transmitter and a model of free-space loss, a receiver can determine the distance traveled by a signal. If three different such signals can be received, a receiver can determine its position in a way similar to the TOA technique. Application of this technique for cellular systems has been investigated by Figuel et al. [11] and Hata and Nagatsu [12]. The main criticism of Caffery and Stüber [7] is about the accuracy of the technique. This is due to transmission phenomena such as multi path fading and shadowing that cause important variation in signal strength.

All these techniques require line of sight propagation between the nodes involved in a signal measurement. Line of sight means that a non obstructed imaginary straight line can be drawn between the nodes. In other words, the accuracy is sensitive to radio propagation phenomena such as obstruction, reflection and scattering. With all the distance-based techniques (i.e. TOA, TDOA, signal strength) three position aware neighbors are required to determine the location of a position unware node, in a two dimensional model (e.g. latitude and longitude are determined), and four neighbors are required in a three dimensional model (e.g. altitude is determined as well). In the sequel, we augment the distance-based techniques with an algorithmic component that can resolve ambiguity in a two dimensional model when only two position aware neighbors are available. The ambiguity can also be resolved using knowledge about the trajectory when the wireless nodes are mobile or using the AOA technique. When the nodes are fixed and the technology required to apply the AOA technique is not available, the algorithm described in this paper can be used. We note that a similar algorithm is also possible in a three dimensional model.

3   Computing the Geographic Location

Consider a set S of n sensors in the plane. Further assume that the sensors have the same reachability radius r. We divide the set S of sensors into two subsets. A subset E of S consists of sensors equipped with GPS devices that enable them to determine their location in the plane. The set U of remaining sensors, i.e., $S \setminus E$, consists of sensors not equipped with GPS devices. (In pictures, the former are represented with solid circles and the latter with empty circles.) In this section, we consider the problem of determining the positions of sensors in a sensor network. In the beginning, we review the well-known *three neighbor algorithm* (3-NA) and conclude with an example illustrating why the algorithm does not necessarily compute the positions of the maximum possible number of nodes. Subsequently, we present an improvement, the *three/two neighbor algorithm* (3/2-NA). Essentially, this is an iteration of the 3-NA followed by an algorithm that uses only two neighbors as well as their distance one neighborhood.

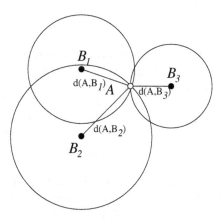

Fig. 3. A sensor at A not equipped with a GPS device can determine its position from the positions of its three neighbors B_1, B_2, B_3.

3.1   Three Neighbor Algorithm and Its Deficiencies

If a sensor at A (see Figure 5) is not equipped with a GPS device, then it can determine its (x, y) coordinates using three neighboring nodes. After receiving messages from B_1, B_2, and B_3, that include their position, node A can determine its distance from these nodes using a distance-based radiolocation method. Its position is determined as the point of intersection of three circles centered at B_1, B_2, and B_3 and distances $d(A, B_1), d(A, B_2)$, and $d(A, B_3)$, respectively.

The well-known 3-NA is as follows. Each node that is not equipped with a GPS device sends a position request message. A sensor that knows or can compute its position sends it to all its neighbors. A sensor that receives position messages from three different nodes, say B_1, B_2, and B_3 can calculate its position as in Figure 3.

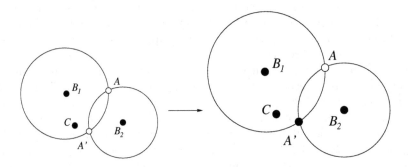

Fig. 4. Application of 3-NA will equip node A' with its (x, y)-coordinates, but it fails to do it with node A.

Computing the position of the maximum number of nodes. The 3-NA does not necessarily compute the positions of the maximum possible number of sensors without GPS devices. We illustrate this with a simple example depicted, in Figure 4. The left side of the picture depicts five nodes A, A', B_1, B_2, and C. Node A' is within communication range of nodes B_1, B_2, and C. Node A is within communication range of nodes B_1 and B_2. Node A is out of the range of both nodes C and A'. Assume nodes B_1, B_2 and C know their (x, y)-coordinates. Application of the 3-NA will equip A' with its (x, y) coordinates (because it will receive messages from all three of its neighboring nodes B_1, B_2, C). This is depicted in the right side of the picture in Figure 4. However, node A will never receive three position messages and therefore will never be able to discover its (x, y)-coordinates. Nevertheless, we will see in the next algorithm that node A can indeed discover its position if additional information (i.e., the distance one neighborhood of its neighbors) is provided.

3.2   Three/Two Neighbor Algorithm

We now consider an extension of the 3-NA for the case where all the sensors have exactly the same reachability radius, say r.

On utilizing the distance one neighborhood. Assume that we have concluded the execution of the 3-NA. For each node P, let $N(P)$ be the set of neighbors of P, i.e., the set of sensors within communication range of P. Suppose we have two nodes B_1 and B_2 (depicted in Figure 5) that know their position. The solid circles are determined by the reachability radius of the nodes. The

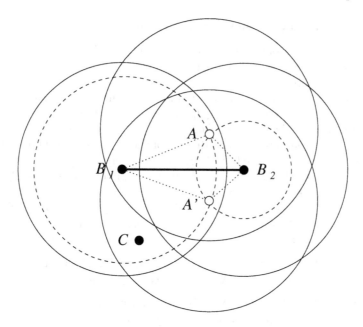

Fig. 5. A sensor not equipped with a GPS device can determine its coordinates using the positions of two neighbors B_1 and B_2 and their neighborhood information.

dashed circles are centered at B_1 and B_2 respectively. After using radiolocation they specify that the inquiring node must be located at one of their points of intersection (in this case either A or A'). Further assume that a given node X receives the positions of nodes B_1 and B_2 by radiolocation. On the basis of this information, X may be located in either position A or A'.

Lemma 1. *Suppose that both nodes A and A' receive from B_1 and B_2 the set $N(B_1) \cup N(B_2)$. If there is a sensor $C \in (N(B_1) \cup N(B_2)) \cap (N(A) \cup N(A'))$ that knows its position then both nodes A and A' can determine their position.*

Proof (of Lemma 1). We must consider two cases. In the first case, assume C is within range of both nodes A and A'. In this case the two nodes will receive position messages from all three nodes B_1, B_2, and C and can therefore determine their position. In the second case, assume C is within range of only one of the two nodes. Without loss of generality assume it is within range of A' but outside the range of A, i.e., $d(C, A') \leq r < d(C, A)$ (see Figure 5). Then A' can determine its position. Also, A can determine its position because it knows it must occupy one of the two positions A or A' and also can determine it cannot be node A' since its distance from C (a node whose position it has received) is outside its range. This completes the proof of Lemma 1. ∎

It is now clear that Lemma 1 gives rise to the following 3/2-NA for computing sensor positions.

3/2-NA (Three/Two Neighbor Algorithm):

1. For each node as long as new nodes determine their position **do**
2. Each sensor executes the 3-NA algorithm and also collects the coordinates of all its neighbors.
3. **If** at the end of the execution of this algorithm a sensor has received the coordinates from only two neighbors, say B_1, B_2 **then**
   a) it computes the distances from its current location to the sensors B_1, B_2 and also computes the coordinates of the two points of intersection A, A' of the circles centered at B_1, B_2, respectively;
   b) it requests the coordinates of all the neighbors of B_1 and B_2 that are aware of their coordinates;
   c) **if** $N(B_1) \cup N(B_1) \neq \emptyset$ then take any node $C \in N(B_1) \cup N(B_1)$ and compute $d(C, A), d(C, A')$; then the sensor occupies the position $X \in \{A, A'\}$ such that $d(X, C) > r$;

For any node P let $D(P; r)$ be the disc centered at P and with radius r, i.e., the set of points X such that $d(P, X) \leq r$. We can prove the following theorem.

Theorem 1. *The Three/Two Neighbor Algorithm terminates in at most $(n-e)^2$ steps, where e is the number of GPS equipped nodes.*

Proof of Theorem 1. First consider the three neighbor algorithm. At each iteration of the algorithm a sensor either waits until it receives three messages (of distances of its neighbors), or else it receives the coordinates of at least three neighboring sensors, in which case it computes its own coordinates and forwards it to all its neighbors. Let E_t be the number of sensors that know their coordinates by time t. Observe that initially $E_0 = E$ and $E_t \subseteq E_{t+1}$. If at any given time no new sensors are not GPS-equipped determines their coordinates (i.e., $E_t = E_{t+1}$) then no new non-equipped sensor will ever be added. It follows that the algorithm terminates in at most $n - e$ steps. After this step no new sensors will be equipped with their coordinates.

Now consider the three/two neighbor algorithm. Concerning correctness we argue as follows. Consider a sensor as above that has received position messages only from two neighbors, say B_1 and B_2. The sensor knows it is located at one of the points of intersection of the two circles (see Figure 5). Since $N(B_1) \cup N(B_2) \neq \emptyset$ and the sensor received no position message from any sensor in $N(B_1) \cup N(B_2)$ after the execution of the 3-NA algorithm it concludes that it must occupy position X, where $X \in \{A, A'\}$ such that $d(C, X) > r$. Since within each iteration at one new sensor computes its position the running time is as claimed. This completes the proof of Theorem 1. ■

Remark 1. There are several interesting issues arising in the algorithm 3/2-NA.

1. The sensors may not know the total number of sensors in the network participating in the positioning algorithm. In this case, they may have to guess an upper bound value n' and use this to execute the above algorithm or even execute the algorithm by increasing incrementally the value n'. The running time stated in Theorem 1 pre-supposes that the sensors know the value $n-e$.

2. The above algorithm will take a lot of messages. To increase efficiency, it may be a good idea that sensors localize their search within limited geographic boundaries.

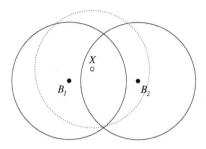

Fig. 6. The termination condition for the 3/2-NA.

3.3   Beyond Distance One Neighborhood

When the 3/2-NA terminates no new node can compute its position. There is an improvement that can be made to the 3/2-NA. In Figure 5, this may happen when for some $k \geq 1$ the node C is at distance k hops from either B_1 or B_2. If the entire distance k neighborhood is being transmitted and C is within A''s range but outside A's range, then the node A can determine its position. This gives rise to the 3/2-NA(k) algorithm which is similar 3/2-NA algorithm except that now the nodes transmit their distance k neighborhood.

We define the distance k hop(s) neighborhood of a node P as follows. First, $N_1(P)$ is defined as $N(P)$. For $k = 2, 3, 4, \ldots$,

$$N_k(P) = \{A | A \in N_{k-1}(P) \vee \exists B \in N_{k-1}(P) \wedge A \in N(B)\}$$

The specific algorithm is as follows:

3/2-NA(k) (Three/Two Neighbor Algorithm):

1. For each node as long as new nodes determine their position **do**
2. Each sensor executes the 3-NA algorithm and also collects the coordinates of all its neighbors.
3. **If** at the end of the execution of this algorithm a sensor has received the coordinates from only two neighbors, say B_1, B_2 **then**
   a) it computes the distances from its current location to the sensors B_1, B_2 and also computes the coordinates of the two points of intersection A, A' of the circles centered at B_1, B_2, respectively;
   b) it requests the coordinates of all the neighbors of B_1 and B_2 that are aware of their coordinates;

c) **if** $N_k(B_1) \cup N_k(B_1) \neq \emptyset$ then take any node $C \in N_k(B_1) \cup N_k(B_1)$ and compute $d(C, A), d(C, A')$; then the sensor occupies the position $X \in \{A, A'\}$ such that $d(X, C) > r$;

4   Simulations

Our approach is sufficiently general and may augment any distance-based geographic location method in use. For example, Capkun et al. [9] propose a GPS-free positioning algorithm for mobile ad hoc networks. In general, their algorithm succeeds in providing a common reference only for a subset of the total number of nodes. In this section, we provide simulations which confirm that our method increases the percentage of nodes that can compute their geographic location in an arbitrary sensor network.

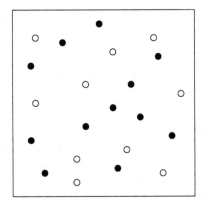

Fig. 7. Sensors dropped randomly and independently over a unit square region. Bolds (respectively, empty) circles denote sensors which are equipped (respectively, not equipped) with GPS devices.

4.1   Connectivity and Coverage in Random Setting

Assume that n sensors (here assumed to be omnidirectional antennas) are dropped randomly and independently with the uniform distribution on the interior of a unit square. For any integer $k \geq 0$ and real number constant c let the sensors have identical radius r, given by the formula

$$r = \sqrt{\frac{\ln n + k \ln \ln n + \ln(k!) + c}{n\pi}}. \tag{2}$$

A network is called k-connected if it cannot be disconnected after the removal of $k - 1$ nodes. Then using the main result of Penrose [16,17] we conclude that this is a threshold value for k-connectivity. Namely, we have the following theorem.

Theorem 2. *Consider sensors with reachability radius r given by Formula 2, and suppose that $k \geq 0$ is an integer and c is a real. Assume n sensors are dropped randomly and independently with the uniform distribution on the interior of a unit square. Then*

$$\lim_{n \to \infty} \Pr[sensor\ network\ is\ (k+1)\text{-}connected] = e^{-e^{-c}},\ and$$
$$\lim_{n \to \infty} \Pr[(k+1)\ is\ the\ \min\ degree\ of\ the\ sensor\ network] = e^{-e^{-c}}.$$
(3)

∎

Thus, for the radius chosen by Formula 2 not only is the network $(k+1)$-connected but within distance r, each node will have $k+1$ neighbors with high probability as indicated by Equations 3.

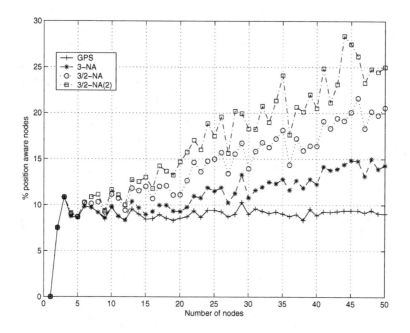

Fig. 8. 9% of the sensors are GPS equipped.

4.2   Experimental Results

A simulation of the algorithms has been conducted. Sensors are spread randomly and independently with uniform distribution on a unit square. The communication range of each sensor is a circle centered at its position and of radius r as defined by Formula 2. Constants k and c are both set to value 1. If the number of nodes equipped with GPS devices is dense (as a proportion to the total),

then we expect that with high probability every node will have three neighbors
that are equipped with GPS devices. Therefore, the standard 3-NA algorithm
is expected to enable all nodes to compute their position, with high probability.
Therefore incremental differences will be more substantial in a sparse setting.
Figure 8 pictures the results of the simulation of one to 50-sensor networks. An
average of 9% of the sensors are GPS equipped and can determine their position
independently of other sensors. Application the 3-NA, 3/2-NA (using one-hop
neighbors only) and 3/2-NA(2) (using one or two-hop neighbors) all make an additional number of sensors aware of their position, up to 5% of the total number
of sensors in each case. The simulation was run for 200 times for each network
size.

Figure 9 pictures the results of the simulation of one to 50-sensor networks.
An average of 28% of the sensors are GPS equipped and can determine their
position independently of other sensors. The simulation was run for 200 times
for each network size.

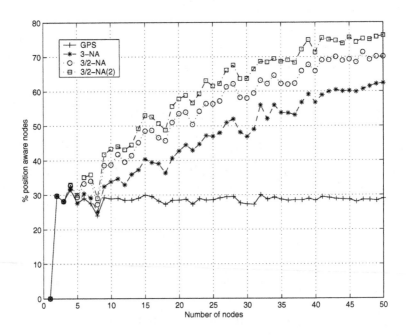

Fig. 9. 28% of the sensors are GPS equipped.

Application of the 3-NA makes the number of sensors aware of their position
up to the double of the number of GPS equipped sensors. This is consistent with
intuition since, with respect to Figure 8, more nodes are available to resolve loci
of position unaware sensors. In addition, up to 10% and 7% more sensors can
resolve their position using respectively the 3/2-NA (using one-hop neighbors
only) and 3/2-NA(2) (using one or two-hop neighbors).

In both cases, the 3/2-NA increase significantly the number of sensors aware of their position.

5   Conclusion

In this paper, we have considered a new class of algorithms for improving any distance based geographic location method. Our technique may augment any existing algorithm by iterating a three-neighbor with a two-neighbor based calculation as well a taking into account the distance k neighbors of the given node. Our simulations show that 5% to 10% more sensors can resolve their position using the 3/2-NA and 3/2-NA(2).

Our analysis focused only on the two dimensional plane. However, it can be easily extended to the three dimensional space. In this case, the intersection of the spheres defined by the distance from four neighbors is required in order to determine the location of a node. The intersection of three spheres alone creates an ambiguity which can be resolved by looking at the distance k neighbors, just like the two dimensional case.

Acknowledgements. The authors graciously acknowledge the financial support received from the following organizations: Natural Sciences and Engineering Research Council of Canada (NSERC) and Mathematics of Information Technology and Complex Systems (MITACS).

References

1. Andrisano, O., Verdone R., Nakagawa, M.: Intelligent transportation systems : The role of third generation mobile radio networks. IEEE Communications Magazine. **36** (2000) 144-151
2. Bahl, P., Padmanabhan, V.N.: RADAR: An in-building, RF-based user location and tracking system. INFOCOM. **2** (2000) 775–784
3. Boone, P., Chavez L. Gleitzky, E, Kranakis, E., Opartny, J., Salazar, G., Urrutia, J.: Morelia test: Improving the efficiency of the Gabriel test and face routing in ad-hoc networks. SIROCCO, Springer Verlag, Lecture Notes in Computer Science. **3104** (2004)
4. Bose, P., Morin, P., Stojmenovic, I., Urrutia, J.: Routing with guaranteed delivery in ad hoc wireless networks. Wireless Networks. **7** (2001) 609–616
5. Briesemeister, L., Hommel, G.: Overcoming fragmentation in mobile ad hoc networks. Journal of Communications and Networks. **2** (2000) 182–187
6. Bachir, A., Benslimane, A.: A multicast protocol in ad-hoc networks: Inter-vehicles geocast. The 57th IEEE Semiannual Vehicular Technology Conference (VTC 2003-Spring). **4** (2003) 22–25
7. Caffery, J.J., Stüber, G.L.: Overview of radiolocation in CDMA cellular systems. IEEE Communications Magazine. **36** (1998) 38–45
8. Caffery, J.Jr., Stüber, G.: Subscriber location in CDMA cellular networks. IEEE Transactions on Vehicular Technology. **47** (1998) 406–416

9. Capkun, S., Hamdi, M., Hubaux, J.-P.: GPS-free positioning in mobile ad hoc networks. 34th Annual Hawaii International Conference on System Sciences. (2001) Pages: 10 pp

10. Cooper, M.: Antennas get smart. Scientific American. July (2003) 49–55

11. Figuel, W., Shepherd, N. Trammell, W.: Vehicle location using signal strength measurements in cellular systems. IEEE Transactions on Vehicular Technology. **VT-18** (1969) 105–110

12. Hata, M., Nagatsu, T.: Mobile location using signal strength measurements in cellular systems. IEEE Transactions on Vehicular Technology. **VT-29** (1980) 245–251

13. Kranakis, E., Singh, H., Urrutia, J.: Compass routing in geometric graphs. 11th Canadian Conference on Computational Geometry. (1999) 51–54

14. Kuhn, F., Wattenhofer, R., Zhang, Y., Zollinger, A.: Geometric ad-hoc routing: Of theory and practice. 22nd ACM Annual Symposium on Principles of Distributed Computing. (2003) 63–72

15. Pahlavan, K., Krishnamurthy, P.: Principles of wireless networks. Prentice Hall PTR, Upper Saddle River, New Jersey. (2002)

16. Penrsose, M. D.: On k-connectivity for a geometric random graph. Random Structures and Algorithms. **15** (1999) 145–164

17. Penrsose, M. D.: The longest edge of the random minimal spanning tree. The Annals of Applied Probability. **7** (1997) 340–361

18. Penrsose, M. D.: Random Geometric Graphs. Oxford University Press. (2003)

19. Venkatraman, S., Caffery, J., You, H.R.: Location using LOS range estimation in NLOS environments. IEEE 55th Vehicular Technology Conference (VTC Spring 2002). (2002) 856–860

20. Wesel, E.K.: Wireless multimedia communications: Networking video, voice and data. Addition-Wesley, One Jacob Way, Reading Massachusetts 01867 USA. (1998)

21. Wylie, M.P., Holtzman, J.: The non-line of sight problem in mobile location estimation. 5th IEEE International Conference on Universal Personal Communications. **2** (1996) 827–831

A Fair Distributed Solution for Selfish Nodes Problem in Wireless Ad Hoc Networks*

Yongwei Wang, Venkata C. Giruka, and Mukesh Singhal

Laboratory of Advanced Networking
Department of Computer Science, University of Kentucky, Lexington, KY 40506
{ywang7,venkata,singhal}@cs.uky.edu

Abstract. In Mobile Ad Hoc Networks (MANETs), nodes depend on each other for routing and forwarding packets. Cooperation among nodes is a key issue in such an environment. However, some of the nodes in MANETs may behave selfishly and may not forward packets to save battery and other resources. Since nodes in MANETs communicate on a peer-to-peer basis, without any central authority (which can monitor selfish behavior of nodes), a centralized solution to stimulate cooperation is not suitable. In this paper, we present a distributed solution to judge, punish and re-admit a selfish node, forcing nodes to cooperate with each other. Unlike previous solutions, we focus on the fairness, i.e., to provide the same chance to all nodes to gain services from the network and to offer services to the network. We also consider the *location privilege* and *counting retransmission* problems which have not been emphasized in any of the previous solutions. We combine our scheme with AODV. To evaluate the performance of our scheme, we conducted simulations using Glomosim. Simulation results show that the proposed scheme improves data forwarding capability substantially in presence of selfish nodes.

Keywords: Mobile Ad Hoc Network (MANET), selfish node, fairness.

1 Introduction

One basic function of a network is to route and forward packets. In MANETs, every node acts as a router to share the information and provide services to the network because of the limited radio transmission range (typically 250m). For resource-poor mobile modes, saving battery is a key issue. As it is a known fact that "packet forwarding requires more power dissipation than packet receiving", nodes can behave *selfishly* and refuse to forward packets for other nodes while gaining services from other nodes for its benefit. If intermediate nodes refuse to forward packets, communication beyond radio range is not possible. Thus cooperation among nodes is necessary to enable communication in MANETs. Specially, in civilian application (for example, ad hoc networks formed at railway

* This work is partly supported by NSF grants IIS-0242384 and ITR-0324836.

I. Nikolaidis et al. (Eds.): ADHOC-NOW 2004, LNCS 3158, pp. 211–224, 2004.
© Springer-Verlag Berlin Heidelberg 2004

stations or airports), where MANET may consist of various users belonging to different authorities, users' cooperation is difficult to enforce.

We define a node as *selfish* if the node wants to gain packet forwarding service from other nodes but avoids rendering the same service to others. Note that a selfish node is different from a *malicious* node. A selfish node is just not cooperative, whereas a malicious node intends to damage other nodes (by launching some kind of attack) and harm the network.

Several solutions proposed in literature [1]- [10] [13]- [15] to enforce cooperation use data packets forwarded by nodes to determine the nodes' behavior. However, nodes in different positions of the network(specially when the network is in two a dimensional area) have different opportunities to serve the network or be served by the network. Some nodes in certain positions (for example, a node in the middle of the network) have obvious privilege over others nodes. We call this unfair utility of network the *location privilege* problem. Generally, routing algorithms select shorter routes (sometimes shortest). Many such routes include more nodes located in the center area than the periphery. Nodes located in the center area have more opportunities to forward packets and this results in the *location privilege* problem. Another important issue is the *counting retransmission* problem. Due to collisions or other reasons, an intended receiver may not receive a packet successfully. The sender may have to retransmit the packet for the maximum number of retransmissions. Thus a forwarding node may spend extra resources. This is critical to virtual money schemes [1] [4] [15] [9], which reimburse forwarding nodes by the number of packets forwarded. If the retransmissions are not counted, a node will not get enough reimbursement for its expense and thus have no incentive to forward packets. Also, it is unfair to ignore the retransmissions. The forwarding node spends extra resources, and the more packets a node forwards, the more resources it consumes, aggravating the *location privilege* problem.

In this paper, we provide a fair scheme to solve the selfish node problem. The term *fair* is used to mean that "nodes in a network have the same chance to serve others and be served by others". The scheme is fully distributed. Nodes in the network monitor, judge, punish and re-admit selfish nodes distributively. Our scheme overcomes the *location privilege* problem and takes care of the *counting retransmission* problem. The rest of the paper is organized as follows. Section 2 reviews the existing work. Section 3 describes our scheme in detail. Section 4 analyzes the performance of our scheme using simulation. Section 5 concludes the paper.

2   Related Work

Several papers in the literatures deal with the problem of nodes cooperation in the MANETS; each solution has its niche of applicability. The solutions can be broadly classified as a) Motivation based protocols and b) Detection based protocols. Motivation based protocols are based on some form of incentive (given to nodes upon serving the network) to enforce nodes' cooperation. Nodes having incentives can use them to gain services from the network. When they don't have any accumulated incentives they cannot take any services from the network. Thus nodes are forced to work to gain services. Detection based protocols, on the other hand, monitor nodes to trace selfish nodes based on observations and take necessary action. We briefly describe protocols under this classification below.

2.1   Motivation Based Protocols

The use of virtual money (specifically called nuglet or credit) to simulate nodes cooperation has been suggested in [1]- [3]. A node earns money by providing forwarding service to others and has to pay to get service from other nodes. To protect the nuglets or credit value from modification and other attacks, they use security modules in each node, which are independent of the node. However, this security module requires tamper-resistant hardware in each node. A certificate for its public key is signed by the manufacturer and stored in the security module. Manufacturers cross-certify each other's public keys, and each security module stores the public key certificates of all manufacturers. All these are not feasible in most cases for mobile nodes. Nodes in the periphery of the network will have fewer packets to forward even if they are willing to do so. So these nodes will have less chance to earn money and will run out of their money soon.

In [4], Fratkin et al. use a virtual money model to establish an APE (Ad Hoc Participation Economy) system. To avoid the tamper-resistant module, they provide a software solution. A trusted third party, called banker node, is used to assure the payment consolidation and its integrity. However, this solution uses a centralized service. In [15], Zhong et al. provide a centralized solution similar to [4], Sprite, for stimulating cooperation among nodes. Every node reports a receipt, the digest of received or forwarded packets, to a central sever Credit Clearance Service (CCS). The CCS then determines the charge and credit to each node involved in the forwarding of the packets. In [9], Anderegg et al. provide ad hoc-VCG, a routing protocol that encourages the intermediate nodes truthfully to reveal their cost of forwarding, and thus gets a cost-efficiency route by paying the intermediate nodes premiums over their cost.

2.2   Detection Based Protocols

In [10], Marti et al. propose two tools, watchdog and pathrater, to mitigate routing misbehavior (including selfish node) in ad hoc networks. Watchdog is used to identify malicious and selfish nodes, while pathrater is used to select a route to avoid them. In their scheme, selfish nodes are not penalized. Instead, they can still use the network while not forwarding packets.

In [6] and [7], Buchegger et al. propose CONFIDANT protocol to monitor the behavior of nodes, evaluate the reputation of corresponding nodes and punish selfish nodes. CONFIDANT suffers from the inconsistent evaluation problem, i.e., different nodes may have different evaluations for the same node and thus, some regard a node as selfish but others do not. Also the location privilege problem, which is opposite to the situation in a virtual economy, still exist in this scheme. This protocol punishes the nodes if they do not forward packets, no matter how they have previously contributed to the network. Thus, nodes situated in the center of the network have to keep on forwarding packets to avoid being selfish. These nodes have to spend much more battery power than those in the periphery of network.

In [5], Michiardi et al. suggest using reputation to measure a node's contribution to the network. The overall reputation is based on a combination of three reputation types: subjective, indirect and functional. In [13], Miranda et al. suggest that a node periodically broadcasts messages stating its view about the neighbors. Also nodes are allowed to publicly declare that they refuse to forward messages for some hosts. This scheme causes heavy communication overhead. In [14], authors introduce security extensions to DSR to detect attacks to the routing process. The scheme relies on neighbors monitoring routing message's context to find the attacker. All the schemes explained so far suffer from the *location privilege* problem. Besides, motivation based schems suffer from the *counting retransmission* problem.

3   Proposed Fair Distributed Solution

As seen from Section 2, each of the broadly classified methods has some plus and minus points. We combine the plus points of both methods to propose a fair solution. We present a distributed scheme to solve the selfish node problem, using a polling mechanism to ensure fairness and cope with false accusations. Also, it solves the *location privilege* problem and the *counting retransmission* problem, and takes into account the scarcity of resources such as battery life.

3.1   Overview

To evaluate selfishness of nodes, we need a criterion to quantify the selfishness of a node. We define the information used to evaluate selfishness as *credit*, and use packet forwarding ratio as *criterion*. Packet forwarding ratio is the ratio of the packets forwarded to the total packets meant to be forwarded. Packet forwarding ratio reflects a node's contribution to the network. Each node must satisfy a preset value of minimum packet forwarding ratio. The minimum forward ratio is based on the battery status of a node. A node may send (its own packet) far more packets than what it forwarded as long as this node always satisfies the required forwarding ratio. This is not possible in motivation based schemes, where a node can only send as many packets as it forwards. The rationale is as follows: every node provides different contribution to the network. Because of location privilege, some nodes can contribute more and others less. However, To be *fair*, as long as they contribute, they should not be regarded as selfish nodes, and the network should provide service to them.

A node's contribution is evaluated by its neighbors from their direct observations. These observations are compared with the node's self-evaluation to justify behavior. In our scheme, nodes broadcast their own credit information and their neighbors verify the credibility according to what they observe. To achieve this, every node calculates its credit and broadcasts it periodically to its 2-hop neighbors. Neighbors monitor the node's behavior and compare the declared value of credit with what they observe. If the deviation between the broadcast self-evaluation and the observed value exceeds a certain threshold, the monitoring node sends out a *warning* message for the monitored node. If more than k nodes accuse the some node, the accused node is considered as a *selfish node* by all nodes in the network. The selfish node is punished by other nodes by dropping packets intended to and originated from such a node. Routes to be established bypass such a selfish node. After some predetermined amount of time, a selfish node is re-admitted to the network. Thus, a selfish node knows that selfishness will do harm to itself, and will be forced to be more cooperative.

3.2   Evaluating Selfish Behavior

Each node maintains *records* to monitor and evaluate the selfishness of 1-hop and 2-hop neighbor nodes. A record has the following fields: *NodeID*, Pkt_{drop}, Pkt_{fwd}, Per_{fwd}, *Batt_stat*, $Credit_{acc}$, *Seq_num*. *NodeID* denotes the identity of the monitored node. Pkt_{drop} and Pkt_{fwd} is the number of packets the monitored node dropped and forwarded, as observed by a monitoring (in promiscuous mode) node. Per_{fwd} is defined as $\frac{Pkt_{fwd}}{(Pkt_{drop}+Pkt_{fwd})}*100\%$. *Batt_stat* represents the battery status of the monitored node. There are four levels: *Normal, Low1, Low2 and Recharged*. *Normal* indicates the node has enough battery. *Low1* indicates that the node does not have enough battery and thus, will not participate in new route discoveries or flooding (if any). *Low2* indicates that the node has very low battery and implies that it will not participate in packet forwarding. *Recharged* indicates the node has been recharged and has enough battery. As discussed later, different battery statuses correspond to different criteria of selfishness. $Credit_{acc}$ is used to record the accumulated net forwarded packets, i.e., $Credit_{acc} = \sum Pkt_{fwd} - \sum Pkt_{drop}$ in each battery life cycle (a cycle is the period during which battery status changes from *Normal* to *Recharged*). *Seq_num* is the latest version of the credit message known to the monitoring node. A node also keeps its own credit information.

Initially, each node sets up an entry for each of its neighbors. For every entry, the initial value of all fields is set to zero, except Per_{fwd}, which is set to 100% (that is, initially all nodes are considered unselfish). Every node maintains its own credit by adding Pkt_{drop} and Pkt_{fwd}, respectively, taking the retransmissions into consideration. Periodically, it sends out Credit message to its neighbors consisting of *NodeID*, Pkt_{drop}, Pkt_{fwd}, *Batt_stat*, *Seq_num*. Each node increases *Seq_num* by one for every credit update message it broadcasts. The credit message is updated upon the expiration of *credit_update_timer* or when the *Batt_stat* is changed (due to energy dissipation).

3.3 Monitoring Selfishness

Nodes monitor neighbors' activity by listening (in promiscuous mode) to all packets within the radio range. If a neighbor forwards a packet, then it increases Pkt_{fwd} for that neighbor, taking the retransmissions into consideration. Similarly, if the neighbor drops a packet, it increases the corresponding Pkt_{drop}. When a node receives its neighbor's *Credit* message, it compares Pkt_{drop} and Pkt_{fwd} data in the message with what it monitored. If the data differs from the observed value beyond a threshold, then the monitoring node suspects that the neighboring node is lying. In practice, there are some cases where it is hard to judge whether the packet has been forwarded or not [10]. So the data declared and the one monitored may not be equal. However, if the originator of the credit message is honest, then the deviation of these data will not be large. If a node finds the deviation is under threshold τ, it adapts the declared value to overwrite its own observed value, and continues monitoring based on this new value. Otherwise, if the deviation exceeds τ, the node sends out *warning* message to accuse the monitored node with the following fields, *AccusedID, AccuserID, Seq_num, Mac*. *AccusedID* and *AccuserID* denote the monitored node and the originator of this message. *Seq_num* is the same as in the latest *Credit* message. *MAC* (Message Authentication Code) is a signed digest over all previous fields with its private key. It is used to provide integrity and non-repudiation.

The conviction is based on voting, i.e., if more than k neighbors accuse the same node, the accused is convicted of being selfish. A selfish node may accuse a normal node as selfish. A larger k can prevent more false accusations and avoid unnecessarily false punishing a good node. However, larger k requirs more time to converge. On the other hand, as the nodes on the periphery of the network have fewer neighbors, a smaller value works well in their situation. Generally a node in an ad hoc network may have 8-10 direct neighbors [14]. So k can be set as 5 or 6, or at least 60% of the neighbors.

3.4 Confirmation of Selfishness

To confirm that a node is selfish, nodes use $Ratio_{\text{fwd}}$ (packet forwarding ratio) and *Batt_stat*. We use three different thresholds, *th1, th2* and *th3* for $Ratio_{\text{fwd}}$, corresponding to battery status. When a node is in *normal* status, its $Ratio_{\text{fwd}}$ should be above *th1*. Otherwise, the monitored node is convicted of selfishness. Similarly, *th2* and *th3* correspond to *Low1* and *Low2* battery status. Note that *th1* > *th2* > *th3*. Also, *th1* should be much higher than *th3* (*th3* is at least 50%) because if the node has enough battery, it should forward packets.

A node may pretend to be in lower battery status to forward few packets to save its energy. To prevent this case, we introduce two strategies. The first strategy is to restrict the benefit a lying node can get. If a node declares its status in *Low1*, then it is free from participating in routing. However, it can not initiate a new route discovery for itself, but it has to participate in packet forwarding. A node in *Low2* status is free from forwarding packets. However, it is allowed to send out its own packets, as many as Pkt_{fwd} - Pkt_{drop}, i.e., its net contribution recorded as credit. After a node is in *Low2* status, we give it a grace period ΔT seconds to get recharged. During ΔT, the node may not send or forward packets. After getting recharged, it issues a new *Credit* message to inform its neighbors. Upon receiving this announcement, the neighbors set $Credit_{acc}$ to $Credit_{acc} + Pkt_{fwd} - Pkt_{drop}$ and resets Pkt_{fwd} and Pkt_{drop} for the originator to zero and $Ratio_{fwd}$ to 100%. The data used for $Ratio_{fwd}$ is based on the new Pkt_{fwd} and Pkt_{drop}. This is because as the node has recharged and has enough battery, it should behave normally and be evaluated normally. On the other hand, its previous contribution should not be ignored. So we use $Credit_{acc}$ to record its previous contribution and the node can use it when in *Low2* status.

Another strategy is to have every node estimate the battery consumption of its neighbors. A node's battery consumption consists of four parts: power used for routine task, power used for working in promiscuous mode, power used for forwarding and sending packets, and power used for receiving packets, denoted by e_R, e_p, e_s, e_r, respectively. Thus, total power consumption $P = e_R + e_p + e_s + e_r$. e_R and e_p is assumed to be the same for all nodes, these two are time dependent quantities. Depending on how long a node has worked during a battery cycle, nodes can estimate e_R and e_p. On the other hand, e_s and e_r are proportional to the number of packets sent or received. Suppose forwarding a packet requires e units of power and every node consumes the same energy for packet forwarding, then we can estimate $e_s = e \star Pkt_{fwd}$ (We ignore the difference of packet length); similarly we can estimate e_r. Since every node can estimate its neighbor's battery consumption, it can verify the credibility of *Batt_stat* in the *Credit* message broadcast by the neighbor. With the above criteria, every node can independently decide if a node is selfish.

3.5   Punishment and Re-admission

Once a selfish node has been identified, it is punished. The neighbors of a convicted node refuse to forward any packets originated from this selfish node. Thus, a selfish node will be excluded from the network. However, we exclude the selfish node temporally (for some predetermined *temp_bypass* time),for the exclusion is not a perfect solution; instead the goal is to force the nodes to cooperate and thus benefit each other. After getting punished, a selfish node is likely to be more cooperative. Besides, as a selfish node is not a malicious node, it's fair to give it a chance to provide service and use the network. However, some nodes may continue to behave selfishly even after being initially punished. In such a case, the *temp_bypass* time is increased exponentially as $2^i \star temp\_bypass$ where i represents the number of times the node has behaved selfishly. Thus, nodes continuing to behave selfishly are punished severely.

3.6   Combining with Routing Protocols

This protocol can be easily combined with routing protocols such as DSR and AODV. In AODV, whenever a source node needs a route to a destination node, it broadcasts a route request *RREQ*. Any intermediate node which knows a route fresh enough to the destination may reply to this request. Otherwise it propagates the request. Upon receiving the request, a destination replies with a *RREP* towards the source along the reverse route. During routing maintenance, upon detecting an inaccessible downstream node, a node sends *RERR* along the active route(s) toward the source(s). Combined with our scheme, a node needs to check the selfish node list whenever it gets routing packets. It drops any packet sent from a selfish node. Thus during route discovery, any known selfish node will be excluded. During routing maintenance, upon convicting its downstream node a selfish node, a node sends *RERR* to the source. The source may initiate a new route request to find a good route. The combination is even simpler for DSR. An intermediate node can exclude selfish nodes as in AODV. Besides, the source itself can check the node list in the Route Reply and drops routes with any selfish node. Similary, selfish nodes are handled during data forwarding by checking route caches.

4   Performance Evaluation

4.1   Simulation Setup and Performance Metrics

We conducted simulations to evaluate the impact of our scheme on a network with selfish nodes. Specifically, our simulation focused on the following metrics,

1. Packet delivery rate: the percentage of total number of packets received by the intended receivers to total packets originated by all nodes, i.e., $D = \sum Packets_{received} / \sum Packets_{sent}$. To be realistic,we consider the packet delivery rate as the comprehensive result of all factors which could affect the packet delivery rate, from the application layer to the physical layer, such as collisions in the MAC layer, rerouting in the network layer, and so on.
2. Overhead: the overall protocol control packets exchanged by nodes in the network, including *Hello* messages and routing packets as *RREQ, RREP* and *RERR*. For our scheme, three additional types of packets (or messages) are introduced: *Credit* message, *Accuse* message and *Black* message. A *Credit* message is periodically broadcast by a sender to provide the credit information. An *Accuse* message is sent whenever a node suspects that its direct neighbor is behaving selfishly based on its observations. Finally, a *Black* message is flooded all over the network whenever a node has been convicted as a selfish node.
3. False conviction: the number of nodes that were convicted selfish when they were cooperative.
4. Missed conviction: the number of nodes that were not detected as selfish when they were selfish.

We use AODV as the underlying routing protocol and have two implementations for it. The original AODV, named *base*, is used for comparison with our enhancement, named *enhance*.

 We model selfish behavior in two ways. First, a selfish node starts behaving selfishly from the beginning. We term this version as *enhance_start* and *base_start*, corresponding to *enhance* and *base* version, respectively. In the second model, the selfish node begins to behave selfishly at some random time during the simulation. Thus a node may drop packets from the beginning of simulation while another node may drop packets from 400 second simulation time. Nodes tend to be selfish after they dissipate more energy, thus we believe this model to be more realistic than the first model. We term the version corresponding to it as *enhance_random* and *base_random*.

Also, we consider the impact of counting the retransmission packets. We implement two versions for calculating credit, one counting packets without retransmissions while the other counting retransmitted packets. The simulation in the current incarnation does not take battery status into account. Unless specified, the data used for performance evaluation is derived from the retransmission version. Furthermore, we use another implementation to detect selfish routing behavior. To study the impact of mobility, we conduct simulations with different maximum mobility speed.

We use Glomosim, a scalable network simulator [16], for simulation. In our simulation, 50 nodes were placed uniformly in an area of 1000m by 1000m. The radio range of nodes is 250m. 802.11 protocol with DCF is used as the MAC protocol. All nodes follow the Random Waypoint mobility model with a speed range of 0 m/s to 10 m/s and a pause time of 30 seconds. Each simulation lasted for 900 seconds simulated real time. To simulate traffic, ten CBR flows were simulated. Each flow sent 4 512-byte data packets per second. Each flow started at 120 seconds and ended at 880 seconds. Data points represented in graphs are averaged over 10 simulation runs, each with a different seed.

4.2   Simulation Results

Packet delivery rate: Figure 1(a) shows the packet delivery rate of simulated systems. Both *enhance* versions have higher packet delivery rate than the corresponding *base* version: *enhance_start* yields as much as a 90% improvement over *base_start* and *enhance_random* shows up to a 36% improvement over *base_random*. The packet delivery rate decreases as the percentage of selfish nodes increases since more nodes drop packets. An important feature is that *enhance_start* performs better than *enhance-random* while *base_start* performs better than *base-random*. This can be explained as follows. Both *base* and *enhance* systems suffer from selfish nodes' dropping packets. However, in the *base* case, no action is taken to bypass these selfish nodes. In *base_start*, all selfish nodes behave selfishly from the beginning and thus they drop all packets passing through them. In *base_random*, selfish nodes only behave selfishly from some random moment. They may behave cooperatively for some time and forward packets. Some selfish nodes even behave well for most of the simulation time. Thus more packets are forwarded in the *base_random* case. However, with a enhanced scheme, selfish behavior will be detected and those selfish nodes will be bypassed. The earlier the behavior is detected, the fewer the dropped packets and thus the higher the packet delivery rate. If selfish nodes behave selfishly from the beginning, they are much easier to be detected. However, if they behave well initially for a period of time and selfishly thereafter, the deviation of credit between the observed and the self-declared may not exceed the threshold or the deviation may not be sharp enough and thus they are hard to be detected. As shown in Fig. 2(b), there are more undetected selfish nodes in *enhance_random* than in *enhance_start*. Finally, in the existence of 30% selfish nodes, the packet delivery is still as high as 37% in *base_start*. This is because a portion of traffic occurs between direct neighbors. Besides, some traffic has short paths, with only one or two intermediate nodes behaving well.

   Overhead incurred: Figure 1(b) depicts the overall protocol control overhead introduced by our enhancement over different evaluating interval vs. percentage of selfish nodes. It is normalized as packets per node per second. *enhance_random_10* and *enhance_random_20* correspond to *enhance_random* with evaluating interval as 10 seconds and 20 seconds, respectively. So do *enhance_start_10* and *enhance_start_20*. The overhead is mainly due to the *Hello* messages, which are broadcast every two seconds by every node, and *RREQ* messages. The periodic *Credit* messages also contribute some overhead. As the evaluating interval increases, the *Credit* messages decrease and thus the overhead decreases. As the percentage of selfish nodes increases, the overhead increases slowly. This is because more selfish nodes will be detected and thus more new *RREQ*s will be initiated. The same reason explains why overhead caused by *enhance_start* is higher than that caused by *enhance_random* although they produce the same number of credit messages. In *enhance_start*, selfish behavior is easier to be detected and thus produces more *RREQ* packets.

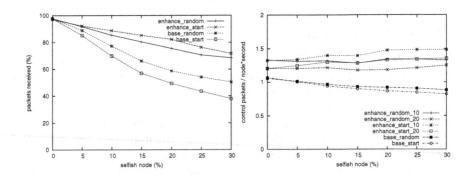

Fig. 1. (a)Packet delivery rate with evaluating interval = 20sec. (b) Overhead incurred

   False conviction: Figure 2(a) depicts the false convictions vs. percentage of selfish nodes. In any case, no more than 0.5 node is falsely convicted. So false conviction is very low. A node will file an accusation as long as the deviation between the observed and the declared exceeds the threshold. Due to radio interference, a node can not always monitor its neighbors accurately and thus convicting a node based on a single accusation is not feasible. A voting system is used to avoid false convictions. However, false convictions still happen in special cases. For example, if a majority of the observed node's neighbors are between the observed node and some other traffic, due to interference from these traffic, these neighboring nodes may misjudge the node as selfish and accuse it. The accused node will be wrongly convicted. We can raise the threshold in the voting system to lower the false convictions. However, it will raise the missed convictions and thus lower the performance (i.e., packet delivery rate). It is a tradeoff between false conviction and missed conviction. Note that false conviction is not sensitive to the evaluating interval since it happens in some special cases.

Missed conviction: Figure 2(b) depicts the missed conviction rate vs. percentage of selfish nodes. *enhance_random* has a high missed conviction rate, averaging about 50% while *enhance_start* has a much lower rate, averaging less than 14% (about 9.7% for evaluating interval = 10s). As discussed earlier, detection of selfish behavior is difficult in the former case and easy in the later one. We can also observe that the evaluating interval affects the missed conviction rate. A larger evaluating interval reduces the missed conviction rate in *enhance_random* while increasing the rate in *enhance_start*. This can be explained as follows. In *enhance_random*, a selfish node may behave well in the beginning and selfishly later, so we need a credit deviation between the observed and the declared sharp enough to detect this behavior. A larger interval helps this deviation. In *enhance_start*, such a requirement is not necessary and a smaller interval is more aggressive in detecting the misbehavior. Of course, too small an interval will blur the credit deviation and thus raise the missed conviction rate.

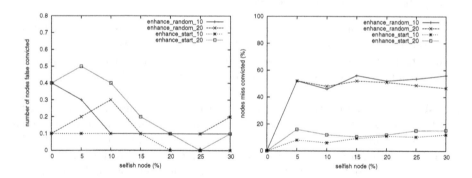

Fig. 2. (a)False conviction for evaluating intervals of 10 and 20 sec. (b)Missed conviction rate for evaluating intervals of 10 and 20 sec.

Effect of mobility: Figure 3(a) and (b) depicts the effect of mobility on packet delivery rate, corresponding to selfish behavior from a random moment and from the beginning, respectively. Four speeds were used in the mobility model - 5m/s, 10m/s,15m/s and 20m/s, keeping the maximum pause time of 30 seconds. In both figures, as speed increases, the packet delivery rate drops. There are two reasons for this. One reason is that higher mobility causes more link breakages. Another reason is that missed convictions increase with higher mobility since it's harder to trace a highly mobile selfish node. Again, the *start* cases perform better than *random* cases.

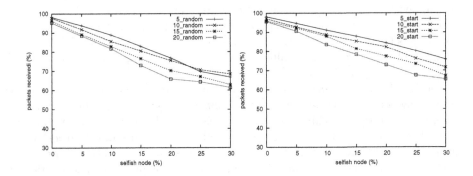

Fig. 3. (a) Packet delivery rate vs. selfish node percentage, with changing speed and behaving selfishly from random time, with evaluating interval = 20 sec. (b) Packet delivery rate vs. selfish node percentage, with changing speed and behaving selfishly from start of simulation, with evaluating interval = 20 sec.

4.3   Discussion and Future Work

We conducted simulations without counting retransmissions. The packet delivery rate in this case is a little lower than when retransmissions are taken into account. This is because counting retransmissions will increase the credit of a node significantly and thus increase credit deviation, which helps in the detection of selfish nodes.

We also simulated the case where selfish nodes do not participate in the routing process. We found the false conviction too high. This is ascribed to the broadcast property of the routing message ($RREQ$). Unlike unicast message, a broadcast does not use RTS/CTS and retransmitting mechanisms. Routing messages ($RREQ$) are flooded in the network and cause collisions and interferences. So the credit counting for routing packets forwarded is much less accurate than that in data forwarding, resulting in high false convictions. From simulations, we infer that purely detection base methods are not suitable for detecting selfish routing behavior.

The solution presented in Section 3 works well when the network size is small. As the network size grows, the performance of the proposed scheme decreases due to its broadcast nature. Also, there is a possibility that all neighbors monitoring a node A can be malicious. In such a case, node A can be falsely accused and punished by all other nodes. An important observation is that, any method that bases decision only on the immediate neighbor monitoring is prone to false accusations. Hence, selfishness evaluation should be based on information from a certain number of nodes spread across the network. Also, as the network grows, the number of monitor nodes for a particular node should grow slowly and should be distributed in all parts of the network. To address the above issues, we plan to investigate the following idea. Each node maintains the *credit* information in a number of nodes (called monitor) distributed throughout the network. Each node selects a set of monitor nodes based on its node ID and is probabilistically different from the set of monitor nodes chosen by any other node. Nodes seeking credibility of a node B can find a monitor node for B solely based on $B's$ ID. Monitor selection is the same as the location server scheme in GLS [17]. Each node N maintains a *credit* record for every node for which N acts as a monitor. Every node sends out its own credit information to its monitors periodically upon expiration of the *credit_update_timer* or when the *Batt_stat* is changed. A monitor can monitor its monitored node in the radio range directly by listening to all passing packets. Also a monitor can randomly poll the nodes out of its radio range as follows. Randomly it generates a "dummy" packet (but indistinguishable from regular packets) to the node being tested and asks it to forward the packet. Upon knowing the successful reception of the dummy packet from the destination, the monitor increases the credit for the tested node, or decreases it otherwise.

5   Conclusion

Selfish nodes pose threats to mobile ad hoc networks. Yet, this problem has not been addressed adequately. In this paper, we provided a fair, distributed solution for the selfish node problem. We use credit as the metric for evaluating a node's contribution to the network. Such metric is different from virtual money and reputation used previously. It can greatly mitigate the *location privilege* problem and the *counting retransmission* problem not addressed in other proposals. Both nodes in the center and those in the periphery of the network have almost the same chance to serve others and be severed by them. We use a voting system to confirm a node's selfishness to avoid the necessity for centralized service. It can also solve the inconsistent evaluation problem. Selfish nodes are re-admitted to the network after being punished by their neighbors for a period to induce a cooperative environment. Our scheme focuses on fairness, which hasn't been emphasized so far. Simulation results show our scheme is effective. It raises the packet delivery rate efficiently, results in few false accusations and has low overhead.

References

1. Levente Buttyan and Jean-Pierre Hubaux, *Enforce Service Availability in Mobile Ad-Hoc WANs*, In proceedings of MobiHOC, 2000
2. Levente Buttyan and Jean-Pierre Hubaux, *Stimulating Cooperation in Self-organizing Mobile Ad Hoc*, Technical Report No. DSC/2001/046, Swiss Federal Institute of Technology, Lausanne, July 2001

3. Jean-Pierre Hubaux, Levente Buttyan, et, al, *Toward Mobile Ad-hoc Wans: Terminodes*, In Technical Report No. DSC/2000/006, Swiss Federal Institute of Technology, Lausanne, July 2000

4. E. Fratkin, V. Vijayaraghavan, Y. Liu, D. Gutierrez, TM Li, and M. Baker. *Participation Incentives for Ad Hoc Networks*, http://www.stanford.edu/ yl314/ape/ paper.ps

5. Pietro Michiardi and Refik Molva, *Prevention of Denial of Service Attacks and Selfishness in Mobile Ad Hoc Networks*, Research Report RR-02-063 - January 2002

6. S. Buchegger, J.Y. Le-Boudec, *Nodes Bearing Grudges: Towards Routing Security, Fairness, and Robustness in Mobile Ad Hoc networks*, In Proceedings of the 10th Euromicro Workshop on Parallel, Distributed and Network-based Processing, Spain, January 2002

7. S. Buchegger, J.Y. Le-Boudec, *Performance Analysis of the Confidant Protocol (Cooperation of Nodes: Fairness in Dynamic Ad-hoc Networks)*, In Proceedings of IEEE/ACM Workshop on Mobile Ad Hoc Networking and Computing (Mobi-HOC), Lausanne, CH, June 2002

8. H. Luo, P. Zerfos, J. Kong, S. Lu and L. Zhang, *Self-securing Ad Hoc Wireless Networks*, 7th IEEE Symposium on Computers and Communications (ISCC'02), July 2002, Italy

9. Luzi Anderegg, Stephan Eidenbenz, *Ad hoc-VCG:A Truthful and Cost-Efficient Routing Protocol for Mobile Ad hoc Networks With Selfish Agents*, In Proceedings of ACM/IEEE International Conference on Mobile Computing and Networking (Mobicom'03), San Diego, September 2003

10. S. Marti, T. J. Giuli, K. Lai and M. Baker, *Mitigating Routing Misbehavior in Mobile Ad Hoc Networks*, In Proceedings of ACM/IEEE International Conference on Mobile Computing and Networking (Mobicom), Boston, August 2000

11. David. B. Johnson and David A. Maltz, *Dynamic Source Routing in Ad Hoc Wireless Networks*, In Internet draft, Mobile Ad Hoc network (MANET) Working Group, IETF, October 1999

12. C. Perkins. *Ad-Hoc on-demand Distance Vector Routing*, In Internet draft RFC, Nov 1997.

13. Hugo Miranda and Luis Rodrigues, *Preventing Selfishness in Open Mobile Ad Hoc Networks*, In Proceedings of 7th CaberNet Radicals Workshop, Bertinoro, Forli, Italy, 13-16 October 2002

14. K. Paul, D. Westhoff, *Context Aware Detection of Selfish Nodes in DSR based Ad-Hoc Networks*, In Proceedings of IEEE Vehicular Technology Conference'02, Vancouver, Canda 2002

15. S. Zhong, J. Chen and Y. R. Yang, *Sprite: A Simple, Cheap-Proof, Credit-Based System for Mobile Ad-Hoc Networks*, Technical Report Yale/DCS/TR1235, Department of Computer Science, Yale University, July 2002

16. http://pcl.cs.ucla.edu/projects/glomosim/

17. J. Li, J. Jannotti, D. S. J. De Coute, D. R. Karger, R. Morris, *A SCalable Location Service for Geographic Ad Hoc Routing*, In Proceedings of ACM/IEEE International Conference on Mobile Computing and Networking (Mobicom), Boston, August 2000

Session-Based Service Discovery and Access Control in Peer-to-Peer Communications

Anand Dersingh[1], Ramiro Liscano[2], Allan Jost[1], and Hao Hu[3]

[1] Faculty of Computer Science, Dalhousie University, Halifax NS B3H 1W5, CA
dersingh, jost@cs.dal.ca
[2] School of Information Technology and Engineering, University of Ottawa, Ottawa
ON K1N 6N5, CA
rliscano@site.uottawa.ca
[3] Intel China Ltd. 2299 Yanan Road, Shanghai, P.R.C. 20000
hao.hu@intel.com

Abstract. Service Location Protocol (SLP) is a standard service discovery protocol proposed by IETF. SLP provides a flexible and scalable
service discovery framework over IP networks. This paper presents an
approach for service discovery and access control under session-based
peer-to-peer communications. In other words, it is an approach providing mechanisms for restricting unauthorized discovery or access to restricted services. The proposed approach integrates and leverages SLP
with Session Initiation Protocol (SIP). Moreover, this approach can be
used to share services from a Wireless Personal Area Network (WPAN)
like Bluetooth across the Internet domain.

1 Introduction

The emergence of Voice over Internet Protocol (VoIP) leads today's Internet
into multimedia data communications. Both wired and wireless networks move
toward supporting multimedia specifically by sharing data during voice conver-
sations. Unsurprisingly, this is one of the goals of 3G and 4G wireless systems.
The nature of multimedia communications is dynamic behavior such that connection parameters are not fixed and contains multiple flows per logical session
which makes multimedia applications differ from traditional applications. This
can be seen with the adoption of Instant Messaging (IM) systems such as MSN,
ICQ, and Yahoo that support multimedia applications like text, voice, and data
sharing.

These approaches all implement proprietary services at the end points and
therefore only operate between clients from the same vendor or with services that
are integrated with the end points. For example, even for simple file sharing it
is not possible to share files between different IM clients and between different
wireless devices because they do not adhere to any set of common service profiles.
Most file sharing is done with the use of email because the SMTP protocol is
so widely in use. The result of this can be a non-integrated communication
procedure that could involve multiple addressing schemes. This is in our view

I. Nikolaidis et al. (Eds.): ADHOC-NOW 2004, LNCS 3158, pp. 225–237, 2004.

where service discovery protocols can have the most impact in communication services. In order to get any form of interoperability between services on different networks it is essential that vendors begin to adopt common service profiles. This is where Bluetooth service profiles are of great value.

The Bluetooth service profiles [1] are designed to support interoperability among wireless devices that form a Bluetooth network which is called wireless piconet. Service discovery is a necessity in a Wireless Personal Area Network (WPAN) like that defined by Bluetooth since devices can easily come and go from the network. As a result, standard Bluetooth provides Bluetooth Service Discovery Protocol (BT-SDP). BT-SDP has strong commercial support. Ironically, it is less flexible in terms of supporting a mechanism of machine readable service templates but gains over other service profiles in that the service specification is very detailed in terms of the expected interaction messages and protocols between the two end points. The result of this is that it is impossible to write a generic Bluetooth service profile that could interpret a service template and instead vendors implement specific service profiles that are highly compatible with the same profiles of another vendor.

Bluetooth wireless technology was developed primarily as a cordless replacement technology and is ideal for creating connections among devices within a short range. One of the characteristics of a Bluetooth network is that the BT-SDP query is limited to devices that can be detected within the range of the Bluetooth radio. The solution proposed in [2] and [3] extends this capability to devices that may exist on a peripheral IP node defined by an existing communication session. This was done by using the Service Location Protocol (SLP) [4] proposed by IETF as a session-based service discovery. The result is that it is possible for a person on a Bluetooth device to view other Bluetooth Services that have been detected by the node at the other end of the communication link.

An up and coming standard that shows great promise of unifying IP-based multi-media communications is the Session Initiation Protocol (SIP). SIP is an application layer signaling protocol for initiating, modifying, and terminating network sessions. This signaling protocol acts similarly to telephone signaling operating over Internet without resource reservation in the network. It can provide variety of multimedia services such as Internet telephony, and multimedia conferencing.

SIP itself does not describe the description for a particular session. In order to manage multimedia services the SIP protocol uses the Session Description Protocol (SDP) to negotiate service parameters when establishing a session. A typical two-party peer to peer SIP session involves a caller and callee. Before a session is established the parties involved in the session are not aware of any of the capabilities of the other end points. It is only through the information in the SDP of the SIP message that the parties get their first view of the service that the session is trying to establish. In addition, SIP and SDP are capable of handling dynamic behavior such that the caller or callee may want to change session parameters on the fly by initiating new INVITE request called RE-INVITE with new description in an SDP message.

Service discovery protocols provide a proactive mechanism by which end points can know a-priori which particular services they support. The advantage of using a service discovery approach over "blind" SIP negotiation approach is that the number of SIP messages can be reduced significantly by knowing beforehand which services are supported. On the other hand service discovery approaches have their own limitations that can be eliminated or reduced by combining them with peer to peer session-based approaches like SIP. For example most service discovery protocols rely on centralized servers for maintaining service information, while under peer to peer communication models they leverage the ability of services to broadcast or multicast their signature. This broadcast approach to the advertisement of services is not scalable to function over the Internet [5] due to the traffic load caused by broadcasting approaches. This is a significant limiting factor in the use and adoption of these service discovery protocols across the Internet.

Another drawback for service discovery protocols is that they simply represent a hypothetical view of the services available. There is no guarantee that the service can actually be accessed due primarily to the fact that the address of the service might be incomplete or simply inaccessible. Moreover, a service discovery protocol like SLP does not provide an access control mechanism in order to restrict services from unauthorized access. However, there is a significant advantage in combining these two approaches.

In this paper, Sect. 2 provides a motivation and example scenario used through out the paper. Section 3 describes a system overview which is mainly based on [2] and [3]. Section 4 and 5 explain the details of session establishment and session-based service discovery. Section 6 describes service accessibility including access control. Section 7 discusses the relation with other works and finally Sect. 8 provides discussion and conclusion.

2   Motivation: Spontaneous Network

The motivation for combining service discovery with session-based communications is illustrated with a scenario of a spontaneous network situation. The scenario involves the sharing of resources between two wireless personal area networks involved in a long distance communication session. This scenario is represented in Fig. 1.

In this example, Jane has Bluetooth-enabled camera and PDA that support the OBEX Object Push Profile. She also has an IP phone that supports Bluetooth networking and the spontaneous session-based service-sharing gateway described later in this paper. A wireless personal area network can be established between Bluetooth-enabled devices and the IP phone on Jane's side. Similarly, Bob has a PC that supports voice communication as well as the session-based service-sharing gateway compatible with Jane's phone, and a Bluetooth-enabled picture frame and PDA that also support the OBEX Object Push Profile. Thus, another wireless personal area network can be established between the PC and Bluetooth-enabled devices on Bob's side. Jane calls Bob on her IP phone and

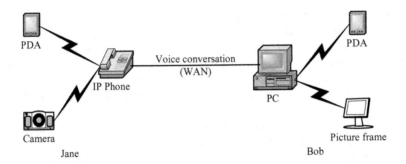

Fig. 1. WPAN peer-to-peer multimedia communications

Bob receives the call on his PC. Since they have established a voice session, Jane's camera can query Bob's WPAN via the phone and request any services it supports. Since Bob's picture frame supports the OBEX Object Push Profile, Jane can request a connection to Bob's picture frame and push a photo to it. In a similar manner Jane is able to push the picture directly to Bob's PDA or another IrDA picture frame if they support the OBEX Object Push Profile. When Jane or Bob ends the conversation, the communications between two WPAN networks terminates as well.

The main issue that comes up with this scenario is controlling access to the advertised services. For example, Bob allows Jane to transfer photos to his picture frame but does not want Jane to discover and access his PDA which is also a device connected to his WPAN. It may be that when Bob communicates in a voice session with a third party, say, Alice, in a similar scenario, he may want to allow Alice access to both his picture frame and PDA. Thus, beyond simply sharing services, access control plays an important role in this type of communication.

3   System Overview

Details of this system have been presented previously [2], [3] and we simply give an overview here in order to detail the new extensions to support session based service discovery and access control that are part of this paper.

The proposed framework solution consists of gateway service collectors located in the end point devices involved in a voice, video and/or Instant Messaging session. These gateways are capable of supporting the creation of a local Bluetooth or IrDA network with other wireless devices in the vicinity. Figure 2 depicts the basic components of the framework.

First of all, the Communication end points must initiate and establish a communication session among themselves, using Session Initiation Protocol User Agents (SIP UA) [5]. Typical communication end points could be a PC device with a SIP user agent, or a SIP IP phone set. A device at one end is able to

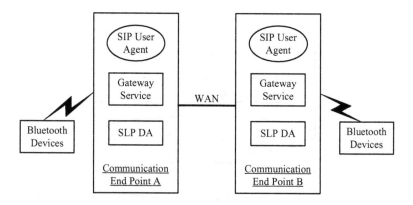

Fig. 2. Components of the session-based service discovery system

discover the available services on the other end via a Gateway Service module [2] that supports a combined IETF Service Location Protocol (SLP) [2] / Bluetooth Service Discovery Protocol (SDP) [1] service discovery model. In this model, the gateway offers the services based on scope that are actually outside in the other end to the inquiring devices as if they provide the services themselves. To enable Bluetooth, IrDA and PC devices to interoperate, we introduced the Bluetooth OBEX profiles to the domain of IrDA and PC devices [2]. A particular OBEX communication between Bluetooth, IrDA or PC devices via the gateway can be fulfilled if the IrDA or PC device conforms to the specific Bluetooth OBEX profile.

The Gateway communication sequence is wrapped around a SIP session as represented in Fig. 3. These messages are the typical SIP messages that are used to commence a session between two end points. These messages are the typical SIP messages that are used to commence a session between two end points. This initialization session is the trigger event for the gateways to commence communication. Establishing a session is also important in order to determine proper IP addresses of the end points that are used as a mechanism to send requests in order to discover services. Communication between the gateways terminates when the overall communication session ends. In other words, it is not possible to discover or access any services between the end points if there is no session active.

4   Session Establishment

SIP is an application-layer protocol for initiating, modifying, and terminating a session. The description of the session is defined by the Session Description Protocol (SDP), embedded in a SIP message. SIP and SDP are capable of adding, deleting, and terminating media in a session. The main focus of this paper is leveraging these capabilities, to handle multiple media which conform to the

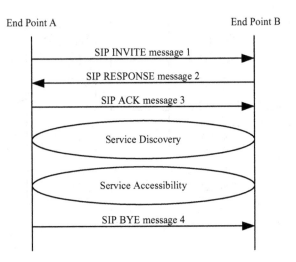

Fig. 3. Session initiation and management of service discovery

standard, to control the communication between end points for service discovery and accessibility of Bluetooth application profiles.

As described in Sect. 3, the underlying communication between two end points is a SIP session. This means, to discover and gain access to the advertised services from the other end the client has to send SIP signaling for negotiation. The main focus in initiating the session in this work is to initiate multiple media in the session. Instead of negotiating only voice media we also combine SLP as another media to the session. This can be done by introducing the SLP media type in the SDP message [3]. The purpose of the SLP media is to let the SIP user agent located in each gateway open the appropriate port, at a time of conversation, for the SLP media in order to exchange service information between the end points. Figure 4 shows the session establishment between two end points including the SLP component.

After establishing a session, the SIP UA at each side records the SIP URI of the other end and Call-ID of this session. This will be described later in Sect. 5. At this time the two end points can perform voice communication as well as service discovery, since the ports used to communicate between the two end points are already negotiated and opened by the user agent during the initiation.

5   Session-Based Service Discovery

A key characteristic of the service discovery model implemented by the gateways is that the gateways periodically exchange service requests while the communication session is active. A gateway will save a local cached copy of the services available at the other end points therefore reducing the time it would take to translate a service request from a wireless end point from BT-SDP to SLP and

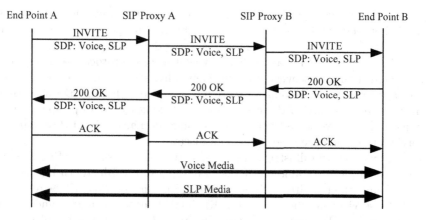

Fig. 4. Messages sequence for session initiation including SLP media

back again at the other end. Bluetooth connection times generally take a couple of seconds to be established and users would not tolerate these delays.

The Gateway Service performs two major functions. It collects local service information from the devices in the WPAN and exchanges service information and re-registration of services when connected to another gateway through a communication session.

5.1   Local Service Collection

The main responsibility of the local service collection is to discover and collect all available services in the local WPAN, then convert them into the format of an SLP service, finally registering these services with a local SLP Directory Agent for advertising.

When the service collection is done, the discovered service information must be translated into the format of an SLP. The SLP service format is called a service template [6] which is the formal description of service types and attributes as information in accessing services. We introduced the Bluetooth OBEX template for SLP as a new service type [3]. The BT-SDP does define service elements (a set of tuples that define attribute names and values) therefore, the translation is not too cumbersome. The BT-SDP/SLP translator simply needs to parse the Bluetooth service information into a basic service element and then reassemble it to conform to the service template.

The challenge is in deciding what particular service address will be used in the SLP description of the service. The natural choice is to use the gateway address and a new unused port number on the hosting end point. Having done that, the gateway maintains a mapping between the new translated SLP service and the original Bluetooth service in order to keep a valid path for future references to the service. At this point the new SLP service is ready to be registered into the SLP directory agent by using a SLP Unicast Service Register method.

Another concern, which was not addressed in the previous publications, [2] [3], is how service discovery and accessibility to end point services are limited. SLP version 2 does provide a mechanism called a scope. Scoping is the administrative method in SLP and other service discovery protocols to group services and control their discovery domain. Generally, specifying a scope in the SLP message reduces the amount of reply messages from the SLP DA such that the SLP DA replies only to those services that belong to the scope specified in the request. In this work we leverage the scoping mechanism in SLP to manage the services that the other end points can discover.

An example will illustrate the point better. Let us say that Bob at one end point has a picture frame and PDA that support the Bluetooth image transfer service and their addresses are 0x1234567890 and 0x1234567899 respectively, and the services are located at BT RFCOMM server channel 5 and 6. This is represented in Bluetooth as a Basic Imaging Profile: SCN 5 for picture frame and Basic Imaging Profile: SCN 6 for PDA. Suppose the gateway IP address is 10.0.0.1 corresponding to Bob's communication end point. During a service collection, Bob's gateway will discover the service of the picture frame and assign it the gateway's IP address 10.0.0.1 as well as dynamically assigned port numbers (say 20001 and 20002). With the service registration, Bob's SLP DA is capable of configuring scope for each service. Let's say Bob specifies the scope of the Bluetooth imaging Profile on his picture frame as 'guest', and 'admin' for the service provided by PDA. These are registered as new SLP services with the SLP DA appearing somewhat like this: Basic Imaging Profile: 10.0.0.1:20001:guest for picture frame, and Basic Imaging Profile: 10.0.0.1:20002:admin for PDA. This applies to all services.

Leveraging scope mechanism does not require any enhancement to the standard SLP since it has already been defined [4]. Later in this paper will describe how the scope mechanism helps prevent unauthorized access to the advertised services.

5.2   Service Exchange and Re-registration

Service exchange involves exchanging services information across the WAN and re-registration is to keep a local copy of services provided from the other end. Recall that service exchange and re-registration are performed while the primary session is still active. After a session is established between two end points, service exchange can take place immediately. This service exchange is performed using the SLP protocol between the two end points. This requires that the gateway support an SLP User Agent (SLP UA) so that it can query the other end points in the session.

It was mentioned in the previous subsection that in registering a service to local SLP DA it is necessary to specify scope to each service in order to limit the discovery of services from the other end. This means that SLP UA has to be configured with scope related to the scopes specified in SLP DA at the other end. Standard SLP version 2 [4] provides three methodologies for configuring SLP UA with scope. The first method is pre-configured by the administrator.

The second method is to obtain it from DHCP options 78 and 79 [4]. Lastly, is to obtain it from SLP DA Advertisement message sending from SLP DA.

The first two methods are not adequate for this work since service discovery in this instance is across a WAN which is located under two different administrative domains. The third method is best suited for our situation. Even though the SIP user agents at each side know the IP addresses of each other it is still necessary for each SLP UA to send a unicast SLP Service Request message specifying the service type as 'directory-agent' to discover the SLP DA at the other end. (This is to guarantee the existence of the SLP DA.) With this request the SLP UA does not know the scope that it belongs to, and therefore this first request is sent without a scope name.

After receiving the first request from the SLP UA, the SLP DA has to respond with an SLP DA Advertisement message. The SLP DA has to specify the scope it wishes the other end to use. It is left to the user to which scopes or services he/she wants to limit the discovery to the other end. This is analogous to the real world scenario where a visitor gets connected to a home network and the owner of a home computer give him/her the privilege to share the content of the drive. In the same manner as in our case, the owner of the WPAN can give the privilege to the other end and use the SIP URI as a key identity.

Figure 5 shows the SLP messages used to obtain scope and perform service discovery. As depicted, once the SLP DA specifies the scope to the SLP UA, subsequent messages have to include scope in the SLP message. Having done this, restriction of the service is achieved based on user perspective.

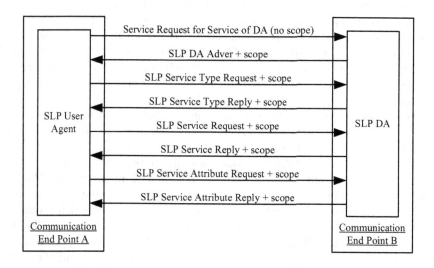

Fig. 5. Service discovery between communication end points with scope mechanism

6   Service Accessibility and Communication

For demonstrational purposes we investigated how Bluetooth profiles that run over the Object Exchange Protocol (OBEX) profile could be exchanged using our approach. The OBEX protocol is a binary protocol similar to HTTP that allows a variety of devices to exchange data simply and spontaneously. The common protocol layer among all the devices and the gateway is the OBEX layer. The gateway replaces Bluetooth RFComm channel identifiers with TCP port numbers and vice versa. No other translations need to occur below the OBEX layer.

The transferring of OBEX data packets is done as part of the existing session, represented as a media under the SIP session. Additional media can be added to the session by sending a RE-INVITE with the new media added in the SDP message. We introduced another media type in the SDP format for transferring OBEX data called Bluetooth-OBEX (BT-OBEX) media [3].

To fully leverage session-based SDP, we also introduced a BT-OBEX SDP profile [3]. The purpose of this profile is to define the payload format in the SDP message in a similar manner as for the RTP/AVP profile [7]. Since OBEX may be used for a variety of applications with somewhat differing requirements, the flexibility to adapt to those requirements is provided by allowing choices in the media, then selecting the appropriate choices for a particular environment and class of applications in a separate profile. Typically, an application will operate under only one profile in a particular BT-OBEX session, so there is no explicit indication within the BT-OBEX protocol itself as to which profile is in use.

In addition to defining BT-OBEX media and BT-OBEX SDP profile, we propose a mechanism controlling access to restricted services. As mentioned in Sect. 3 that after establishing a call, the SIP UA at each end point records the SIP URI of the other end and the Call-ID of the session. Call-ID is one of the header fields in a SIP message, and is a globally unique identifier for a session. These two fields will be validated during the negotiation before accessing the services.

For example, consider Jane transferring an image file from her camera to Bob's picture frame. After discovering and registering the service to Jane's gateway, the IP address and port number are known to Jane's gateway. However, while setting up the BT-OBEX session, this information (destination IP address and port number) cannot be sent in the form of an SDP message. The IP address and port number in the SDP message belongs to the source, not the destination. By defining a BT-OBEX SDP profile, Bob's gateway can tell which profile Jane would like to access after receiving a RE-INVITE from Jane with the BT-OBEX media added to the SDP message. The SIP UA at Bob's gateway determines the RE-INVITE message by matching the SIP URI of the sender and the Call-ID previously recorded during the initialization of the session. If they do not match, that means it came from an unauthorized user not participating in this call. If so, we compare these values with the scope previously given and the service (profile) that Jane is trying to access. If the requested service belongs to the same scope provided by Bob then Bob's gateway can establish the media at the proper IP address and port number, based on the profile, which are initially determined during local service registration.

By enhancing the SIP UA and keeping records of the SIP URI, Call-ID, and scope of an existing session, we increase the level of security to restricted services. This achieves access control when the other end is trying to access to services not belonging to its scopes. Figure 6 shows the messages sequence during negotiation for accessing a service.

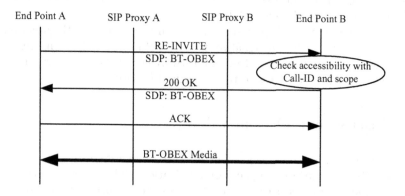

Fig. 6. Messages sequence for accessing a service

The termination of a media can be done by sending a new RE-INVITE by setting the port number of that media to zero which will automatically inactivate that particular media. Moreover, the entire communication relies on the session. Termination of the session terminates all communications.

7   Related Work

This work addresses principally the extension of service discovery across the Internet by leveraging session-based communications and service discovery protocol, with an example that uses Bluetooth profiles. In terms of previous work in the domain of large scale service discovery it is worth noting Cascella's thesis [8], in particular the section on the Service Peer Discovery Protocol (SPDP). In this work he leverages the SIP event model by defining SPDP as an extension to the SIP event framework and embeds it in the SIP messaging. The user has to be able to package location and context based services as an information registration to a context server in order to be reachable. He does not consider at all the use of these services. Another important contribution in this area is the Mesh-enhanced Service Location Protocol (mSLP) [9]. In this protocol SLP directory agents exchange service registration information and form fully-meshed peering according to the registered services. Finally, Collaborative Peer Groups

[10] forms peer groups based on a template that provides peer's information. However, they do not address communication sessions at all.

In terms of previous work that has investigated the extension of Bluetooth service profiles across the Internet most existing approaches in this domain have primarily considered how to integrate existing IP-based service discovery protocols into the Bluetooth framework [12] [11]. In our work we do not impose an IP-based service discovery protocol onto the Bluetooth stack but rely more on a gateway between Bluetooth and SLP and the creation of a new SLP service profile to support Bluetooth profiles. We see similar approaches with the mapping of UPnP [13], Jini [14], and Salutation [15].

The standard SLP [3] provides security in the form of authentication, principally by adding a digital signature to the SLP message. This ensures that the SLP UA receives information from a trustworthy SLP DA. However, it does not prevent unauthorized access to the advertised services.

8   Discussion and Conclusion

This framework is a commencement for further investigations of session-based service discovery in peer-to-peer communications. This leverages session information and service discovery protocol in order to perform service discovery and accessing services effectively which adds another security level to the system from the user perspective. In particular, this approach does not require opening ports that belong to the services all the time. Ports are opened only after a successful negotiation and closed after termination of the media or session. Moreover it also provides an access control mechanism for restricted services. However, there are a number of outstanding issues that are important and require further investigation. These are:

- Investigating how other service discovery approaches would fit into the framework
- Considering multi-party communication sessions
- Managing Firewalls in session-based service discovery

This work was a continuation of [2] and [3], attempting to extend service discovery and access control beyond the boundaries of a Wireless Personal Area Network such as that defined by Bluetooth. The approach presented leverages the existing standards SIP, SDP, SLP, and BT-SDP to embed service discovery and accessing services in a SIP session. The authors believe that serious consideration should be given to the use of Bluetooth profiles beyond the realm of Bluetooth networks particularly over session-based communications links.

Acknowledgement. The authors would like to acknowledge financial and technical support from the Strategic Technology Group at Mitel Networks located in Kanata, Ontario, Canada.

References

1. Bluetooth Special Interest Group, Specification of the Bluetooth System v1.1: Protocol, Service Discovery Protocol, (Bluetooth Special Interest Group, 2000)
2. R. Liscano, H. Hu, A. Jost: SIP Session-based Access to Bluetooth Application Profiles. Proc. Wireless and Optical Communications, Banff, Alberta, July 2003, pp. 349–354
3. R. Liscano, A. Jost, A. Dersingh, H. Hu: Session-based Service Discovery in Peer-to-Peer Communications, Canadian Conference on Electrical and Computer Engineering (CCECE'04), Niagara Falls, Ontario, May 2004
4. E. Guttman, C. Perkins, J. Veizades, and M. Day: Service Location Protocol, Version 2, IETF, RFC 2608, June 1999
5. J. Rosenberg, H. Schulzrinne, G. Camarillo: SIP: Session Initiation Protocol, IETF, RFC 3261, June 2002
6. E. Guttman, C. Perkins, J. Kempf: Service Templates and Service: Schemes, IETF, RFC 2609, June 1999
7. H. Schulzrine, S. Casner: RTP Profile for Audio and Video Conferences with Minimal Control, IETF, RFC 3551, July 2003
8. R. Cascella: Reconfigurable Application Networks through Peer Discovery and Handovers, Master's thesis, Royal Institute of Technology KTH, Department of Microelectronics and Information Technology, June 2003
9. W. Zhao, H. Schulzrinne, E. Guttman: Mesh-enhanced Service Location Protocol (mSLP), IETF, RFC 3528, April 2003
10. V. Sunderam, J. Pascoe, R. Loader: Towards a Framework for Collaborative Peer Groups, Proc. 3rd Int. Symp. on Cluster Computing and the Grid, Tokyo, Japan, May 12 - 15, 2003. pp. 428-433
11. S. Baatz, M. Frank, R. Göpffarth, D. Kassatkine, P. Martini, M. Schetelig, and A. Vilavaara: Handoff Support for Mobility with IP over Bluetooth, Proc.25th Annual Conf. on Local Computer Networks (LCN'00), Tampa, FL, USA, Nov. 2000, pp. 143-154
12. M. Albrecht, M. Frank, P. Martini, M. Schetelig, A. Vilavaara, and A. Wenzel, IP services over Bluetooth: leading the way to a new mobility, Proc. 24th Conf. on Local Computer Networks (LCN'99), Lowell, MA, USA, October 1999, pp. 2-11
13. A. Ayyagari, S. S. AbiEzzi, W. M. Zintel, and T. M. Moore: Proxy-bridge connecting remote users to a limited connectivity network, U.S. Patent US 2001/0033554 A1, October 2001
14. L. Marchand: Ad hoc network and gateway, International Patent WO 01/76154 A2, October 2001
15. B. Miller and R. Pascoe: Mapping Salutation Architecture APIs to the Bluetooth Service Discovery Layer, white paper; available online at http://www.salutation.org/whitepaper/BtoothMapping.pdf

A Controlled Mobility Protocol for Topology Management of Ad Hoc Networks

Amit Shirsat

Department of Computer Sciences,
250 N. University Street,
Purdue University, West Lafayette,
IN 47907-2066, USA.
ajs@cs.purdue.edu,
http://www.cs.purdue.edu/homes/ajs

Abstract. In this paper we present a new approach towards self configuration in mobile ad hoc sensor networks. In the proposed scheme sensor
nodes are location aware and are informed with the region of the sensor
field that needs to be covered. The sensor nodes then occupy calculated
positions to enhance network connectivity and coverage. These points
referred to as grid points are computed dynamically and in a decentralized fashion. These grid positions are *live* -meaning a power replenished
sensor node is stationed at each grid point. The geometric graph induced by the nodes occupying the grid positions has the following nice
properties: 1)it has a subgraph which is a sparse spanner and 2)planar.
3)it forms a connected dominating set over all the connected nodes in
the sensor field. 4)It is a bounded degree graph. Property 1 ensures the
existence of efficient routes which are constant factor off the Euclidean
distance between the nodes. Property 2 is required to perform greedy
routing such as GPSR [1] to find those routes. Property 3 states that
the nodes occupying the grid points form a forwarding backbone for the
sensor field while the remaining nodes periodically enter into the idle
state for power savings. Property 4 implies that the interference between
the nodes occupying the grid points is limited to only a constant number
of nodes.

1 Introduction

The overall design goal for mobile ad hoc networks is to extend its lifetime while
maintaining sustained coverage. The nodes however are constrained with limited
battery power. Therefore effective power management and topology control of
the nodes is imperative to prolong the network lifetime. Power management in ad
hoc networks spans all layers of the communication protocol stack. At the MAC
layer the power management techniques react to local traffic changes whereas at
the network layer optimizations respond to local topological transitions. Among
the MAC protocols such as PAMAS [2], IEEE 802.11 and STEM [3] significant
power savings are obtained especially by putting a node into sleeping state. The
protocols differ in their mechanisms that indentify the nodes that power off their

I. Nikolaidis et al. (Eds.): ADHOC-NOW 2004, LNCS 3158, pp. 238–251, 2004.

radios. The focus for optimization at the network layer shifts to the distibuted computation of the connected dominating set (CDS) of the underlying unit disk graph (UDG) induced by the placement of the nodes. Since the minimum CDS problem is NP-Complete, the approach is to find heuristics to better approximate the mininum CDS. [4],[5],[6] describe approximate algorithms for MCDS in the centralized and distributed setting. Among the topology management protocols, Geographical Adaptive Fidelity (GAF) [7] conserves energy by identifying nodes that are equivalent from a routing perspective and then turning off unnecessary nodes, keeping a constant level of routing fidelity. In SPAN [8] a node decides to wakeup based on a local count of neigbors that would benefit by it being awake and amount of energy available to it. Similarly, a node could put itself into sleep state if it finds itself redundant for routing purposes. The Low Energy Adaptive Clustering Hierarchy (LEACH) [9] protocol selects rotating cluster-heads to collect local information and send the aggregated information to a wireless base station. Another relevant problem in ad hoc networks is to compute efficient routes in ad hoc networks. [1] describe Greedy Perimeter Stateless Routing (GPSR) which is an optimal routing protocol on planar graphs. [10] describe a spanner graph called as the Restricted Delaunay Graph (RDG) which has provably good topological and euclidean stretch factors (within a constant value from the optimal). Since the decisions for node wakeup and sleep at the MAC and network layers are intertwined recent work such as [11] point towards a cross layer design approach. From a pure theoretical perspective [12] show that the network a.s. percolates (i.e contains an unbounded connected component) if the node densities derived from a point Poisson process exceed a threshold λ. For a unit disc $\mathcal{D}$ containing n nodes distributed uniformly and independently the network a.s. percolates if $\pi r^2(n) = \frac{\log n + c(n)}{n}$ and $c(n) \to \infty$ as $n \to \infty$. In contrast to the statistical arguments above, it is easily seen that there exist sparse deterministic spanners which can cover the region with much lesser transmission radii. For e.g. the unit disc $\mathcal{D}$ can be covered by nodes placed on $\sqrt{n} \times \sqrt{n}$ square grid with inter-grid spacing $\frac{1}{\sqrt{n}}$. Contrast this with the value of r derived from a random Poisson process $r = \Omega(\sqrt{(\frac{\log n + c(n)}{n})})$. Our work aims to construct a grid-like topology of $O(n)$ nodes from a random Poisson distribution of $\Omega(n \log n)$ nodes. We assume that nodes are position and power aware and collectively strive to enhance connectivity and preserve energy. A subset of nodes form a forwarding backbone for the rest of the network and those nodes which occupy the forwarding backbone are called grid points and connectors. In contrast to the free nodes which are *reactive* to the network state, the grid points are more *proactive* and take measures to improve connectivity and coverage. The connectors are gateways between grid points; they could be reactive or pro-active based on the connections they offer to the network. This paper attempts to address the following key question: *Can nodes leverage their mobility to form an almost unstructured grid-like network while maintaining sustained coverage?*. The primary advantage is to keep the nodes occupying the topological grid points "live" while the rest of the nodes hibernate or forward traffic through a neighboring grid node. Since the nodes cooperate to build an ad hoc network, the MAC layer can take advantage of this infrastructure for further optimizations. We further motivate the problem and define it in the next section. We give a distributed protocol to solve it in section 3. In section 4 and 5 we prove the properties of the topological graph and the routing graph respectively. Section 6 deals with the optimizations involved in maintaining the topology. In section 7 we conclude with future directions to this work.

2   Motivation and Problem Definition

In fixed topology networks such as the mesh, one can guarantee number of network properties such as existence of efficient routing paths, bandwidth, redundancy in the network etc. If we restrict fixed topology networks to an ad hoc setting then we are limited to positioning nodes at points within communication distance. Ad hoc networks are however not deployed to form a fixed grid like structure for variety of different reasons. One primary difficulty is to maintain the structure in a distributed setting. Sometimes the structure itself could be inviable due to terrain constraints. However the structured grid model can serve as a reasonable test bed to analyze our protocols and also motivate the desired features in the unstructured grid creation. We enlist a few desired features in a structure grid such as the mesh or the hexagonal mesh. These grids form a connected dominating set (CDS) over the points of the region they cover and the size of this set is a constant factor from the optimal. Secondly, the structure gives a very good foundation to design greedy and optimal routing protocols. The mesh also insures network connectivity despite a few node failures. By this we mean that despite the failure of few random nodes in the network, by and large most of the network would still be connected. Furthermore there are multiple paths between two distant nodes that are within a constant (stretch) factor from the optimal. Assume a set of identical n sensor nodes with unit transmission radii distributed over a region Γ by means of a Poisson point process with density $\Omega(\log n)$ the problem then is to design a protocol that creates a topology with the properties as described for a fixed topology network.

2.1   Grid Point Election Protocol

Before we even describe how the grid topology is generated, we describe a generalized arbitration protocol for occupying a grid point. In a typical scenario there could be number of nodes willing to be co-ordinators for a single grid point. This protocol underlines the mechanisms for nodes to participate in grid point election and subsequent occupation of the grid point. In the ensuing scenario assume that a node u initiates the Grid Point Election Protocol for a point $p \in D(u) \bigcap \Gamma$ by proposing itself as the co-ordinator for the grid point. In order to be eligible for the election u must be less than α from the grid point p, where $0 < \alpha \leq \frac{1}{2}$ is a parameter to be fixed later. $\alpha = \frac{1}{2}$ for the fixed grid case and $\alpha = \frac{1}{3}$ for the unstructured grid algorithm which we describe later. Also u must not change its position while it is a candidate for the election.

1. The grid point p is in the intermediate state, if there is already a candidate bid for that grid point. There is a timeout period set, beyond which the leading bidder is declared as a winner for that grid point.

2. If there is an occupied grid point q within $1 - \alpha$ of p, the co-ordinator for that grid point will broadcast a packet in unit distance radius making the election for the grid point void. Since all contestants for the grid point p are within α of p all of them can listen to the packet.

3. A successive bid for the grid point must be made within the timeout period and be atleast a constant fraction better than the earlier bid. The tradeoff here is the time and energy required to bid for a better candidate as opposed to the time and energy required for the existing leading bidder to occupy the grid point.

4. If the leading bidder is not challenged in the timeout period, it assumes that it is the co-ordinator for the grid point and moves to the position to occupy the grid point. The state for the grid point is intermediate until the node gets to the grid point to occupy it. Upon occupancy the co-ordinator broadcasts beacons (that declare its id and co-ordinates) at regular intervals advertising itself as *live*.

5. If due to communication failures, there are more than one contenders for a grid point co-ordinator then all but the first are annulled.

2.2   Analysis of Grid Election Protocol(GEP) for Mesh Topology

We analyse the GEP in the following abstract setting. Assume there is a virtual grid (mesh) of size $\sqrt{n} \times \sqrt{n}$ that covers the sensor field Γ. The neighboring grid (lattice) points are at a unit communication distance apart. Let us assume there are cn nodes distributed uniformly at random in Γ. The problem is then to assign one node to each virtual grid point in a distributed fashion so that the backbone network is established. Each node has an apriori knowledege of the virtual grid points in the network or in other words the topology of the grid. We first consider the Voronoi cell associated with each virtual grid point. The Voronoi cell for a grid point p is a region in Γ whose points have p as their closest grid point. Note that for the mesh the Voronoi region takes the simple form as shown in Figure 1. If we run the GEP with parameter $\alpha = \frac{1}{2}$ for the virtual grid points then every grid point will be occupied provided the voronoi cell for that point has atleast one node inside it. Consider a natural lexicographic labelling of cells as shown in Figure 1. Let $X_{i,j}$ be the random variable that takes the value 0 if no node occupies cell (i, j) and 1 otherwise. Let $X = \sum_{i,j} X_{i,j}$.

$$\Pr(X_{i,j} = 0) = \left(1 - \left(\tfrac{1}{n}\right)\right)^{cn} < e^{-c}$$
$$\Pr(X_{i,j} > 0) \ \geq \left(1 - e^{-c}\right)$$
$$E[X] = \sum_{i,j} E[X_{i,j}] \geq n(1 - e^{-c})$$

We also note that if $c = \ln n + O(1)$ then all grid points are a.s. occupied.

$$\Pr(\bigwedge_{i,j} X_{i,j} > 0) = 1 - \Pr(\bigvee_{i,j} X_{i,j} = 0)$$

$$\geq 1 - e^{-c}$$

$$= 1 - \left(\tfrac{1}{n}\right)^{O(1)}$$

However only $O(n)$ nodes are actually required for occupying the grid points. The rest of the nodes increase the redundancy in the network. By better load distribution through controlled mobility, one can extend the lifetime of the network by leveraging these redundant nodes.

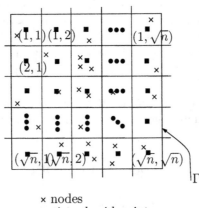

× nodes
■ virtual grid points

Fig. 1. Voronoi region for a regular grid

3   Construction of the Grid Graph

In this section we present the decentralized protocols which determine which points in Γ would serve to maintain the sparse connected forwarding backbone for the mobile ad hoc network.

3.1   Topology Creation Protocol

1. At any instant of time the network maintains a graph of co-ordinating grid points and connectors called $GridG$. Each node u which is not a grid point first finds out the live grid points in its unit distance neighborhood set $N(u)$. Let $D_r(u)$ denote the disk of radius r centered at u. For $0 < r \leq 1$ let us denote the set of grid points within distance r to u as $NGP_r(u)$. Let $NGP(u) \doteq NGP_1(u)$ and $\forall r, 0 < r \leq 1, NGP_r(u) \subseteq N(u)$. The gridpoint u

also finds if there is an election in progress for a grid point in its neighborhood. If u is a better candidate for that grid point it proposes itself as the candidate for that grid point otherwise it backs off for a random (bounded) amount of time and upon wakeup runs this step again. If there is no election in progress and if $\forall v \in NGP(u), d(u,v) > \frac{2}{3}$ then it initiates the *grid election protocol* (GEP) with parameter $\alpha = \frac{1}{3}$ proposing itself as a grid point. This pre-condition for running the GEP is referred to as *distance invariance*. Each grid point v receives neighborhood information from the set $NGP(v)$ in form of periodic beacons. The node which occupies and controls a grid position is considered to be *live* and such a node is called the co-ordinator for that grid point. For the purpose of rest of the paper whenever we use the term grid point in context of a node it refers to the co-ordinator of the grid point.

2. v keeps track of its neigboring grid points and potential connectors to its one-hop grid points. It does so by maintaining the information it receives from $u \in N(v)$. A grid point co-ordinator maintains the following local invariant. Let C_v''' be a circular arc of radius $\frac{7}{6}$ centered at v. It ensures that no arc of C_v''' of central angle greater than $\arccos\left(\frac{13}{14}\right) \approx 21 \cdot 78°$ is left uncovered by some grid point disc $D(u)$ for some $u \in D(v) \setminus D_{\frac{2}{3}}(v)$ or by a connector disc $D_{\frac{1}{2}}(c)$ centered at some connector $c \in D(v) \setminus D_{\frac{5}{6}}(v)$. This property is referred to as the *coverage criteria*. Note that the grid points are elected autonomously where as the connectors are nominated (assigned) by existing grid points to cover the arc C_v''' described above. Also the criteria for coverage of C_v''' is different for a connector and a grid point inside $D(u)$. See Figure 2. The requirement for the gridpoint is that it must be atleast $\frac{2}{3}$ away from every other grid point in its neighborhood while the connector has to be atleast $\frac{5}{6}$ away from its nominating grid point. Also the radii for coverage computation of C_v''' is unit for a grid point whereas it is $\frac{1}{2}$ for a connector. When two connector disks (of radii $\frac{1}{2}$) overlap then the two connectors are directly connected and thus are their nominating grid points. A grid point v chooses a connector and its position in a way to minimize node movement and maximize the coverage of C_v'''. If a connector assigned to a point upon reaching that point satisfies the invariants for a grid point it will run the grid point election protocol and if elected gets promoted to a grid point. The grid point co-ordinator also annuls redundant connector nodes when new grid points are created. Also two grid points can share a connector for optimizing purposes. A connector moves to its assigned position only when there is no election in progress in its visible neighborhood.

4   Properties of the Grid Graph

In what follows, we refer the circles with center u and radii $\frac{2}{3}, \frac{5}{6}, 1, \frac{7}{6}$ as C_u', C_u'', C_u, C_u''' respectively and their corresponding discs as D_u', D_u'', D_u, D_u''' respectively.

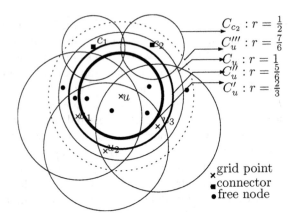

$$C_{c_2} : r = \tfrac{1}{2}$$
$$C_u''' : r = \tfrac{7}{6}$$
$$C_u : r = 1$$
$$C_u'' : r = \tfrac{5}{6}$$
$$C_u' : r = \tfrac{2}{3}$$

× grid point
■ connector
● free node

Fig. 2. Connectors and Grid Points covering C_u'''

Lemma 1. Boundedness. *The Grid graph has node degree bounded by a constant.*

Proof. Consider a grid point $u \in GridG$. Let $CP(u)$ be the connectors assigned bu u in $D_u \setminus D_u''$. We need to prove $\mid NGP(u) \cup CP(u) \mid \le c$. By the protocol invariant, if u is a grid point there cannot be any grid point inside D_u'. Consider a sector aub of angle $38 \cdot 94°$ centered at u. See Figure 3. And consider the region of the sector in $D_u \setminus D_u'$. No two points in this region are less than $\tfrac{2}{3}$ apart. So there cannot be two grid points in a sector with central angle $38 \cdot 94°$. This gives an upper bound of $\lfloor \tfrac{360}{38 \cdot 94} \rfloor = 9$ on the number of neighboring grid points within unit distance. A connector is placed so as to cover an uncovered sector of angle greater than $\theta = \arccos(\tfrac{13}{14}) \approx 21 \cdot 78°$. Since a connector covers a sector angle of size atleast 2θ. This gives a simple upper bound on the number of connectors to be $\lceil \tfrac{2\pi}{3\theta} \rceil = 9$. In practice the number of connectors in $D_u \setminus D_u''$ depends on the number and placement of grid points in $D_u \setminus D_u'$ and the upper bound is expected to be much less.

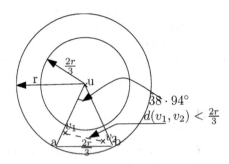

$$\tfrac{2r}{3}$$
r
u
$$38 \cdot 94°$$
$$d(v_1, v_2) < \tfrac{2r}{3}$$
$$\tfrac{2r}{3}$$

Fig. 3. Degree of a node in GridG is bounded

Lemma 2. *Overlap.* *Grid points within a distance of $\frac{7}{3}$ are a.s. connected.*

Proof. We say that an arc of a circle is of length θ if the angle that it subtends at its center is θ. The disc $D_{u'}$ of a grid point u' or the disc $D_{\frac{1}{2}}(c')$ of a connector c' which are neighbors u in GridG cover an arc of C_u''' of length atleast $2\theta = 2\arccos(\frac{13}{14})$. The value for 2θ comes from the arc of C_u''' cut by a disc of radius $\frac{1}{2}$ of a connector placed at distance $\frac{5}{6}$ from u and is the least such value that follows from the distance invariance. Note that the coverage requirement is met a.s. if we assume that the node density is $\Omega(\log n)$ with a Poisson distribution. Consider a disk D_u''' and its closest overlapping disk D_v'''. For the purpose of this proof we will also require the following geometric fact. If a circle radii r_2 overlaps an arc of length θ of a circle with radii r_1 then the distance between the two circle centers d (using the cosine rule) satisfies $r_1^2 + d^2 - 2r_1 d \cos(\frac{\theta}{2}) = r_2^2$. So we can get a lower bound for d if we know that the overlap must be less than θ. We refer to this fact as the separation condition. Let α be the arc of length 2θ in C_u''' about the line connecting the centers of the two discs as drawn in red with maximum thickness in the Figure 4. The corresponding labels for the arc in C_v''' be γ. Let $(\alpha_i, \alpha_{i+1}), 1 \leq i \leq 2$ be the equi-partition of α and $(\gamma_i, \gamma_{i+1}), 1 \leq i \leq 2$ the corresponding one for γ. Let β_1 be the midpoint of arc (α_1, α_2) and β_2 be the midpoint of the arc (α_2, α_3). Let β be the arc (β_1, β_2). Let δ_1 be the midpoint of arc (γ_1, γ_2) and δ_2 be the midpoint of the arc (γ_2, γ_3). Let δ be the arc (δ_1, δ_2). p_1, p_2 are points on C_u''' at a distance $\frac{1}{2}$ from α_2. i.e. if we place connectors on p_1 or p_2 that would barely cover α_2. Since β is of length θ atleast one among β_1 or β_2 and one among δ_1 or δ_2 would be covered by a connector or a grid point neighbor of u. We have the following two cases:

case 1: no portion of β is covered by a grid point $\in D_u \setminus D_u'$ and no portion of δ by a grid point in $D_v \setminus D_v'$

Therefore some connector disc $D_{u'}$ of radius $\frac{1}{2}$ must overlap a portion of β. Let u' be a connector that overlaps β and in particular covers β_2. The overlap with C_v''' must be less than θ so by the separation condition $d(u', v) > 1.5944$. Using cosine rule to $\triangle u p_1 v$ we get $d(p_1, v) < \sqrt{\frac{7}{3}^2 + \frac{5}{6}^2 - 2(\frac{7}{3})(\frac{5}{6})\cos\theta} = \frac{\sqrt{91}}{6} \approx 1.589 < 1.5944$. Therefore the connectors placed by u must lie outside the shaded lune of radius $d(p_1, v)$ in D_u''' as seen in Figure 4. This is because a connector for u placed in its lune is less than distance $d(p_1, v)$ from v and hence overlaps an arc of C_v''' of length greater than θ. Now since $d(p_1, \alpha_2) = \frac{1}{2}$ for a connector u' outside the lune $d(u', \alpha_2) > \frac{1}{2}$ therefore no connector for u would cover α_2 without violating previously assumed conditions. This implies that the points α_1, α_3 have to be covered by some connector in D_u and likewise γ_1, γ_3 by connectors in D_v (to satisfy the coverage invariant). wlog assume u' covers β_2, α_3 and v' covers δ_1, γ_1 then the connector disc that covers α_1 intersects the disc of v'. Since the two connector disks of radii $\frac{1}{2}$ overlap they are connected and so are their respective grid points u and v.

case 2: some portion of β say β_1 is covered by a grid point $u_1 \in D_u \setminus D_u'$ i.e $C_{u_1} \cap \alpha \neq \phi$.

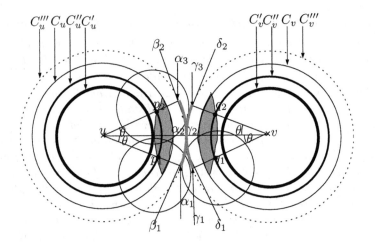

Fig. 4. Connectors satisfying the coverage criteria

Here again we consider two subcases $d(v, \alpha_2) > \Delta \approx 0.07$. As seen in the Figure 5, Δ is an approximate threshold such that if $d(v, \alpha_2) > \Delta$ then the grid point in the neighbor of u, say u_1 that covers β_1, is closer to v than u. The intuition is that if the disk D_{u_1}''' covers the point α_2 then some point of D_{u_1}''' is definitely closer to v than α_2 which is the closest point in D_u''' from v. This implies $d(u_1, v) < d(u, v)$. As before, let $\alpha_1, \beta_1, \alpha_2, \beta_2, \alpha_3$ be equi-distant points on arc $\alpha \subset C_u'''$ that is closest from v. Let γ be the mirror reflection of α about the tangent at the closest point of approach in α to v i.e α_2. Then the arc $\gamma \subset D_v'''$. if $d(v, \alpha_2) > \Delta$. As seen in the case (i) in Figure 5,

$$\theta = \arccos(\frac{13}{14}) \approx 21.786°$$

$$\Delta = (\frac{(\frac{7}{6})^2 \cdot (1 - \cos\frac{\theta}{2}) - \frac{1}{72}}{(\frac{7}{6}) \cdot \cos\frac{\theta}{2} - 1}) \approx 0.007$$

Therefore,

$$d(v, \alpha_2) > \Delta \Rightarrow d(v, \beta_1) - \frac{1}{6} < d(v, \alpha_2)$$

$$d(u_1, v) < d(u_1, \beta_1) + \frac{1}{6} + d(v, \beta_1) - \frac{1}{6}$$

$$< d(u, \alpha_2) + d(v, \alpha_2) = d(u, v)$$

Thus, $D_{u_1}''' \cap D_v \neq \phi$ and u_1 is more closer to v than u. Let $u_0 = u$. Now we could repeat the same argument as above for u_1. Continuing this way, there exists connectors or grid points of the form u_i such that u_i is connected to u_{i+1} and $\bigcup_i D_{u_i} \cap D_v$ increases with each i. This process cannot continue beyond a constant number of iterations as u_i, u_{i+1} are separated by atleast $\frac{2}{3}$ and their discs of radii $\frac{7}{6}$ overlap a finite region i.e. D_v. This implies is a there would be some intermediate grid point or a connector that will connect directly to v. if $d(v, \alpha_2) \leq \Delta$ then we have the same argument as before but we cannot necessarily have succesive discs closer to v. But now the discs D_{u_i} are considerably close to v as seen in case(ii) in the Figure 5. They would cover a substantial portion of $D_v \setminus D'_v$ so that they would contain a grid point or connector placed by v.

case(i): $d(v, \alpha_2) > \Delta$ case(ii): $d(v, \alpha_2) \leq \Delta$

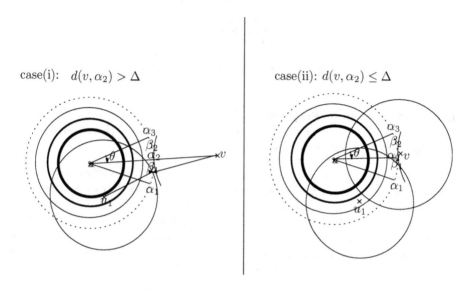

Fig. 5. Grid Points satisfying the coverage criteria

Lemma 3. Connectivity. *If the underlying unit disk graph (UDG) defined by the distribution of nodes in Γ is connected then GridG is connected. In other words, vertices of GridG forms a connected dominating set.*

Proof. Suppose there are atleast two disconnected components in GridG but the underlying geometric graph induced by the nodes is connected. Then there must be two nodes which are less than a unit distance apart but in separate components in GridG. Let u and v be the closest nodes that are in disjoint components of GridG but $d(u, v) < 1$. Note that neither u nor v cannot be grid points since otherwise they are in the same component in GridG. This implies there exist grid points u' and v' such that $d(u, u') < \frac{2}{3}$ and $d(v, v') < \frac{2}{3}$. See Figure 6. By triangle inequality, we have $d(u, v) < d(u, u') + d(u', v') + d(v, v') < \frac{2}{3} + 1 + \frac{2}{3} = \frac{7}{3}$. But by Lemma 2 this implies u, v are connected. This contradicts the assumption that nodes less than a unit distance are in separate components in GridG. Since the underlying geometric graph in connected GridG is connected. Note that we can have an a.s. guarantee of the connectivity of the geometric graph by assuming the Poisson density of the nodes to be $\Omega(\log n)$.

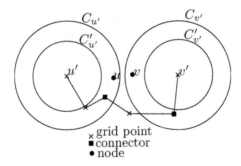

Fig. 6. GRIDG forms a connected dominating set

Lemma 4. *Spanning Property.* *If the underlying unit disk graph (UDG) defined by the Poisson distribution of nodes in Γ with density $\Omega(\log n)$ then GridG is a topological and euclidean spanner.*

Proof. Based on the proof of Lemma 2, between any two discs whose centers are separated by distance no more than $\frac{7}{3}$ there exists a path in GridG that contains atmost a constant number of nodes. Consider the shortest path $\mathcal{P}$ between any two nodes u_0, u_n in G. By Lemma 3 any two adjacent nodes in $\mathcal{P}$ are contained in some discs whose centers are no more than $\frac{7}{3}$ apart. So a.s. there are paths of constant length (euclidean and topological) in GridG between two adjacent nodes in $\mathcal{P}$. Thus GridG is a constant spanner of UDG.

5   Routing Graph

A Routing graph is a subgraph of the underlying connectivity graph used for forwarding purposes. The motivation for maintaining a routing graph is to guarantee efficient routing algorithms in the network. We consider a route to be efficient if the stretch factors (both euclidean and topological) are within a constant factor from the optimal. The graph induced by the grid points and connectors (GridG) is sparse. However, to ensure efficient routing strategies such as the Greedy Perimeter Stateless Routing (GPSR) [1], it is required that the routing graph be planar, i.e. no two edges in the connectivity graph (GridG) cross each other. Furthermore the routing graph must also be a good spanner. It is well known that Delaunay triangulation is a constant factor Euclidean spanner for the complete graph. Dobkin et al. [13] have proved that for any two nodes x, y there exists a path in the Delaunay triangulation that lies entirely inside the circle with x, y as the diameter and the path length is atmost $\frac{(1+\sqrt{5})\pi}{2} \cdot d(u,v)$. The Delaunay triangulation on the vertices of GridG may have long edges and it is not easy to compute locally as seen by the following fact: An edge u, v is Delaunay iff there exists an empty circle passing through u, v which contains no other points in its interior. Gao et al. [10] propose the following modification to Delaunay graph called as the Restricted Delaunay Graph (RDG) which could be computed locally and preserves the spanning properties for the Delaunay Graph(DG). RDG is a subset of DG and contains all short edges i.e. edges with length less than one. To compute short Delaunay edges each node v of the grid graph computes a Delaunay triangulation $T(v)$ of $\tau(v)$ where $\tau(v)$ are the neighbors of v including itself in the grid graph. It sends $T(v)$ to each of its neighbors in $\tau(v)$. Now an edge $e = (u, v) \in T(v)$ is a short Delaunay edge if it is short (length < 1) and $\forall w \in \tau(u) \bigcap \tau(v), e \in T(w)$. Since each v has constant number of neighbors and $T(v)$ is constant size the RDG $\subseteq$ GridG. Furthermore RDG is connected if GridG is connected since the minimum spanning tree (EMST) is always a subgraph of the RDG. We summarize this section with two lemmas concerning Planarity and Spanning Property and refer the reader to the paper by by [13] and [10] for further details. These lemmas imply that a greedy routing strategy such as GPSR can perform well on GridG if the nodes maintain an RDG at all times.

Lemma 5. *RDG is a topological spanner and Euclidean spanner of GridG.*

Lemma 6. *RDG is a planar graph.*

6 Maintaining the Grid Graph

6.1 Topology Variations

In this section we discuss how the network adjusts itself to various scenarios such as node failures, node movements etc. It should be noted that grid points and connectors share the responsibility of maintaining the grid graph and hence are not expected to move too often once they assume their respective positions.

Graceful Termination or Movement

1. When a grid point v is low on power or decides to move it would request a nearby node u (which is not a grid point) to initiate the grid point election protocol at a position which will minimize the movement of the replacement node and preserve the local visibility graph (GRIDG) seen at v (as much as possible). The natural way to achieve this is to compute the region in $S = \bigcup_{i \in NGP_{\subseteq}(v)} (D(v) - D_{\frac{2}{3}}(i))$ and choose a node in this region where $NGP_{\subseteq}(v)$ is the largest subset of $NGP(v)$ (the neighboring grid points of v) which yields an eligible node. If no such node exists, it would request one of its connectors to run the GEP at some point in S.
2. v would then inform $NGP(v)$ and connectors about its graceful termination. It would then ready the chosen replacement (if any) to initiate the GEP.
3. The connectors to v would inform the grid points at the other end about the termination of the link to v. The grid points connected directly or indirectly to v would update their link status and correct their local topology accordingly.

Node Failures or Abrupt Movement. The nodes failures or abrupt movements of grid nodes will be detected either by neighboring live nodes or ordinary nodes. The live nodes update their link status at regular intervals often by piggybacking their status with data or sending beacons that contain their coordinate positions. The ordinary nodes sleep for a random interval based on the node density and power capacity. We would discuss the backoff issues in the next section.

1. if an existing connector or free node detects that it is eligible for election it initiates the GEP.
2. If there is no election in progress in their neighborhood, the neighboring grid points would initiate their topology correction to ensure their coverage criteria is satisfied. This would involve moving a node minimally to maximize the coverage. The node moved could be free or a redundant connector. The node can move only if no election is in progress in its visibility neighborhood. It can also participate in an ongoing election if it is eligible. Since atmost one node per grid point in the neighborhood of the failed grid point moves at any instant the effect of the failure is restricted only to the neighboring grid point co-ordinators. Furthermore since motion is restricted only when there is no election in progress, this allows the network to stabilize.

6.2   Cross Layer Optimizations

The GEP is a bidding style protocol. A crucial issue in such protocols is to select the timeout intervals for the election process. Since elections have to be completed in a timely fashion a timeout value is fixed at T which is atleast as large as the expected (or worst case) sleep time of a free node. The other sleep timers take randomized values. Although a node listens to the control beacons before it contests the elections, contentions could still occur. These contentions are resolved using randomized backoff timers as in [8]. The critical factor in selecting the backoff duration is determining the utility of a node. It is computed based on its own power capacity and the utility of its neighboring grid points. Each grid point will compute the utility factor of its neighborhood based on the utilities of the nodes it cached. Thus the wakeup interval for a free node or a redundant connector will depend on a random variable whose range is decided by the utility factor for that node. All non-redundant connectors can synchronize between their end points, since they usually communicate between only few end points. Thus based on the topology and power constraints nodes can gain significant savings by optimizing with the MAC layer protocols. At the network layer, a region with high utility factor can initiate transfer of nodes to a nearby region with low utility factor there by balancing the utility in the network.

7   Conclusion and Future Work

In this paper we present a power aware topology management protocol which uses controlled mobility to enhance connectivity and coverage in the network.

We investigate the theoretical aspects to ensure a constant degree forwarding backbone which is connected as long as the underlying UDG of nodes are connected. We suggest toplology based cross layer design optimizations between the MAC and the network layers in the protocols. We propose to incorporate mobility costs and effects into our analysis in the future. We would also be simulating the protocols in near future.

References

1. Karp, B., Kung., H.: GPSR: greedy perimeter stateless routing for wireless networks. In: ACM/IEEE Symposium International Conference on Mobile Computing and Networking (Mobicom). (2000) 243–254
2. Singh, S., Woo, M., Raghavendra, C.S.: PAMAS: Power aware multi-acess protocol with signalling for ad hoc networks. ACM SIGCOMM Computer Communication Review **28** (1998) 5–26
3. Schurgers, C., Tsiatsis, V., Ganeriwal, S., Srivasatava, M.: Topology management for sensor networks: Exploiting latency and density. In: Third ACM International Smposium on Mobile Ad hoc Networking and Computing. (2002)
4. Guha, S., Khuller, S.: Approximation algorithms for connected dominating sets. Algorithmica **20** (1998) 374–387
5. Das, B., Sivakumar, E., Bhargavan, V.: Routing in ad hoc networks using a virtual backbone. In: 6th International Conference on Computers and Communication Networks (IC3N). (1997) 1–20
6. Wu, J., Li, H.: On calculating connected dominating set for efficient routing in ad hoc wireless networks. In: 3rd International Workshop on Discrete Algorithms and Methods for Mobile Computing and Communications. (1999) 7–14
7. Xu, Y., Heidemann, J., Estrin, D.: Geography-informed energy conservation for ad hoc routing. In: 7th Annual ACM/IEEE Mobile Computing and Networking (MobiCom). (2001)
8. Chen, B., Jamieson, K., Balakrishnan, H., Morris, R.: SPAN: Energy-efficient coordination algorithm for topolpogy maintenance in ad hoc wireless networks. In: ACM/IEEE Mobicom. (2001)
9. Heinzelman, W.R., Chrandrakasan, A., Balakrishnan, H.: Energy-efficient communication protocols for wireless microsensor networks. In: Hawaaian International Conference on Systems Science. (2000)
10. Gao, J., Guibas, L., Hershberger, J., Zhang, L., Zhu, A.: Geometric spanner for routing in mobile networks. In: 2nd ACM Symposium on Mobile Ad Hoc Networking and Computing (MobiHoc-01). (2001) 45–55
11. Zheng, R., Kraverts, R.: On-demand power management for ad hoc networks. In: IEEE InfoComm. (2003)
12. Meester, R., Roy, R.: Continuum Percolation. Cambridge University Press, Cambridge, Massachusetts (1996)
13. Dobkin, D.P., Friedman, S., Supowit, K.J.: Delaunay graphs are almost as good as complete graphs. In: 28th Annual IEEE Symposium on Foundations of Computer Science. (1987) 20–26

Efficient Data Collection Trees
in Sensor Networks with Redundancy Removal

Shoudong Zou, Ioanis Nikolaidis, and Janelle J. Harms

Computing Science Department
University of Alberta
Edmonton, Alberta T6G 2E8, Canada
{szou,yannis,harms}@cs.ualberta.ca

Abstract. In sensor networks, overlapping sensed areas result in the collection of partially redundant data. Interior sensor nodes along a data collection tree can remove redundancies and reduce the required bandwidth for the aggregate data. Assuming that the sensed information is proportional to the sensed area, we propose heuristic algorithms to build data gathering trees in order to reduce communication costs. The algorithms are completely distributed and are based on finding shortest hop paths and employing particular tie-breaking mechanisms. The mechanisms can be divided into two categories: those that use simple knowledge of the cardinality of a neighbor's neighbor set and those that rely on knowledge of a distance–monotonic metric to a neighbor. We compare five heuristics based on communication cost and also on energy and processing costs. For smaller configurations, the performance of the heuristics are also compared to the optimal. With the proposed approaches, it is possible to build a cost efficient data gathering tree which takes advantage of the data redundancy among closely located sensors.

1   Introduction

In a Wireless Sensor Network (WSN), dispersed sensors are usually battery–powered and difficult or impossible to recharge. Constrained by the limited energy supply, collecting sensed data efficiently from each sensor to the sink is becoming one of the most important research challenges. Sensor nodes are often densely deployed in order to, (a) guarantee the connectivity of the WSN, (b) collect enough information from the entire surveyed area to satisfy the application requirements, and (c) to cope with multiple failures of deployed nodes. A direct implication of dense deployment is that sensed data between nearby sensors tend to be correlated to the extent that they correspond to intersecting sensed areas. This correlation can be viewed as an indication of redundancy in the collected data. If the data collected from each sensor is sent to the "sink" without any form of processing to remove redundancies, substantial energy (and bandwidth, due to unnecessarily long data transfers) is wasted without good reason. We do not address specifically how the redundant data are removed. The interested reader should review correlation and compression techniques, e.g., see [1], for particular correlated data gathering algorithms.

I. Nikolaidis et al. (Eds.): ADHOC-NOW 2004, LNCS 3158, pp. 252–265, 2004.

For the sake of exposition, we adopt the convention that the information collected by a sensor is proportional to the area it senses. Hence, the redundancy between the sensed data of two sensors is proportional to the overlap between their corresponding sensed areas. If the sensor nodes are capable of removing the redundancies, then the particular tree along which data collection is performed (rooted at the sink) will impact how "early" (i.e., how close to a leaf node) and how effectively data redundancies are removed, resulting also in different communication costs. As an example, consider the three nodes of Figure 1. Suppose that the sensed area of node i is A_i and the amount of information sensed is proportional to the size of A_i, i.e., $\gamma|A_i|$ for some constant γ. If the data collection involves nodes x and y forwarding their sensed data to node z. Node z transmits the combined sensed data of x and y, along with its own to a final destination (sink). Then, the total data aggregated at node z are $\gamma|A_x \cup A_y \cup A_z|$ and the total transmission cost, including the transmissions from x to z and from y to z is $\gamma(|A_x|+|A_y|+|A_x \cup A_y \cup A_z|)$. Naturally, the above example is a simplification to illustrate the point. Depending on how the tree is structured the total data collection communication cost is different. Clearly, the overlap of the sensed areas is usually not as simple as in this example, as certain areas can be covered by several sensors.

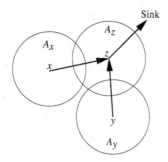

Fig. 1. Three overlapping sensing areas.

The purpose of this paper is to suggest ways in which the data collection tree (rooted at the sink) should be formed in WSNs where nodes are capable of eliminating redundant data. The objective is to minimize the total (over the entire tree) communication cost because it is well known that, compared to the energy cost of computation, the dominant energy cost in a WSN is due to the wireless transmission [5]. What makes the particular solution approaches interesting is that they achieve good performance without undue demands for particular hardware (for example hardware providing location information) or global information (all presented schemes operate on local information). Calling for redundancy removal as part of the assumptions may seem rather extravagant at first, given the limited capabilities of sensors. However, depending on the application, simple techniques to detect similarities between sensed data may be sufficient. For example, if the sensed data are expressed as *name–value* pairs, then identifying names that are associated with the same value (plus/minus a small error epsilon) across sensors could be sufficient to detect and remove redundancies.

The most complex hardware suggested (for certain of the schemes) is nothing more than signal strength monitoring. That is, being able to detect whether other nodes are either close or far on the basis of measuring the received signal strength of their transmissions (but *without* the ability to translate signal strength to precise distance or location). Without much information to go by, we rely on well known techniques for constructing spanning trees. The one considered in our heuristics is the union of shortest paths (in terms of hops) to the sink, i.e., the hop-based Shortest Paths Tree (SPT). However, as we will shortly see, the optimization criteria for the (communication) optimal data collection tree with redundancy removal are different from the criteria used for SPT.

Even though SPT is an approximate solution, it is simple to implement and possesses certain advantages in the particular setting. Namely, in a dense and uniform deployment of sensors, the data sensed by a sensor heavily overlap the data sensed by many of its neighbors. When it comes to deciding which neighbor to forward one's own data (for subsequent redundancy removal) a discriminator as to which neighbor to select is whether the neighbor is in any way "closer" to the eventual destination/sink, hence the value of hop–based SPTs. As it turns out, in a dense sensor network, each node also has many neighbors that are the same (minimum) number of hops away from the sink. Forming the SPT to route the sensed data requires that we break ties between next–hop neighbors that have the same minimum cost to the sink. Remarkably, what our results indicate is that a clever way of breaking ties (within the confines of our assumptions on hardware complexity) makes a lot of difference.

The remainder of this paper is organized as follows: In Section 2, the problem formulation is provided, quantifying redundancy in sensed data and specifying two cost metrics and their related optimization problems (total communication minimization and transmission energy minimization) as well as one approximate metric for the processing cost associated with redundancy removal. Section 3 outlines five heuristics targeting the communication minimization subtopic. All heuristics operate with local information and follow a baseline distance vector protocol, on top of which particular tie–breaking schemes are defined. The 4th section presents and contrasts some of the simulation results produced from applying the presented heuristics on small and large (sparse and dense) WSNs. An exhaustive search algorithm capable of determining the optimal communication or energy cost is applied to the small configurations as well. Finally, in Section 5, we provide a summary of our findings, describe future directions and put certain of our assumptions in perspective.

2   Problem Formulation

First we introduce some notation. We consider a sensor network of N homogeneous sensor devices deployed in 2–dimensional space. We use index $i = 0$ to denote the sink, and $i = 1, \ldots, N$ to denote the sensing nodes. The sink does not perform any sensing, it merely collects the data. Let $V = \{0, 1, \ldots, N\}$ denote the node/vertex set, inclusive of the sink. The area sensed by sensor i is denoted by A_i which is approximated by a circular disk of radius R_s. Assume that each sensor is capable of communicating within a radius of R_t and assume that $R_t \geq 2R_s$, thus ensuring that two sensors with overlapping sensing areas are capable of communicating directly with each other. The area sensed by a subset of sensors is the union of their corresponding A_is. In addition, $d_{i,j}$ is the Euclidean distance between any two nodes i and j. We use $\mathcal{N}_i$ to denote the set of all neighbor nodes of i within its transmission radius, i.e., $\mathcal{N}_i = \{j \mid d_{i,j} \leq R_t, i \neq j, j \in V\}$.

Assume G is the graph defined as $G(V, E(R_t))$ where $E(R_t) = \{(i,j) \mid d_{i,j} \leq R_t, i \neq j, i,j \in V\}$, i.e. the edges are between nodes close enough to allow direct communication between them. Without loss of generality we assume that G is connected. We note that the objective of any data collection algorithm in a sensor network is to forward sensed data along a spanning tree subgraph of $G(N, E(R_t))$ with the root of the tree at vertex 0. For notational convenience, we denote as $c(i)$ the set of children of node i in the data collection tree, and $p(i)$ the parent of node i in the data collection tree (as a convention $p(0) = 0$). Each node i in the data collection tree defines a set, T_i, of all the nodes (including i) that belong to the subtree rooted at i (note that $T_0 = V$, while in the case of leaf nodes $T_i = \{i\}$). Let S_i denote the sensed area covered by all the nodes in subtree T_i, i.e., $S_i = \cup_{j \in T_i} A_j$.

Communication Cost Metric. A node in the tree removes redundancy in the sensed data that exists between its children's subtrees as well as redundancy between its own sensed data and those of its children subtrees. It forwards up the tree the resulting (non-redundant) information. Assuming the information content is proportional (by a constant factor γ) to the sensed area (after removal of redundancies), then the total communication cost of collecting data along the tree is $\sum_{i=1}^{N} \gamma |S_i|$. Therefore the task of minimizing the communication cost is in essence:

$$\text{minimize } \sum_{i=1}^{N} \gamma |S_i| \tag{1}$$

where of course γ is irrelevant to the optimization, and the real impact on finding the optimal is in the way the tree is structured, which determines the T_is and, consequently, the S_is.

Energy Cost Metric. If the transmission power is the same for each transmitted bit and it is the same regardless of the node, then minimization of the energy cost metric for the entire tree is equivalent to the minimization of the communication cost (Equation 1), differing only in the constant factor γ. However, a more refined energy cost can be considered whereby, once the data collection tree has been determined, the transmission radius, r_i of each node is adjusted to just enough (but always less than R_t) to reach its parent node in the tree. Thus, $r_i = d_{i,p(i)} \leq R_t$. Assuming the loss exponent capturing the attenuation of transmissions with distance is α, then the total energy cost is $\sum_{i=1}^{N} \gamma' |S_i| d_{i,p(i)}^{\alpha}$, thus the corresponding energy optimization problem is:

$$\text{minimize} \quad \sum_{i=1}^{N} \gamma' |S_i| d_{i,p(i)}^{\alpha} \qquad (2)$$

for a constant γ'. We note that we generally have no expectation that sensors are capable of fine transmission power adjustments to arbitrarily set r_i. Therefore, the results of optimizing this metric should be seen only as an indication of how much better the energy consumption would have been if the hardware complexity could allow for transmission power adjustments.

Processing Cost Metric. We also provide a very approximate metric for the processing cost at each node in the data collection tree. We consider the process of redundancy removal to be solely expressed by the cost of identifying cross-correlations between data sets. At its simplest, cross–correlation calculation involves quadratic processing cost (more specifically, cost on the order of the product of the data set cardinalities). Thus, let $c(i)$, the set of children of node i, be expressed as $c(i) = \{i_1, i_2, \ldots, i_m\}$ indicating, in an arbitrary order, the set of the indices of the m children of node i in the tree ($|c(i)| = m$). We define the auxiliary sets $f_{i,j}$, recursively: $f_{i,0} = A_i$ and $f_{i,j} = f_{i,j-1} \cup S_{i_j}$. Hence, the total computation cost can be approximated by:

$$\sum_{i=1}^{N} \sum_{j=1}^{c(i)} (|f_{i,j-1}| \, |S_{i_j}|) \qquad (3)$$

We refrain from considering the optimization problem which minimizes Equation 3 because on top of determining T_is it will also have to establish an order of the elements of $c(i)$'s (can no longer be arbitrary) to minimize the total cost of the summation in Equation 3. While this is conceivably doable, the cost expression supplied is highly artificial because it will eventually depend on the particular application. Indeed, the processing cost presented here assumes: (a) determining cross-correlation is the dominant process cost, (b) cross-correlation is performed in a pair–wise fashion, while one could opt for simultaneous cross–correlation among more than two sequences at a time, and (c) nodes possess no location information, so they have to indirectly (via cross–correlation) discover redundancies. We remark that, because sensors are typically *not* mobile, the cross–correlation computation may be performed only once and its results (expressed as the elements from the data sets to eliminate as redundant) can be used for all time afterwards. Thus, determining the cross–correlation can be a non–recurring cost, in contrast, for example, to the communication cost.

3   Distributed Algorithms

We consider several (five) distributed heuristic approximation algorithms for solving the problem of efficiently building the data gathering tree in sensor networks with redundancy removal, i.e., for finding an approximate solution of the optimization in Equation 1. The algorithms are a combination of the Distance Vector routing (DV) protocol (see e.g. [11]), but simplified because the only destination of interest is the sink, plus a tie-breaking scheme that characterizes the particular algorithm. As we noted earlier ties are common in a densely deployed network so the effect of tie-breaking rules is significant. The result of the DV run is that each node i determines the minimum number of hops, h_i, to the sink.

The algorithms rely only on local information and are therefore scalable as the network grows. None of them uses location information. Three of the algorithms (`Random`, `MaxDegree` and `MaxSubset`) are based on a simple-to-determine graph property, namely the cardinality of the set of neighbors $\mathcal{N}_i$. Nodes announce to their immediate neighbors the cardinality of their corresponding neighbor sets. Two of the algorithms (`MaxDist` and `MinDist`) are based on the existence of a monotonic distance-dependent metric (e). A realistic example of such a metric is the received signal strength $e_{i,j}$ (from transmissions of j as received at node i). Specifically if, by convention, transmissions (e.g. certain control traffic transmissions) are at the same transmission power level regardless of the sensor emitting them, a low received signal strength (averaged over a sufficiently long period of time to account for channel imperfections) implies a longer distance from the transmitter to the receiver. Hence, if $d_{i,j_1} \leq d_{i,j_2} \leq \ldots \leq d_{i,j_m}$ (where $j_k \in \mathcal{N}_i$) then $e_{i,j_1} \geq e_{i,j_2} \geq \ldots \geq e_{i,j_m}$.

Specifically, the studied approximation heuristics are:

Minimum Distance to Next Hop (MinDist). In this algorithm, a tie is broken by choosing the "nearest" neighbor (in the sense of largest $e_{i,j}$) out of all the neighbors which are the same (minimum) distance away from the sink. Thus, if $\mathcal{N}_i^\star$ is the set of neighbors of i that have the minimum hops to the sink, i.e., $\mathcal{N}_i^\star = \{j \mid h_j = \min_{m \in \mathcal{N}_i} h_m , \ j \in \mathcal{N}_i\}$ then, the tie-breaking is:

$$p(i) = \{j \mid e_{i,j} = \max_{k \in \mathcal{N}_i^\star} e_{i,k} , \ j \in \mathcal{N}_i^\star\} \qquad (4)$$

The intuition behind `MinDist` is that the most overlap will be with the sensed area of the closest neighbor. That is, of course, as long as the closest neighbor does not divert us away from the path to the sink.

Maximum Distance to Next Hop (MaxDist). The idea behind `MaxDist` is the opposite of `MinDist`, we attempt to take the longest "stride" forward along the path to the sink, but we do so at the expense of sacrificing opportunities for better redundancy removal. Symbolically:

$$p(i) = \{j \mid e_{i,j} = \min_{k \in \mathcal{N}_i^\star} e_{i,k} , \ j \in \mathcal{N}_i^\star\} \qquad (5)$$

*Random Next Hop (*Random*).* This algorithm breaks ties by uniformly randomly selecting the next hop among all the candidate sensors with minimum hops to the sink. This approach is introduced here for illustrating that a more "educated" selection, in the vein of MinDist, does have some impact. Indeed, one would expect Random to perform in-between the performance of MinDist and MaxDist (it turns out that this is the case). Assuming uniform($\mathcal{A}$) is a function denoting the random uniform selection of an element of set $\mathcal{A}$, then the tie-breaking rule is simply:

$$p(i) = \{j \mid j = \text{uniform}(\mathcal{N}_i^\star) \} \tag{6}$$

*Maximum Degree of Next Hop (*MaxDegree*).* MaxDegree's tie-breaking selects as the next hop the neighbor that possesses the maximum cardinality neighbor set among all minimum hop-count neighbors. The intuition behind it is that the node with the most neighbors is bound to have several overlaps with its neighbors. Choosing that kind of neighbor as the next hop will greatly improve the opportunity of getting rid of data redundancy among many nodes as soon as possible.

$$p(i) = \{j \mid |\mathcal{N}_j| = \max_{k \in \mathcal{N}_i^\star} |\mathcal{N}_k| , \ \ j \in \mathcal{N}_i^\star\} \tag{7}$$

*Maximum Subset of Next Hop (*MaxSubset*).* MaxSubset also looks at the neighbor set of the neighbors of node i to break ties. Consider j, a neighbor node of node i. We observe that some nodes included in the neighborhood of node j, are not going to select j as their parent because their minimum hops are less than node j and our routing is based on minimum hop shortest path. Sometimes, MaxDegree does not distinguish these nodes as not being part of the potential "fan-in" towards node j, and misses better choices. MaxSubset is based on the "partial degree" of node j, which is the number of nodes within j's transmission radius whose minimum hops to sink are *greater* than the minimum hops for node j itself. Only those nodes have the possibility to choose node j as their parent and should be counted as likely to have their redundancies removed by j. We use $\mathcal{N}_j'$ to denote the set of all neighbor nodes of j within its transmission radius except those nodes whose minimum hops is less than or equal to node j, i.e., $\mathcal{N}_j' = \{t \mid h_t > h_j, t \in \mathcal{N}_j\}$. Then,

$$p(i) = \{j \mid |\mathcal{N}_j'| = \max_{k \in \mathcal{N}_i^\star} |\mathcal{N}_k'| , \ \ j \in \mathcal{N}_i^\star\} \tag{8}$$

4   Numerical Results

To evaluate the performance, we generate random configurations and compare the algorithms based on communication cost, energy cost and processing cost. The communication cost metric was the main design criterion for our heuristics, however, it is of interest to see how the algorithms perform for the other two metrics as well. The configurations are generated as follows. A sink node is placed at the origin of a 2-dimensional coordinate system. Experiments were made with other placements of the sink node but there were no significant differences in relative performance. Sensor nodes are placed uniformly randomly within a square area defined by the origin $(0,0)$ and (f, f). Ten independent instances (node placements) are generated and the results for communication cost, energy cost and processing cost are averaged for each algorithm. The maximum and minimum values (that is, the range) over the 10 instances are also recorded. The loss exponent, α, for the energy metric is set to 2.

In the first set of experiments we consider small configurations where it is computationally feasible to find optimal results. We construct an omniscient exhaustive search algorithm over all spanning trees [9,10] which determines the optimal spanning tree under either the criterion of minimizing communication cost from Equation 1 or that of minimizing energy cost from Equation 2. This is done by calculating the union of sensed areas via Voronoi tessellations [8]. However, the exhaustive search is computationally very costly to perform and assumes knowledge (exact location information) that we lack in the presented heuristics. Due to its computation load, we were unable to determine the optimal solution on configurations of more than 13 nodes. The set up for the small configuration experiments have area 25 x 25 (that is, $f = 25$), transmission range $R_t = 10$ and sensing range $R_s = 5$.

To illustrate the behavior of the optimal algorithms and the heuristics, we examine the sink trees for a particular instance of the 13 node case in Figure 2. The optimal trees for communication and energy are shown in Figures 2(a) and 2(b) respectively. Note that Node 9 in the minimum communication cost sink tree connects to Node 1; while in the minimum energy cost sink tree, Node 9 conserves energy by connecting to Node 11 which is closer. In Figures 2(c) and 2(d) the sink trees for the heuristic algorithms MaxSubset and MaxDegree respectively are shown. Unlike the optimal cases, both heuristic cases connect Node 7 to Node 1 which has fewer hops from the sink than Node 4. Node 8 has a choice of connecting to Node 4 or Node 7 and this tie is broken differently for the two heuristics. In the MaxDegree case it connects to Node 7 which has more neighbors; while in the MaxSubset case it connects to Node 4 which has more neighbors further from the sink. In this example MaxDegree has lower communication cost than MaxSubset but higher energy cost.

We next consider results averaged over the 10 iterations for the small configurations. With the given area and transmission range, fewer than 8 nodes results in many cases where paths to the sink do not exist for some nodes (underlying graph is not connected). Therefore, we vary the number of nodes from 8 to 18 for the heuristic algorithms with the optimal results computed for up to 13 nodes. In Figure 3(a), we show the averaged communication costs for the optimal communication cost sink trees and the heuristic sink trees. As can be seen in this figure, the heuristics are very close to the optimal. This indicates that forming shortest hop trees produces close to optimal results for these small cases. Figure 3(b) shows that the energy costs of the heuristics are more different from optimal. Reducing energy was not the main design criterion of these algorithms, neither would it have been possible because knowing the $d_{i,j}$ values (possibly

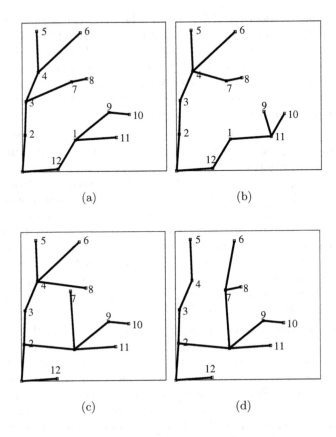

Fig. 2. Small network example (a) optimal (communication) tree, (b) optimal (energy) tree, (c) `MaxSubset` tree, (d) `MaxDegree` tree.

via knowledge of their exact location) is not within the assumptions about the sensors hardware complexity. Moreover, for both cost metrics, the tie-breaking heuristics do not greatly affect the results since in small networks there are few ties to break.

In the next set of experiments we look at larger scenarios with many more nodes. In these cases we have a 100x100 grid ($f = 100$) with nodes of transmission range $R_t = 20$ and sensing range $R_s = 10$. First consider a particular instance of finding a sink tree for 300 nodes. For these large networks, there are clearer patterns in the sink trees produced by the heuristics. The sink tree for `MinDist` in Figure 4(a) shows that this heuristic produces trees with small line segments strongly oriented towards the sink. The fan-in, which indicates opportunity for merging redundant data, is good. The fan-in improves for the heuristics of `MaxSubset` and `MaxDegree` as seen in Figures 4(b) and 4(c) respectively. These algorithms seek explicitly to improve the fan-in by choosing parents with high potential for a large number of children. `MaxSubset` has the

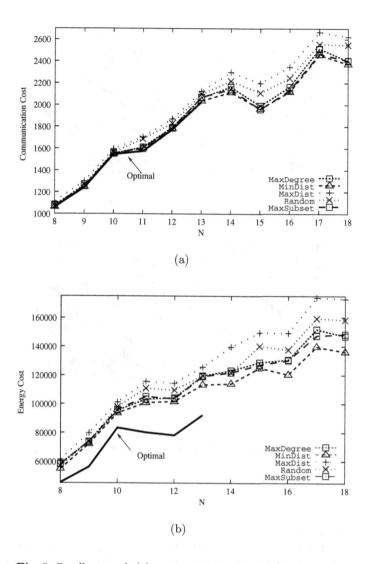

Fig. 3. Small network (a) communication cost, (b) energy cost.

largest fan-in. These designs, however do result in longer segments than the MinDist heuristic. Finally MaxDist, which can be seen in Figure 4(d), has the least-ordered appearance with very long segments because it will choose as a parent the minimum hop neighbor that is furthest away.

We next compare the communication cost, energy cost and processing cost of the heuristic algorithms for a range of 100 to 600 nodes. In general, for all of the metrics, as the number of nodes increases, the costs increase as well. This can be explained by the fact that leaf nodes are not able to merge data until they reach

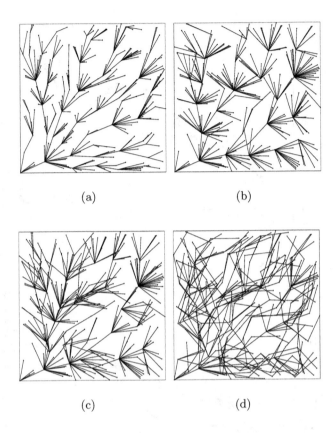

(a)   (b)

(c)   (d)

Fig. 4. Large network example (a) `MinDist` tree, (b) `MaxSubset` tree, (c) `MaxDegree` tree, (d) `MaxDist` tree.

their parent and as you increase the number of nodes you potentially increase the number of leaves. In Figure 5(a), the communication cost of the heuristic algorithms is found for different numbers of nodes. The error bars on the graphs represent the range (min and max values) over the 10 iterations. As the number of nodes is increased, the differences in the heuristic tie breakers becomes more significant. This occurs because there are more chances for ties as the network becomes denser. `MinDist` and `MaxSubset` have the lowest communication cost; while `MaxDegree` has only slightly higher cost. All of these tie-breaking algorithms seek to increase the amount of overlapping data: `MinDist` by connecting to close nodes and `MaxDegree` and `MaxSubset` by connecting to nodes with more potential to combine data. Their communication costs are significantly lower than random tie-breaking indicating that intelligent tie-breaking is important. `MaxDist` has the highest communication cost because it reduces the amount of overlap by connecting to nodes that are further away.

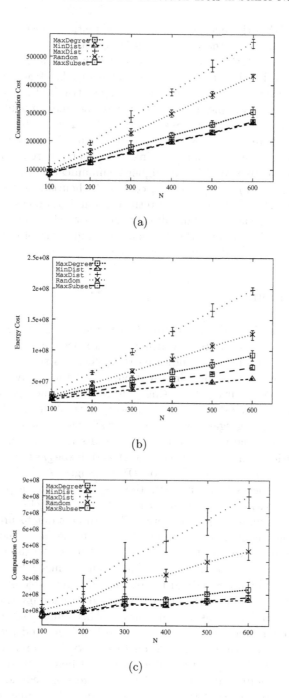

Fig. 5. Large network (a) communication cost, (b) energy cost, (c) computation cost.

In Figure 5(b), the energy cost of the heuristic algorithms is shown. The relative performance of the 5 heuristics is similar to the communication cost results except that MinDist now has lower energy cost than MaxSubset. This is reasonable since by choosing the closest neighbor, the MinDist algorithm will reduce the squared distance factor of the energy metric (at least locally). This can also be seen in the examples of Figure 4 where the line segments are noticeably longer for MaxDegree and MaxSubset than for MinDist. MaxDist has the longest segments and therefore highest energy cost because it connects a node to the furthest minimum hop neighbor. In doing so, it defers the removal of redundancy to further up the tree resulting in higher communication and processing costs as well. In Figure 5(c), the processing cost of the heuristic algorithms is shown. The relative results are very similar to the communication cost metric results of Figure 5(a). The general method already chooses a minimum hop path so that the ability to combine data is the dominating requirement of the tie-breakers for both communication and processing cost metrics.

5   Conclusions

In this paper, we have described five distributed heuristic algorithms for constructing data collection trees, which take into account the removal of data redundancy. Our results have indicated that when restricted to heuristics that operate using exclusively local information, and if the hardware complexity of the sensors is low (equivalently: using sensors that do not possess location information, neither by GPS, nor by any other means), it is still possible to effectively form the aggregate data tree in a way that exploits the reduced data traffic from sensor data aggregation. The results are particularly encouraging when one considers the fact that MaxSubset and MaxDegree require no complexity beyond the piggybacking of $\mathcal{N}_i$ on DV messages. At little added hardware cost MinDist appears to outperform the rest, suggesting a possible potential for greedy approaches, i.e., communicate with the node closest to you, since it represents the highest degree of overlap. This directs our future efforts to consider balancing locally greedy (communicate with the closest node) with global direction–preserving (move closer to the sink) schemes. We also examined particular instances of data gathering trees from our algorithms in both small and large configurations to help understand the effect of different tree building schemes.

Under our assumptions, the heuristics use no location information, while the exhaustive search does. This may appear as rather unfair, but we note the need to quantify redundancy in order to compare heuristics and the optimal which leads inevitably to location information (necessary for Voronoi tessellations to be performed). Avoiding location information, yet being able to quantify redundancy, would require a different model for expressing redundancies devoid of the geometric properties of the sensed areas. It is debatable whether a sensible redundancy model, disconnected from geometry, can be derived. Finally, in future work we will consider cases where $R_t >> R_s$ as well as the impact of non–uniform density of nodes on the data aggregation tree.

Acknowledgements. This work has been supported in part by grants from the Natural Sciences and Engineering Research Council (NSERC) of Canada.

References

1. R. Cristescu, B. Beferull-Lozano, and M. Vetterli. On network correlated data gathering. In 23rd Conf. of the IEEE Communications Society (INFOCOM), 2004.
2. C.-F. Huang, and Y.-C. Tseng. The coverage problem in a wireless sensor network. In 2nd ACM Int'l Conf. on Wireless Sensor Networks and Applications (WSNA), pages 115–121, 2003.
3. S. Slijepcevic, and M. Potkonjak. Power efficient organization of wireless sensor networks. In IEEE Int'l Conf. on Communications (ICC), pages 472–476, 2001.
4. F. Ye, G. Zhong, S. Lu, and L. Zhang. PEAS: A robust energy conserving protocol for long-lived sensor networks. In Int'l Conf. on Distributed Computing Systems (ICDCS), pages 28–37, 2003.
5. K. Barr, and K. Asanovic. Energy aware lossless data compression. In 1st Int'l Conf. on Mobile Systems, Applications, and Services (MOBISYS), 2003.
6. I.F. Akyildiz, W. Su, Y. Sankarasubramaniam, and E. Cayirci. Wireless sensor networks: a survey. Computer Networks (Elsevier), vol.38, pp.393-422, 2002.
7. K. Sohrabi, J. Gao, V. Ailawadhi, and G.J. Pottie. Protocols for Self-Organization of a Wireless Sensor Network. IEEE Personal Communications, October 2002.
8. F. Aurenhammer. Voronoi diagrams - a survey of a fundamental geometric data structure, ACM Comp. Surveys 23:345 – 405, 1991.
9. A. Shioura, and A. Tamura. Efficiently scanning all spanning trees of an undirected graph, J.Oper. Res. Soc. Japan, 38(1995), pp. 331 – 344
10. A. Shioura, A. Tamura, and T. Uno. An optimal algorithm for scanning all spanning trees of undirected graphs. SIAM J.COMPUT. Vol. 26, No. 3, pp. 678 – 692, June 1997
11. D. Bertsekas, and R. Gallager. *Data Networks*, Prentice-Hall, 1992.

Cross-Layer Optimization for High Density Sensor Networks: Distributed Passive Routing Decisions

Primož Škraba, Hamid Aghajan, and Ahmad Bahai

Stanford University, Stanford CA 94305, USA

Abstract. The resource limited nature of WSNs requires that protocols implemented on these networks be energy-efficient, scalable and distributed. This paper presents an analysis of a novel combined routing and MAC protocol. The protocol achieves energy-efficiency by minimizing signaling overhead through state-less routing decisions that are made at the receiver rather than at the sender. The protocol depends on a source node advertising its location and the packet destination to its neighbors, which then contend to become the receiver by measuring their local optimality for the packet and map this into a *time-to-respond* value. More optimal nodes have smaller *time-to-respond* values and so respond before less optimal nodes.

Our analysis focuses on the physical layer requirements of the system and the effects of different system parameters on per hop delay and total energy used. Some examples of mappings are examined to obtain analytical results for delay and probability of collision.

1   Introduction

Wireless sensor networks (WSN) are a special case of ad-hoc wireless networks where the constraints on resources are especially tight. With a small power supply and limited processing power, any overhead equates to performance degradation. However, in any network system there is overhead in every layer of the network model. Therefore, improved performance may be achieved by reducing overhead through cross-layer optimization. This paper proposes Distributed Passive Routing Desicions (DPRD), a novel combined MAC/routing scheme based on geographic routing which achieves energy efficiency through minimization of control overhead. Reduction in overhead is achieved by utilizing implicitly distributed information of each node's location to make routing decisions at the receiver. Optimization and analysis of the protocol is done with physical layer constraints in mind. It is shown that with proper choice of design parameters, a low upper bound on delay can be achieved.

Communication in WSNs is not between specific nodes, but rather based on some attribute that the nodes may possess. Perhaps the best example of this is where a user wishes to know what is being sensed in a particular area. In this case, it is irrelevant which node the information comes from but only that the

I. Nikolaidis et al. (Eds.): ADHOC-NOW 2004, LNCS 3158, pp. 266–279, 2004.

data quality and accuracy are sufficient. Also, the large number of nodes in the network makes tracking the network state impractical. For these two reasons, geographic routing is used in this protocol. To further reduce overhead to a minimum, the MAC layer is utilized to make the routing depend only on locally known information.

This paper begins by presenting relevant previous work. A description of the proposed protocol and relevant parameters follows. Then the performance metrics are explained and the analysis is presented. Finally, resolving deadlocks in the protocol is discussed, followed by the conclusions and future work.

2 Related Work

Several research results have been reported on designing routing algorithms and MAC layer schemes specifically for WSNs. In [1], it is shown that in a wireless environment, the unstable nature of the topology of a wireless network requires on-demand protocols. [2,3] showed that routing decisions made with respect to remaining battery power can dramatically increase the lifetime of a network.

In addition to geographic routing, several other data-centric routing schemes have been proposed. [4,5] describe variations of directed diffusion, which is based upon sink nodes advertising interest to sensing nodes to create gradients that packets can follow. Directed diffusion is designed for locally "pulling" information, since it uses a form of limited flooding. Geographic routing is more practical when data needs to be sent to or retrieved from a distant geographic area (i.e. querying the state of an area, notifying an information sink of an event). Greedy geographic routing was first proposed when GPS became available. It is the simplest algorithm and in most situations finds a near optimal path. The greedy algorithm was extended in [6] to include perimeter routing, thus allowing geographic routing to make its way around voids and to escape dead-end local maxima. Geographic routing, however, assumes that nodes are location aware[1].

In the MAC layer, the S-MAC protocol [7] deals with networks where nodes are on a low duty cycle and propose how to synchronize the sleep schedules without incurring too much overhead. The concept of a low duty cycle is key for WSNs, because they are intended to last over such long periods with small batteries.

Developed independently, two schemes have been proposed recently which embody the same idea as the protocol presented here. Geographic Random Forwarding (GeRaF) [8,9], and Implicit Geographic Forwarding (IGF) [10], both propose receiver contention schemes. The main difference lies in how the receiver is chosen and the emphasis of the analysis. The analysis of GeRaF stressed the energy savings achieved mostly through sleep schedules, while IGF is developed from a more algorithmic point of view, with a specific implementation proposed

---

[1] The cost of location awareness can be quite large and is a topic of research in its own right. However, in any sensing application of WSN's, it is crucial that each node have some spatial knowledge of it's own location

in the paper. The contribution of this paper is that the constraints of the physical layer are examined in the analysis.

3   Protocol Description

This protocol is based on geographic routing, which uses distributed information to achieve state-less operation. This work attempts to further utilize this distributed information to minimize overhead in the lower layers of the network, most notably in the MAC layer. The dynamic nature of the environment results in an unstable network topology. Thus, routes must be created in an on-demand fashion. As described above, given a destination location, geographic routing finds a route on a hop by hop basis, always routing towards the destination, thereby using locally optimum nodes to route a packet. However, in a dense network, if local parameters other than node location (i.e. remaining battery power, link quality) are also taken into account, it quickly becomes impractical to exchange this information with all neighboring nodes. By using the broadcast nature of wireless communication and moving the routing decision to the receivers, this protocol allows this information to be incorporated without having to gather the data at the sending node.

The decision is made through "receiver contention." The sending node sends out a request to send a data packet (RTS). The packet includes its own location and the location of the final destination of the packet. Based on this information, each receiving node calculates its own optimality and maps this into a delay, τ. After this delay, the receiving node transmits a CTS packet to the sending node, unless another node has responded before it. Then the less optimal node does not transmit anything and turns off its radio to prevent overhearing.

This mapping from a measure of optimality to delay will be refered to as the "delay function." Consider some function $f(\theta) \to \tau$ where θ represents all relevant state information. To ensure that an optimal or near optimal node is selected, the delay function should map the state to delay in a monotonically decreasing fashion with respect to optimality (i.e. the more optimal the state of the node, the smaller the delay). To give good delay characteristics, $f(\theta)$ should be bounded. In real implementations, having a purely deterministic scheme without centralized control is impractical due to the possibility of deadlocks. However, the more random a protocol is, the larger the variance in performance will be. In a tightly constrained environment, worst case performance is more important than average performance, making protocols with a smaller amount of randomness more attractive. As will be shown later, however, some randomness is still needed to break deadlocks. Before the performance of different delay functions can be investigated, a quantative metric of optimality must be found.

3.1   Optimality in Geographic Routing

The choice of a metric representing optimality depends on the goals of the system. To simplify the analysis, the smallest set of paramters is chosen. For any one hop using geographic routing, the location of source node, the location of the candidate receiver node, and destination of the packet must be known. With a fixed transmission range, the optimal node is given as the neighbor of the source node that is closest to the destination of the packet as shown in Fig. (1.a). The metric of optimality is taken as the distance from a receiving node to the destination, D, normalized over the range $[L - R, L]$, where L is the distance from the source node to the destination and R is the transmission radius.

Given a specific topology, there is always a locally optimum next hop node. If a node knew that it is optimal, it should respond to the source node immediately. Since this would require knowing the full network state, a node must estimate the likelihood it is the optimal node. The likelihood is exactly equal to the probability that there are no nodes in the shaded area, as shown in Fig. (1.b)[2]. The area can be found as

$$A(L, R, r) = \int_{L-D}^{x'} \sqrt{D^2 - (x - L)^2}\, dx + \int_{x'}^{R} \sqrt{R^2 - x^2}\, dx \qquad (1)$$

where $\quad x' = \dfrac{L^2 - D^2 + R^2}{2L}$

and $\quad D = \sqrt{L^2 + r^2 - 2Lr\cos(\theta)}$.

A closed form expression for $A(L, R, r)$ can easily be found from (1). A two dimensional Poisson process is consistent with a "random" network and is used as the network model to find the probability of optimality. It results in a uniform distribution for a given number of nodes in a fixed area. For a Poisson process, the probability that there are no nodes in the shaded area in Fig. (1.b) is the probability of first arrival which can be written as

$$\mathbb{P}\{n_j \text{ is optimal}\} = e^{-\rho A(L, R, r)} . \qquad (2)$$

Note that the probability is given by the network density, ρ and A. Implementing the complex calculation for A might not be feasible in a sensor node, but it is possible to approximate with a much simpler function of $\{L, R, D\}$. For the purposes of analysis, the assumption is that A or some estimate of A is known.

4   Physical Layer

Several assumptions are made about the characteristics of the underlying physical layer. First, is the absence of power control. This is a fair assumption because the energy saving over short distances that are envisioned for a WSN (on the order of ≈ 10 m) would be negligible. Without this assumption, the optimality criterion is no longer valid and to minimize the total energy, the transmission energy must be jointly optimized with the number of hops. Any model of the

---

[2] The assumption of the circular transmission area is an idealization. A probabilistic transmission area would not change the analysis significantly.

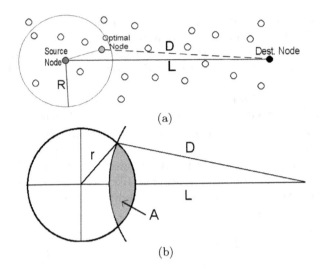

Fig. 1. (a) Example of the one-hop optimal node in a network, the node transmission range is R. (b) Geometric interpretation of receiver position in geographic routing.

number of hops has too much variance to be useful in the optimization, making this problem much more intractable. For the purposes of analysis, links are considered lossless. Link quality/probability of packet loss could easily be incorporated into the metric of optimality but is not considered here due to space constraints.

Distributerd Passive Routing Decisions (DPRD) is a CSMA/CA based scheme, and so the hardware must be available to do a carrier sense. The first consideration is the time it takes to perform the carrier sense[3], which will be referred to as δ_{CS}. The longer the carrier sense is done, the smaller the probability of a miss or false alarm. However, a long δ_{CS} introduces several constraints on the system that are detrimental to performance.

Another important consideration in designing this protocol is that all the nodes that can hear the source node may not hear the receiver node. To solve the resulting hidden terminal problem, [8,9] introduced a dual-tone system. The solution this paper suggests is more flexible in terms of trading off performance characteristics. Two thresholds are introduced for the carrier sense SNR, γ_{CS} for carrier sense and $\gamma_{RECEIVE}$ for data transmission. If we assume a simple path loss model,

$$P_R = P_T K \left[\frac{d_0}{R} \right]^{\alpha} \tag{3}$$

the SNR and receiving range R, can be found to be

---

[3] Most previous work assumes the carrier sense to be instantaneous.

$$\text{SNR} = \frac{P_R}{N_0} \geq \gamma$$

$$R \leq d_0 \sqrt[\alpha]{\frac{P_T K}{N_0 \gamma}} \tag{4}$$

where K, α, and d_0 are channel constants and P_T and P_R are the transmitted power and received power, respectively. Then the ratio of R_{CS} to $R_{RECEIVE}$ becomes

$$\frac{R_{CS}}{R_{RECEIVE}} = \sqrt[\alpha]{\frac{\gamma_{RECEIVE}}{\gamma_{CS}}}. \tag{5}$$

The ratio should take a value between 1 and 2. A value of 1 corresponds to a successful carrier sense being equivalent to a successful reception, while 2 is the most energy efficient value, since it ensures that no interferers can be present within 2R. While this does reduce the network capacity, generally WSNs are used in low data rate applications and so the traffic is assumed to be light. It will be assumed from this point on that the value is near 2 and so the only collisions that occur are a result of two nodes responding to the RTS at the same time. This is the most conservative configuration in terms of energy usage. Thus for nodes neighboring the source node but out of receiving range, it is assumed they would be able to sense the original RTS and so shut down to avoid overhearing.

The introduction of duty cycles can be dealt with by stating that the network density is simply the density of the currently awake nodes. Thus if the true node density is λ, then $\rho = \lambda p$, where p is the probability of a node being awake. However, in this case a dual-tone system should be used to avoid a node waking up and transmitting. Alternatively, this analysis assumes the source wakes up all of its neighbors for a short period of time with an out of band wake-up scheme. Both methods have their advantages and disadvantages, but a quantative comparison is beyond the scope of this paper and does not affect its results.

5   Framework

The general framework of this paper for analyzing the performance of the protocol uses several "types" of delay functions based on certain physical layer parameters. First an analysis of energy efficiency is given, followed by a formulation to find the probability of collision, and finally an analysis of the average one-hop delay will be presented. It will be shown that the latter two are the most crucial parameters affecting energy efficiency.

5.1   Energy Efficiency

A node in a WSN consumes energy in the communication tasks of transmitting, receiving and listening. The energy spent in these activities is denoted $E_T, E_R,$ and E_L respectively. It has been found in [11] that when communicating over a short range, the energy spent in receiving data is of the same magnitude as of the transmit energy, and the energy expended in listening often dominates the total energy consumed by a radio.

Thus, to send a packet from a source with N neighbors to a receiver, assuming no collisions occur, the total energy expended (assuming no overhead with reporting each node's routing information) will be

$$E_{packet} = (N+1)E_{RTS} + 2E_{CTS} + 2E_{DATA} + E_L \ . \tag{6}$$

It is assumed that the energy used to transmit the actual packet is fixed as a physical layer parameter (i.e. Energy/bit, Number of bits/Packet). So all that remains is to minimize the energy consumed in listening.

Using this protocol, it can be shown that once the RTS is sent, the neighboring nodes must listen for $\delta_{CS} + \delta_{PROC}$ before they transmit. δ_{PROC} represents the processing delay of the sending node. If a node is optimal, there will be no other node transmitting before it, so no listening time is required. Assuming a node is not optimal it must detect either the optimal node's CTS or the actual data transfer. As shown in Fig. (2.a), since its delay, τ_1, will be greater than the optimal node's delay, τ_0, the only way to miss the "optimal" node is if τ_1 falls between the optimal CTS and the data transfer. This is exactly $\delta_{CS} + \delta_{PROC}$. This represents an upper bound on the receiving nodes' listening time. Since no node knows a priori whether it is optimal, it must always listen before transmitting. Once a node has detected that it is not optimal, it would turn its radio off for the duration of the transmission to prevent overhearing.

The listening time of the source node is from the RTS until the optimal node responds. This is exactly the expected value of the delay function. So the expected amount of energy is

$$\overline{E_L} = \left(\overline{f(\theta)} + N(\delta_{PROC} + \delta_{CS}) \right) P_L \tag{7}$$

where P_L is the power consumed in listening mode. The formulation above assumes that the transmission is successful. Once collisions or lost packets due to channel conditions are considered, the formulation changes to

$$\overline{E_{total}} = P_S E_{packet} \sum_{n=0}^{\infty} n P_F^n = \frac{1}{P_S} E_{packet} \tag{8}$$

where P_S is the probability of success and P_F is the probability of failure. It is assumed that the physical layer is relatively robust and so the only failures coincide with collisions. So it can be seen that to reduce energy consumption, probability of collision and per hop delay must be reduced.

5.2   Probability of Collision

The probability of a node being chosen is given in (2) and is solely a function of A and node density (which is considered known). Hence, it is natural to make

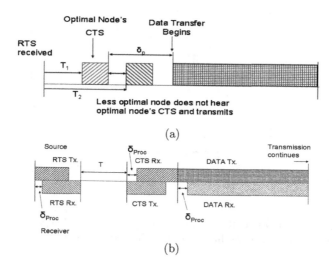

Fig. 2. (a) Example of how a node must listen for the turnaround time of the source node to ensure that it does not miss the optimal node's CTS (b) Example of successful RTS-CTS-DATA transmission. δ_{Proc} is the processing delay.

the delay function a function of A. This is becuase A encapsulates all relevant spatial information about the probability of a node being optimal. As stated in Sect. 3.1, A is considered to be known, or more precisely, each node is able to calculate or estimate it upon receiving the information from the source node's RTS packet. Note that only considering the spatial information is a special case of the formulation and other local parameters may be added as required.

The problem can the be stated as of finding the probability of collision given a delay function. The only source of collisions that will be considered will be when the CTS packets from two receiving nodes collide. As discussed in Sect. 4, the carrier sense threshold is set so that not only can all relevant nodes hear the first CTS, but also it ensures that the neighboring nodes outside the range of the source node do not transmit, because they would sense either the RTS, CTS, or the data transmission.

To find the probability of collision, note that $f(A) : A \to \tau$, where τ represents the delay from the end of the processing of the RTS as shown in Fig. (2.b). A collision will occur if two nodes calculate two values of τ that are separated by less than $\delta_{CS} + \tau_{PROP}$, where δ_{CS} represents the time required for a carrier sense and τ_{PROP} represents propogation delay. Based on this, we can find the following condition to ensure that no collisions occur:

$$\tau_i + \delta_{CS} + \tau_{PROP} < \tau_j \tag{9}$$

where τ_i represents the optimal node and τ_j represents the next closest node. In (9), τ_{PROP} may be negligible due to the short range of the transmission links. δ_{CS} is a parameter determined by the physical layer. The longer the carrier sense

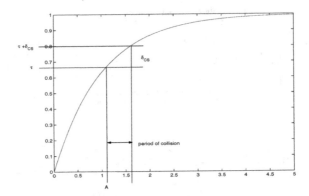

Fig. 3. Mapping from vulnerable period of collision to spatial separation in terms of A with respect to some general function $f(A)$, here shown as a concave function.

is done, the smaller the probability of miss and probability of false alarm will be. However, the shorter it is, the smaller will be the probability of collision. Optimizing this depends on the hardware available and is beyond the scope of this paper. The effects of different δ_{CS} will however be explored. Taking a general function that fulfills the criteria stated above, we wish to obtain an expression for a collision in terms of A rather than in terms of time. As shown in Fig. (3), for a given A and δ_{CS}, the interval in terms of A is given by

$$g(A, \delta_{CS}) = f^{-1}(f(A) + \delta_{CS}) - A \ . \tag{10}$$

By the independent increment property of the Poisson process, the probability of a collision is given by

$$\mathbb{P}\{\text{coll.}|A\} = 1 - e^{-\rho g(A, \delta_{CS})} = 1 - \frac{e^{-\rho f^{-1}(f(A) + \delta_{CS})}}{e^{-\rho A}} \ . \tag{11}$$

5.3 One-Hop Delay

Before investigating the performance of some test functions, the delay characteristics with respect to protocol parameters and energy efficiency are presented. The delay is characterized as

$$\tau_{TOTAL\ DELAY} = \# \text{ of hops} \cdot E[\text{One Hop Delay}] \ . \tag{12}$$

The number of hops can be found to be roughly linear with distance and the variance decreases with node density. Since the actual number of hops that will result is more a function of the specific scenario topology, it is not considered a descriptive statistic of the network performance. As long as the number of hops taken by geographic routing with perfect next hop knowledge versus the proposed algorithm is equal, the algorithm is near optimal.

The expected value of one hop delay can be found simply by finding $E[\tau]$, which is equivalent to finding the expected value of the delay function:

$$E[f(A)] = \int_0^{A_{MAX}} f(A)e^{-\rho A}dA \tag{13}$$

where $A_{MAX} \approx \frac{\pi R^2}{2}$. Thus the total delay is

$$\tau_{TOTAL\ DELAY} = N_{hops}\left(\int_0^{\frac{\pi R^2}{2}} f(A)e^{-\rho A}dA + \tau_{DATA}\right). \tag{14}$$

6 Delay Functions

Two classes of delay functions are examined, exponential and linear. The exponential model was chosen becuase it best matches the model of the distribution of nodes and the linear model was chosen for its simplicity. For any delay function

$$f(A) : [0, A_{MAX}] \rightarrow [0, T_{MAX}]. \tag{15}$$

T_{MAX} represents the upper bound on one contention period, after which the source node knows that there are no closer nodes than itself to the packet destination or that a collision has occured (but for the purposes of this analysis, it is assumed that a collision would be detected).

6.1 Exponential Model

To ensure that it is increasing and bounded, the function and its inverse must be of the form

$$f_1(A) = \frac{T_{MAX}\left(1 - e^{-sA}\right)}{1 - e^{-s\frac{\pi}{2}}} \tag{16}$$

$$f^{-1}(\tau) = \frac{1}{s}\ln\left(\frac{T_{MAX}}{T_{MAX} - \tau(1 - e^{-s\frac{\pi R^2}{2}})}\right). \tag{17}$$

Using this we can find the one hop delay

$$E[f_1(A)] = \int_0^{\frac{\pi R^2}{2}} f_1(A)e^{-\rho A}dA = T_{MAX}\left(\frac{s\frac{1-e^{-\rho\frac{\pi R^2}{2}}}{1-e^{-s\frac{\pi R^2}{2}}} - \rho e^{-\rho\frac{\pi R^2}{2}}}{\rho(\rho + s)}\right). \tag{18}$$

Trivially, (18) has a minimum at $s = 0$. This is because s must take a positive value (so that the resulting delay is positive). There is a tradeoff between delay and probability of collision, which is computed next. In this analysis, we use C which constitutes the delay over which we wish to study the probability of

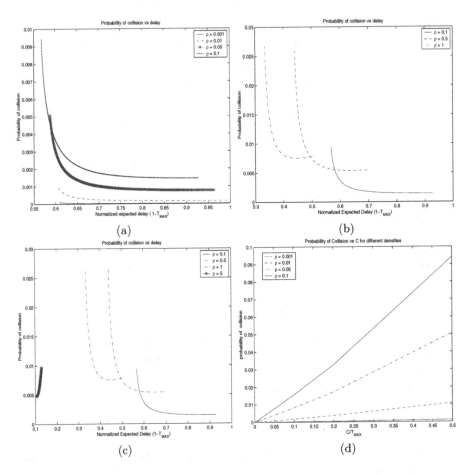

Fig. 4. $\mathbb{P}\{\text{coll.}\}$ vs. Delay for (a) low node densities, (b) higher node densities, and (c) the highest node densities. (d) $\mathbb{P}\{\text{coll.}\}$ vs. $\frac{C}{T_{MAX}}$ for a few node densities

collision. So in all cases presented here, it can be assumed that $C = \delta_{CS}$. By substituting (17) and (16) into (11) and integrating over all possible values of A, the probability of collision in terms of s and C, is given by

$$\mathbb{P}\{\text{coll.}\} = \frac{1 - e^{\frac{\pi \rho R^2}{2}}}{\rho} - \int_0^{\frac{\pi R^2}{2}} \left(e^{-sA} - \frac{C}{T_{MAX}}(1 - e^{-s\frac{\pi R^2}{2}}) \right)^{\frac{\rho}{s}} dA . \quad (19)$$

As Figs. (4.a), (4.b), and (4.c) show, by changing the parameter s, increased delay can be traded off with lower probability of collision. The graphs show the operating curves of the protocol using an exponential delay function under different node densities. Figure (4.a) shows that at low densities, an increase in density only marginally improves delay. In Fig. (4.b), the improvement in expected delay becomes more dramatic as the density increases. Figure (4.c) shows that at extremely high densities, it is possible to minimize both delay and probability of collision, due to the large probability that a "very" optimal node exists. From the graphs, it can be seen that the performance of the protocol improves with density, provided that C is small enough. As Fig. (4.d) shows, at high densities the probability of collision increases quickly with C.

6.2 Linear Model

For linear functions, there is only one possible form of the delay function

$$f_2(A) = \frac{T_{MAX}}{A_{MAX}} A \approx \frac{2T_{MAX}}{\pi R^2} A \tag{20}$$

$$f_2^{-1}(\tau) = \frac{A_{MAX}}{T_{MAX}} \tau . \tag{21}$$

The analysis for this function is straightforward. The delay is given by

$$E[f_2(A)] = \int_0^{\frac{\pi R^2}{2}} f_2(A)e^{-\rho A} dA = \frac{T_{MAX}}{\pi R^2} \left(\frac{2 - e^{-\rho \frac{\pi R^2}{2}}(2 + \pi R^2 \rho)}{\rho^2} \right) . \tag{22}$$

The probability of collision can be found in the same way as for the exponential function. Again, (20) and (21) are substituted into 11 to find $\mathbb{P}\{\text{coll.}|A\}$. This is integrated over all A to find that the probability of collision is given by

$$\mathbb{P}\{\text{coll.}\} = \left(1 - e^{-\rho \frac{\pi R^2}{2} \frac{C}{T_{MAX}}} \right) \left(\frac{1 - e^{\frac{\pi \rho R^2}{2}}}{\rho} \right) . \tag{23}$$

For a linear system, it can be seen that the probability of collision increases with the node density and C, while the expected delay decreases with node density. Since there is no free parameter, the effect of the different ratios of C and T_{MAX} at different node densities can be seen in Figs. (5.a) and (5.b). Fig. (5.b) especially shows that for increasing values of C/T_{MAX}, the probability of collision can increase rapidly with node density to unacceptably high levels. Fig. (5.c) shows how the expected delay deceases as the node density increases. Fig. (5.d) shows the operating curves using the linear delay function at different node densities. It is interesting because it shows that for a fixed collision window C, by choosing an optimal T_{MAX}, the protocol performs better in low node densities. This result does not consider that a the lower density implies a larger probability of a void, which results in a breakdown of greedy geographic routing. This implies longer delays and higher energy usage. It is an open problem to find the optimal density that maximizes the probability that another node is closer to the destination but minimizes the probability of collision.

7 Resolving Deadlocks

The protocol as described is deterministic. If a collision does occur, it will occur every time a packet will have the same destination. This implies that a deadlock

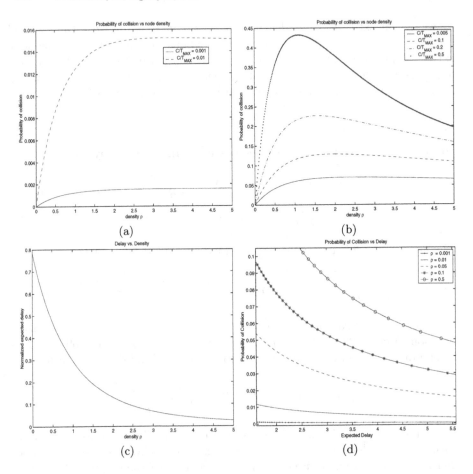

Fig. 5. (a) $\mathbb{P}\{\text{coll.}\}$ vs. node density for $C \ll T_{MAX}$ (b) $\mathbb{P}\{\text{coll.}\}$ vs. node density for $C < T_{MAX}$ (c) Delay vs. node density for a linear delay function (d) $\mathbb{P}\{\text{coll.}\}$ vs. Delay for some node densities, varying T_{MAX} assuming $C = 1$.

can exist between two nodes. A random term can be added to the delay function. Specifically, this term should take on some value $\pm k\delta_{CS}$ for $k = 0, 1, 2 \ldots$. Since deadlocks will be relatively rare and ideally δ_{CS} will be small, a value of $k_{MAX} = 2$ should suffice to resolve most deadlocks. In addition, as more local parameters are added, especially the remaining battery power, more randomness will inherently be added to the system which will help prevent deadlocks.

8 Conclusion and Further Work

This paper has proposed and analyzed a novel combined routing/MAC scheme. It was shown that physical layer parameters introduce limits on network parameters such as δ_{CS} and T_{MAX}. The purpose of this paper has been more to highlight that the available system parameters such as network density and physical layer constraints must be considered in designing an efficient routing protocol. The analysis of two delay functions was presented, illustrating that in high node densities a more complex function may perform better. This is especially important for geographic routing because it performs best in high node densities. Much further research remains be done on how to add local parameters, such as battery power, to the delay function without altering the basic system behaviour. Furthermore, since network node density is not uniform over space or time (e.g. local low densities may be casued by node failures), this protocol allows for the possibility of using an estimate of local node density to obtain optimal operating efficiency. While many questions remain open, the protocol has many of the desirable properties of deterministic MAC schemes while minimizing the control overhead of making routing decisions.

References

1. Broch, J., Maltz, D.A., Johnson, D.B., Hu, Y.C., Jetcheva, J.: A performance comparison of multi-hop wireless ad hoc network routing protocols. In: Mobile Computing and Networking. (1998) 85–97
2. Singh, S., Woo, M., Raghavendra, C.S.: Power-aware routing in mobile ad hoc networks. In: Proceedings of the 4th annual ACM/IEEE international conference on Mobile computing and networking, ACM Press (1998) 181–190
3. Toh, C.K., Cobb, H., Scott, D.: Performance evaluation of battery-life-aware routing schemes for wireless ad hoc networks. In: Proceedings of IEEE International Conference on Communications, ICC 2001. Volume 9. (2001) 2824–2829
4. Schurgers, C., Srivastava, M.: Energy efficient routing in wireless sensor networks. In: Proceedings of MILCOM 2001. Volume 1., IEEE (2001) 357–361
5. Intanagonwiwat, C., Govindan, R., Estrin, D.: Directed diffusion: a scalable and robust communication paradigm for sensor networks. In: Mobile Computing and Networking. (2000) 56–67
6. Karp, B., Kung, H.T.: GPSR: greedy perimeter stateless routing for wireless networks. In: Mobile Computing and Networking. (2000) 243–254
7. Ye, W., Heidemann, J., Estrin, D.: An energy-efficient mac protocol for wireless sensor networks. In: Proceedings of INFOCOM 2002. Volume 3. (2002) 1567–1576
8. Zorzi, M., Rao, R.: Geographic random forwarding (geraf) for ad hoc and sensor networks: multihop performance. IEEE Transactions on Mobile Computing **2** (2003) 337–348
9. Zorzi, M., Rao, R.: Geographic random forwarding (geraf) for ad hoc and sensor networks: energy and latency performance. IEEE Transactions on Mobile Computing **2** (2003) 349–365
10. Blum, B.M., He, T., Son, S., Stankovic, J.A.: Igf: A state-free robust communication protocol for wireless sensor networks. Technical Report CS-2003-11, University of Virginia CS Department (2003)
11. Min, R., Chandrakasan, A.: Energy-efficient communication for ad-hoc wireless sensor networks. In: Proceedings of 35th Asilomar Conference on Signals, Systems, and Computers. Volume 1. (2001) 139–143

Adaptive Probing and Communication in Sensor Networks

Iftach Ragoler, Yossi Matias, and Nimrod Aviram

School of Computer Science, Tel Aviv University
{ragoleri, matias, aviramni}@post.tau.ac.il

Abstract. Sensor networks consist of multiple low-cost, autonomous, ad-hoc sensors, that periodically probe and react to the environment and communicate with other sensors or devices. A primary concern in the operation of sensor networks is the limited energy capacity per sensor. As a result, a common challenge is in setting the probing frequency, so as to compromise between the cost of frequent probing and the inaccuracy resulting from infrequent probing.
We present adaptive probing algorithms that enable sensors to make effective selections of their next probing time, based on prior probes. We also present adaptive communication techniques, which allow reduced communication between sensors, and hence significant energy savings, without sacrificing accuracy. The presented algorithms were implemented in Motes sensors and are shown to be effective by testing them on real data.

1 Introduction

Sensor networks consist of multiple low-cost, autonomous, ad-hoc sensors, that periodically probe and react to the environment and communicate with other sensors or devices. A primary concern in the operations of sensor networks is the limited energy capacity per sensor. There has been extensive research aimed at reducing energy consumption in the application itself, network level, Datalink level and physical level. A few examples are works on network aggregation, improved routing algorithms, efficient communication, synchronizing wake-up time between nodes, data compression and efficient tracking algorithms.

This paper presents two primitives for energy saving in the service level: *adaptive probing* and *adaptive communication*. These primitives assume very little about the system settings or the behavior of the measured data, and thus many applications may gain significant energy savings simply by using the new primitives instead of raw probing and communication.

Adaptive Probing

In current settings, the probing rate of sensors is typically fixed, and is set as a compromise between the cost of frequent probing and the inaccuracy resulting from infrequent probing. In contrast, *adaptive probing* is based on variable rate

I. Nikolaidis et al. (Eds.): ADHOC-NOW 2004, LNCS 3158, pp. 280–293, 2004.

probing so that the base rate is lower than the typical one, and is increased as required when the probed measurements change more rapidly, or are approaching the vicinity of a predefined threshold. The *adaptive probing* primitive enables sensors to make effective selections of their next probing times, based on what they have learned so far. This results with reduced energy consumption while improving accuracy.

The energy saving obtained when using adaptive probing is three fold: (1) reduced probing: as probing is the most frequent operation in sensor networks and thus reduction in the average probing rate could be significant; (2) reduced communication, as a result of reduced probing - the communication step is typically the most expensive operation in sensor networks; and (3) reduced wake-up time: the average sleep time is increased resulting with potentially reduced energy consumption.

The problem of devising a suitable adaptive probing can be formalized as an optimization problem with respect to cost versus approximation error: given a stream of data, our goal is to probe the stream so as to obtain as good approximation as possible, with minimum number of probes. We denote by *Approximate Probing* (*AP*) a method for online scheduling of probes with the goal of learning the measured value with minimum error and minimum cost.

Approximate probing is suitable for settings in which the entire pattern of a measured value over time is of interest. There are other situations of interest in which the sensor network has to alert when certain values exceed a predefined threshold. Rather than having a constant rate probing, it is more effective to have an adaptive probing approach: the next probe should be set according to the approximate time to reach the threshold. We denote by *Threshold Probing* (*TP*) a method for efficiently alerting on a value reaching a threshold in minimum detecting latency and at a minimum cost.

Adaptive Communication

The second optimization primitive is *Adaptive Communication* (*AC*), which is a method for efficiently adapting the communication to the actual values of the data stream.

In *Adaptive Communication*, a sender and its receiver maintain each a synopsis of the stream history and compute each a prediction for the next value. When a node needs to send the next value, it sends a message only in case the value deviates significantly from its prediction, which is also shared (approximately) by its receiver. Otherwise this can be thought of as an *Implicit Send* operation, in which no value is actually transmitted and its receiver will compute the next value by itself based on its own prediction. In this setting, we show additional significant energy savings, as in many cases communication is a dominant factor in energy consumption.

In the heart of all our methodology is computing an instant, low cost model of the changes in the environment and issuing a tiny prediction from this model for the next desired probe or send time.

The requirement to obtain an instant prediction implies using a simple model to compute that prediction, so that it can be computed efficiently. There is extensive literature about prediction computations such as adaptive filters which include Weiner and Kalman filters (e.g., [4]), Box-Jenkins modeling (e.g., [1]) and time series analysis (e.g., [16]) to name a few. Those methods usually involve complex calculations that can be computed on a powerful server such as a base station, but are too costly to be computed on the sensors themselves. In addition, those methods typically compute the expected value in a particular time, rather than the expected time to get to some value as our method requires. It is possible to find an approximation to this inverse function using doubling and binary search, but this will be even more costly. We further elaborate on related works in Section 6. In contrast, our method is built on computing a tiny model in the severe constraint of the sensor itself and not in a powerful base station. Adding to this constraint the fact that the calculation should be done in real-time for the next probing or sending epoch, limits the possible complexity of the prediction. Our model makes no assumptions about the model of the environment (e.g., sinusoid or any other periodic model), thus it can serve as a basic primitive in a large number of applications.

To fully exploit the advantages of adaptive probing and adaptive communication, a tiny low cost non-volatile memory is sufficient. In particular, about a dozen of registers that remain non-volatile in sleep period, and for which storing and retrieval is low-cost, would be sufficient to hold the history synopses.

We devise basic concrete algorithms and validate their potential impact with experimentation. The algorithms were tested by simulation on data gathered from the Great Duck Island (GDI) [9]. We also implemented the algorithms on Mica motes and tested the performance of these implementations. For instance, at the GDI case we obtain up to 55% improvement in the error for the same number of probes (AP), up to 95% improvements in latency of threshold detection (TP), and up to 75% reduction in the number of messages needed for the same error (AC). In the Mica motes experiments we got up to up to 48% improvement in the error for the same cost (AP) and up to 70% reduction in the number of messages needed for the same error (AC).

2   Basic Prediction Techniques

We begin by presenting basic low-cost prediction techniques that will later be used by both the adaptive probing and adaptive communication algorithms. Given a target value τ, one of our objectives is to predict the time in which the measured value will be τ, based on the recorded information. The prediction techniques, presented in increasing complexities, are used for computing the next probing time based on the recorded history of the measured stream, and also estimate the value of the stream in a future time.

Distance prediction. This technique is based on the distance between the current measured value and a given threshold. While the value is far from the threshold, the sensing node can probe at a slow rate, but as the value advances towards the threshold, the node should increase the probing rate.

Let Δt_{i+1} be the time between probe i and probe $i+1$, and let x_i be the value measured at time t_i. Our objective is to predict the time t_{i+1} at which the measured value will satisfy $x_{i+1} = \tau$. The time until the next probe $\Delta t_{i+1} = t_{i+1} - t_i$ will be taken as linear in the distance from the threshold: $\Delta t_{i+1} = \Theta(\tau - x_i)$

There are no memory requirements for calculating the distance prediction, except for τ, as the next probing time is calculated using the current value only.

Linear Prediction. This technique adapts the local probing rate not only to the distance from the target value, but also to the rate of change in the measured values. Intuitively, while the probed value is changing fast, the node should probe at a high rate, but when the change is slow the node can probe at a slower rate and still detect the essential properties.

Computing the estimated value at different points is based on linear extrapolation. Given two probe points (t_{i-1}, x_{i-1}) and (t_i, x_i), the predicted value x_{i+1} in time t_{i+1} is computed as :

$$x_{i+1} = x_i + (t_{i+1} - t_i) \cdot \frac{x_i - x_{i-1}}{t_i - t_{i-1}} \tag{1}$$

Based on this prediction, the time t_{i+1} in which the value is expected to get to x_{i+1} is hence:

$$t_{i+1} = t_i + (x_{i+1} - x_i) \cdot \frac{t_i - t_{i-1}}{x_i - x_{i-1}} \tag{2}$$

We can then compute Δt_{i+1} by substituting x_{i+1} with τ. If the computed Δt_{i+1} is not positive, then we can probe as far in the future as desired, setting Δt_{i+1} to be as large as allowed by considering the adjustment parameter (see Section 3.3).

The memory requirement consists of two additional registers, holding t_{i-1} and x_{i-1}, which are updated at each epoch.

Quadratic Prediction. This technique extends the linear prediction by accounting for the acceleration / deceleration in the measured value; i.e., to its second derivative. Intuitively, while the measured function is accelerating significantly, the node should correspond by accelerating the probing frequency, whereas while the acceleration is moderate, the prediction can be done using linear prediction.

Computing the estimated value at different points is based on quadratic extrapolation. Given three probe points (t_{i-2}, x_{i-2}), (t_{i-1}, x_{i-1}) and (t_i, x_i), computing the value x_{i+1} in time t_{i+1} can be computed using the Lagrange Equation (LE) for three points:

$$LE = x_{i-2} \cdot \frac{(t_{i+1} - t_{i-1})(t_{i+1} - t_i)}{(t_{i-2} - t_{i-1})(t_{i-2} - t_i)} + x_{i-1} \cdot \frac{(t_{i+1} - t_{i-2})(t_{i+1} - t_i)}{(t_{i-1} - t_{i-2})(t_{i-1} - t_i)} \tag{3}$$
$$+ x_i \cdot \frac{(t_{i+1} - t_{i-2})(t_{i+1} - t_{i-1})}{(t_i - t_{i-2})(t_i - t_{i-1})}$$

We provide an improved progressive computation, by leveraging on previous computations: let $x_i' = \frac{x_i - x_{i-1}}{t_i - t_{i-1}}$, $t_i' = \frac{t_i - t_{i-1}}{2}$ be the values as computed for linear prediction, x_{i-1}' and t_{i-1}' be these values as computed in the previous epoch. The *progressive estimation* (PE) is computed as:

$$PE = x_i + x'_i(t_{i+1} - t_i) + \frac{(x'_i - x'_{i-1})(t_{i+1} - t_i)(t_{i+1} - t_{i-1})}{t'_i - t'_{i-1}} \quad (4)$$

The number of basic arithmetic operations in the progressive estimation is 13, compared to 23 in the Lagrange computation.

It can be proven using basic arithmetic that LE as computed by the Lagrange Equation (Eq. 3) equals to PE as computed by the progressive estimation (Eq. 4). The detailed proof can be found in the extended paper.

Given the next desired value $x_{i+1} = \tau$, the time t_{i+1} can be computed by solving the quadratic equation. There are a few possibilities for the equation results, depending on the number of solutions that specify a time in the future. We take the smallest t_{i+1} such that $t_{i+1} > t_i$. If no such solution exists, we can probe again very far in the future, such that Δt_{i+1} is very large. Quadratic prediction requires two more non-volatile registers for x'_i and t'_i, bringing the total number of required registers to four.

Higher Dimension Prediction. It is possible to extend to higher prediction levels using the general Lagrange equation. However, the additional cost may not be justified by the possible improvement, which is expected to be diminishing.

3   Adaptive Probing

We define two types of adaptive probing: *Approximate Probing* (AP) (Section 3.1) and *Threshold Probing* (TP) (Section 3.2).

3.1   Approximate Probing (AP)

The basic function of sensors is to approximate their environment as they probe it in a periodic manner. Our goal is to improve this approximation, so that we probe as infrequently as possible with minimum approximation error.

The AP algorithm uses extrapolation for predicting future values and determining the right time for the next probe. It may utilize any of the prediction techniques, described in Section 2. We present algorithms with linear and quadratic predictions. The *Naive* algorithm can be viewed as probing in equal *time* differences between probes, therefore the changes in the probed environment do not influence its execution. In contrast, the AP algorithm will aim to probe in equal *value* differences between probes, adapting to the changes in the environment. At each epoch the algorithm should decide *when* the next probe should take place.

For a given desired *value* difference c between the last probe and the next probe, the goal of the algorithm is to decide the *time* difference, Δt until the next probe.

Recall that x_{i-1} is the probed value at time t_{i-1} and x_i is the value in the current time t_i. By setting $c = x_{i+1} - x_i$ and using Eq. 2 for linear prediction, we get the next probing time:

$$t_{i+1} = t_i + c \cdot \frac{t_i - t_{i-1}}{x_i - x_{i-1}} \quad (5)$$

For quadratic prediction, substituting $c = x_{i+1} - x_i$ in Eq. 4 gives:

$$c = x_i'(t_{i+1} - t_i) + \frac{(x_i' - x_{i-1}')(t_{i+1} - t_i)(t_{i+1} - t_{i-1})}{t_i' - t_{i-1}'} \tag{6}$$

Calculating quadratic equation for both c and $-c$ (for increasing values) may result with four possible solutions. The selected solution is taken as discussed for the quadratic prediction of Section 2.

The Error Metric: Let E be the error, N be the number of probes, RV_i be the real value, and AV_i be the value estimated by the algorithm at epoch i. The mean absolute error and the mean relative error are computed as: $E_{abs} = \frac{1}{N} \sum_{i=1}^{N} |RV_i - AV_i|$, $E_{rel} = \frac{1}{N} \sum_{i=1}^{N} \left| \frac{RV_i - AV_i}{RV_i} \right|$, respectively. Higher moments of errors can be defined in a similar manner.

3.2 Threshold Probing (TP)

In many cases, the function of a sensor network is to alert when the measured value reaches a certain level. That would be the case, for instance, in fire detection. In such cases, we do not want to approximate an entire measured signal, but rather to alert when reaching the desired threshold, with minimum cost and minimum latency error.

The algorithm for TP is based on computing *estimate time of arrival (ETA)*: the predicted time in which the threshold value is expected to be reached, based on the stream history.

Let τ be the threshold value; we desire to predict the time t_{i+1} in which the measured value will become $x_{i+1} = \tau$. The TP algorithm based on linear prediction, and using Eq. 2, gives:

$$t_{i+1} = t_i + \frac{(\tau - x_i) \cdot (t_i - t_{i-1})}{x_i - x_{i-1}} \tag{7}$$

Eq. 7 might return time t_{i+1} in the past (i.e., $< t_i$), when the value is getting farther from the threshold. In that case the next probe can be determined using a predetermined bound, as discussed in Section 3.3.

The TP algorithm based on quadratic prediction, using Eq. 4, gives:

$$\tau = x_i + x_i' \Delta t_{i \to i+1} + \frac{(x_i' - x_{i-1}')\, \Delta t_{i \to i+1} \Delta t_{i-1 \to i+1}}{t_i' - t_{i-1}'} \tag{8}$$

As for AP, solving the quadratic equation might return two possible results, in order to be conservative and thus not lose important events, the algorithm should take a point in time t, such that t is the minimal solution that is larger than the current time t_i. If the equation does not have any result, we use the adjustment parameters as discussed in Section 3.3.

The Error Metric: The goal of TP is to detect an event as soon as possible, therefore the error measure is the delay between the time in which a real event occurs and the time in which the algorithm detects this event.

Let T_e the time of an event e, in which the measured value reaches a predefined threshold, and let T_a be the detection time based on the TP algorithm. Then, the *latency error* is defined as: $E_{TP} = T_a - T_e$

Note that for any algorithm, but especially for the *Naive* algorithm, the error in each run is influenced by the starting time of the probing. The average error in the *Naive* case over all possible starting times, is half of the time between probes. For the TP algorithm, the starting point is less important as the algorithm adjusts itself to the value changes.

3.3   Realization

We consider the realization of the AP and TP algorithms. The basic algorithms, as described so far, only treat the anticipated case where a very small synopsis of history perfectly predicts the future. We utilize two methods to address inherent prediction errors: (1) using adjustment parameters for computing Δt; (2) incorporating a fall back methodology in which the algorithm monitors its own performance and transforms to a *Naive* algorithm in case it performs poorly. Due to space limitation, we only describe briefly the method of adjustment parameters and leave the discussion on the fall back method to the full paper.

Let Δt_{i+1} be the estimated time difference so that $x_{i+1} = \tau$, when letting $t_{i+1} - t_i = \Delta t_{i+1}$, as computed by a prediction method. To address the imperfection of this prediction, we conservatively enforce the next probe to be earlier, based on two *adjustment parameters*: $\alpha \geq 0$, and $\beta \geq 0$:

$$\Delta t_a = \frac{\Delta t_{i+1}}{(1+\alpha)} - \beta \tag{9}$$

The parameter α is multiplicative and it would mostly affect the mean relative error; the parameter β is additive and it would mostly affect the mean absolute error.

We also use two additional *adjustment parameters*: $\overline{\Delta}$ and $\underline{\Delta}$. The upper-adjustment parameter $\overline{\Delta}$ provides an upper bound on the time difference between two probes, and hence on the latency error. The lower-adjustment parameter $\underline{\Delta}$ provides a lower bound on the time difference, to avoid costly probing that is beyond the required granularity. The waiting time for the next probe is computed as a function of the adjusted prediction Δt_a and the two adjustment parameters:

$$\Delta t = Min\left(\text{Max}\left(\Delta t_a, \underline{\Delta}\right), \overline{\Delta}\right) \tag{10}$$

4   Adaptive Communication (AC)

One of the most common uses of communication in sensor networks is the transfer of probed values of a sensor node to other nodes. The value can be passed as raw data or aggregated along the communication path. In either case, a node transfers data to its parent node, which in turn transfers data onwards. The base assumption for our methodology is that the sequence of values are taken from a signal that has some structure.

4.1   AC Protocol

The sender maintains a synopsis that represents the stream of values sent so far, and a prediction of the value that is to be sent next, based on the synopsis. Similarly, the receiver maintains a synopsis that represents the stream of values received so far, and a prediction of the value that is to be received next, based on the synopsis. The synopses and the corresponding predictions can be computed using the prediction techniques of Section 2.

If the next probe is as predicted, within a small allowed error, then the sender does not send it. If the receiver does not receive a value at a designated time, it assumes that it equals its predicted value. Thus we manage to avoid transmission and still transfer the appropriate information. This event is therefore called *implicit send*.

In cases in which the time of transmission is unknown, as is the case for probing with a variable rate, the receiver does not compute any prediction. Its synopsis is then only used for the purpose of extrapolation for reporting values. The detailed protocol is described fully in the extended paper.

4.2   AC Implementation

In many installations, the base station needs all the probed information from each node in a multi-hop network of nodes (e.g., [9]). In such scenarios the *AC* protocol can be applied between each node and the base station. It has two advantages: (1) each message saving implies savings along all the routing to the root; and (2) the base station might be a powerful node with 'unlimited' memory, energy and computation power for issuing the *Receiver* protocol.

There are cases in which the user is interested only in aggregation of results or the aggregation is desired from an energy saving perspective (e.g., [8]). In such scenarios the *AC* protocol is applied between each node and its parent, and the *Receiver* saves k states for its k children. If this is too much for its constrained memory, it is used for the maximum number of children which can use the protocol, while other children will continue propagating all their values as before.

In sensor networks there might be extensive loss of communication due to collisions, synchronization issues, lack of resources and other interferences. There are cases in which the communication has an acknowledgement mechanism. In that case our protocol continues to work as is.

When the communication is unreliable, we propose transmitting redundant information. In that situation, the sender should send the subsequent probe after each sent probe, in hope that at least one of the two sent probes will be received. Thus, we increase the cost of *AC*, but the cost will still be less or equal to the *Naive*, which sends all of the probes.

5   Experimental Results

In the experimental study, we demonstrate the efficiency of the algorithms proposed in this paper. For each type of algorithm (i.e., *AP*, *TP* and *AC*) we conducted the following experiments: (1) running the algorithm with an ad-hoc simulator over real nodes traces, which were taken from the Great Duck Island (GDI) project [9]; (2) running the algorithms on Mica motes [12] using the TinyOS operating system [15]. Due to place limitations, we present here a subset of the experiments; the full experiments can be found in the extended paper.

5.1   Framework

Traces from GDI: The algorithms were implemented using an ad-hoc simulator written in Java. All our experiments were held with the adjustment parameters. We plugged the real motes traces taken from the GDI project [2] into the simulator. The traces include a few attributes from each mote, and consist of temperature probes taken about once every 5 minutes. The *Naive* algorithm was simulated using different time intervals between probes. For each time interval, the error and cost (number of probes) were recorded. For the adaptive algorithm, the time interval was used in order to generate $\overline{\Delta}$ and $\underline{\Delta}$, which were division and multiplicity of the *Naive* rate by a constant factor. For example, in most experiments we used $\overline{\Delta}$ as thrice the *Naive* rate, and $\underline{\Delta}$ as a third of the *Naive* rate.

Mica Motes: We implemented the algorithms on Mica2 Motes [12] using the Nesc language and over the TinyOS operating system [15]. Our test bed includes two motes in which one was attached to the base station in order to present the results, and the second ran the *Naive* algorithm and adaptive algorithm in parallel. We chose this setup in order to measure the same values using the same sensing calibration. The executed *Naive* algorithm used very high probing frequency, serving two purposes: 1) obtaining a highly accurate approximation of the measured signal; and 2) enabling the simulation of several *Naive* algorithms with different probing rates, by considering only sub-sequences of probes. A *Naive* algorithm with probing rate that is $1/k$ fraction of the actual probing rate was obtained by taking only one every k probes. We used the light sensing devices, typically measuring time interval around sunset or sunrise.

5.2   Approximate Probing (AP)

As is evident from the results for the GDI case (Fig. 1a), the AP is superior over the *Naive*, especially if a low error rate is desired or a higher cost is allowed. It can be seen that for the same cost we obtained up to 54% improvements in the error (the horizonal line in the figure, which compares between the *Naive* and AP with desired value difference between probes of 5). For the same error we obtained up to 43% reduction in probes (the vertical line in the figure, which compares between the *Naive* and AP with desired value difference between probes of 5). We also computed the errors in various other metrics for the same experiments and got similar results; details can be seen in the extended paper.

For the real Motes experiments (Fig. 1b), we ran the algorithms a few times. For each experiment we obtained a few results of the *Naive* algorithm that are represented as a plot and one AP result which is presented as a point with the same shape and color. We can see that for the same cost we get up to 48% improvement in the error (vertical line in the figure, which compares between the *Naive* and *AP* on the 25th of May). For the same error we get up to 51% improvements in cost (horizontal line in the figure, which compares between the *Naive* and *AP* on the 17th of May).

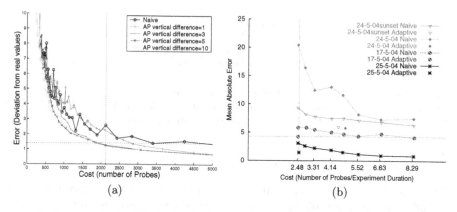

Fig. 1. Error vs. Cost. (a) *AP* algorithm simulated over the measured data of GDI of node 101. The adjustment parameters $\underline{\Delta}$ and $\overline{\Delta}$ are 1/3 and 3 times the naive rate. The displayed plots are for vertical difference of $1°, 3°, 5°, and 10°$ Celsius (b) *AP* algorithm run over Mica Motes. For each day there are a few *Naive* experiments displayed on a single plot, and one adaptive experiment displayed as a dot with the same symbol.

5.3   Threshold Probing (TP)

We tested the TP algorithm by setting a temperature threshold for measurements from the GDI trace data. For node 101 of the GDI, around time $2,030,000$ there is a single instance in which the value climbs over $37°$, staying over this value for some 3900 seconds (reaching even to $39°$). We defined $\tau = 37°$ as the Threshold. The results for the comparisons are shown in Fig. 2. It is evident that TP with first derivative (linear prediction) is quite superior to the *Naive*. Positive values of α and β improve the results, which can reach in some cases up to 90-95% improvements for the same cost.

It is notable that the TP algorithm is much more stable in its results, whereas the *Naive* results are somewhat arbitrary; this is due to the fact that the *Naive* is much more vulnerable to the starting time relative to the threshold crossing time. In contrast, the TP algorithm is adapting to the measured signal and thus converges in a timely fashion to high probing frequency when getting near the threshold.

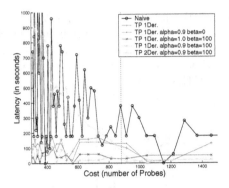

Fig. 2. Latency vs. Cost. TP algorithm simulated over node 101 of GDI. The Threshold was set to 37° Celsius. The adjustment parameters $\underline{\Delta}$ and $\overline{\Delta}$ are 1/3 and 3 times the naive rate, respectively . TP with several adjustment parameters α,β are tested.

5.4   Adaptive Communication (AC)

We ran both the *Naive* communication algorithm and the *AC* algorithms with an underlying *Naive* probing algorithm; that is, the probed signal is taken with equal time intervals. The *Naive* communication algorithm is always sending the probed values while the *AC* algorithm is sending on an as-needed basis only.

For the GDI data (Fig. 3a), we ran simulations with different time intervals between probes, and displayed a plot showing the obtained error vs. the cost measured as the number of probes. We can see that for the same cost we get up to 75% error improvements compared to the *Naive* (see vertical line, which compares between the *Naive* and *AC* with $\varepsilon = 0.5$). For the same error we get up to 52% cost saving compared to the *Naive* (see horizontal line, which compares between the *Naive* and *AC* with $\varepsilon = 1$).

For the Mica Motes experiments (Fig. 3b), we can see that for the same cost we get up to 70% error improvements compared to the *Naive* (see vertical line, which compares between the *Naive* and *AC* on the 29th of May). For the same error we get up to 70% reduction in the number of messages needed for the same error compared to the *Naive* (see horizontal line, which compares between the *Naive* and *AC* on the 28th of May).

6   Related Work

There is rich literature on forecasting and adaptive techniques. We highlight a few adaptive techniques that were considered in various scenarios, and relevant works on sensor networks.

Adaptive Filters theory is a well researched domain with substantial literature (e.g., [4]). The idea is to apply filters on a signal for filtering, smoothing and prediction. For instance, Weiner filters are used over stationary data, and Kalman filters are used for ongoing updated data where the computation is done incrementally. The basic prediction technique in adaptive filters is to find the

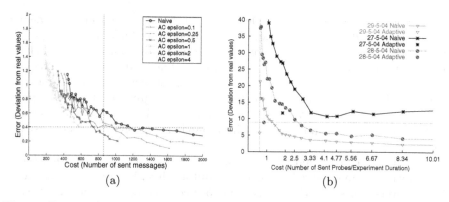

Fig. 3. Error as function of the Cost. (a) *AC* algorithm simulated over the measured data of GDI over node 101. Each plot describes a different ε value. (b) *AC* algorithm run over Mica Motes. For each day there are a few *Naive* experiments and one adaptive experiment

next expected value based on n previous values, assuming the value is changing linearly. Box-Jenkins modeling [1] is a mathematical modeling of time series used for forecasting. It involves identifying an appropriate so-called ARIMA process, fitting it to the data, and then using the fitted model for forecasting. It has three iterative processes: Model selection, Parameter estimation and Model checking. Recent improvements in the process [10] added two more steps of data preparation and forecasting. All those methods are too computational and memory intensive for our limited devices and online computation involved in our goal.

In [13] the authors propose an automatic environmental model construction, but their sensors ("PDA-like devices") are order of magnitude more powerful than the sensors considered in our work (e.g., Mica Motes). They are trying to construct a global model for future forecast whereas we are forming a local model for forecasting the very near future.

In [11] the authors propose to adapt the sampling rate based on an a priory model such as sinusoid. In contrast, our model is changing over time and adapts to the changing measured stream. Thus, we do not assume an a priory model which makes our solution more general and adequate for a larger set of applications. As far as we know, except for this work, existing probing algorithms use constant rate probing frequency. In [3] the authors propose to create a temporal model at the base station. Each sensor in this case sends a message only if the value deviates from this model. In our *AC* algorithm, the sender maintains its own model. Thus, the base station does not need to issue and send a model which is a costly communication step on its own.

The notion of a single node that uses only its own resources for event detection was considered in [7,5]. However, [7] defines an event as a value over some predefined threshold but continued to probe in constant rate. Paper [5] proposes to put a special hardware in designated nodes, which has cost and need a priory knowledge of the topology.

The idea of balancing the tradeoff between accuracy and lifetime was considered independently in [6]. They use this tradeoff in building a synopsis that represents the data for queries of aggregation purposes, while our approach is to use this tradeoff for reducing the number of probing and basic communication. In [14] the authors propose the idea of sending messages on a changing basis; however, this method is proposed in the context of aggregations only, and in case it deviates from previous values. In contrast, AC deals with any communication and sends in case the value deviates from previous prediction and not just from the previous value.

The questions considered in the above works is typically computing the expected value of an observed sequence at a future time. We are mostly interested in the inverse question of when is the expected time to get to a certain value. It is possible to find an approximation to this inverse function using doubling and binary search but this will be rather costly in our context. Note that our adaptive methods should be computed every epoch in runtime manner towards the next epoch; therefore it should be done very cheaply and efficiently.

7   Conclusions

In this paper we present the first study of adaptive probing and adaptive communication for improved utilization of sensor networks. We present basic algorithms that effectively realize the adaptive methods. The potential impact of our approach is demonstrated using experimentations. There are many open questions with regard to adaptive methods in sensor networks. Below we mention a few.

It would be interesting to consider various objective functions for approximation and their corresponding error metrics. We described some basic mathematical models that guide the adaptive algorithms. It would be interesting to further develop mathematical models for measured functions that provide a good approximation for real data and at the same time are manageable with low overhead, given the limited computing and communication resources available in sensor networks.

Prediction of future patterns based on past information is at the heart of the proposed adaptive methods. There may be significant promise in incorporating methods from other disciplines that address related questions. We expect that due to the computational constraints in sensor networks, many known techniques would require considerable adaptation in order to be useful here. Some of them might be derived from fields like Adaptive Filters and Box-Jenkins modeling.

This work focuses on adaptive methods for single sensors, or for pairs of sensors. Applying more advanced adaptive methods for the entire network may be a rich area for research with a potentially significant impact.

Acknowledgments. We thank Christos Faloutsos for helpful pointers to relevant literature in the forecasting domain. Research supported in part by the Israel Science Foundation founded by The Academy of Sciences and Humanities, by the Sackler Fund, and by an Intel fund.

References

1. G. Edward, P. Box, and G. M. Jenkins. *Time Series Analysis: Forecasting and Control.* Prentice Hall PTR, 1994.
2. Habitat monitoring on great duck island. http://www.greatduckisland.net/.
3. S. Goel and T. Imielinski. Prediction-based monitoring in sensor networks: Taking lessons from mpeg. In *ACM Computer Communication Review, Vol. 31, No. 5,* 2001.
4. S. Haykin. *Adaptive Filter Theory.* Prentice Hall, Upper Saddle River, NJ, 3rd edition, 1996.
5. J. M. Hellerstein, W. Hong, S. Madden, and K. Stanek. Beyond average: Toward sophisticated sensing with queries. In *2nd International Workshop on Information Processing in Sensor Networks (IPSN '03),* March 2003.
6. K. Kalpakis, V. Puttagunta, and P. Namjoshi. Accuracy vs. lifetime: Linear sketches for approximate aggregate range queries in sensor networks. available as umbc cs tr-04-04, February 11, 2004.
7. J. Liu, J. Liu, J. Reich, P. Cheung, and F. Zhao. Distributed group management for track initiation and maintenance in target localization applications. In *2nd International Workshop on Information Processing in Sensor Networks,* 2003.
8. S. Madden, M. J. Franklin, J. M. Hellerstein, and W. Hong. Tag: a tiny aggregation service for ad-hoc sensor networks. In *THE MAGAZINE OF USENIX and SAGE April 2003 volume 28 number 2,* page 8, 2003.
9. A. Mainwaring, J. Polastre, R. Szewczyk, D. Culler, and J. Anderson. Wireless sensor networks for habitat monitoring. In *ACM International Workshop on Wireless Sensor Networks and Applications (WSNA'02),* Atlanta, GA, Sept. 2002.
10. S. Makridakis, S. Wheelwright, and R. J. Hyndman. *Forecasting : Methods and Applications.* John Wiley & Sons., 1998.
11. A. D. Marbini and L. E. Sacks. Adaptive sampling mechanisms in sensor networks. In *London Communications Symposium,* 2003.
12. Berkley mica motes.
    http://www.xbow.com/Products/Wireless_Sensor_Networks.htm.
13. S. Papadimitriou, A. Brockwell, and C. Faloutsos. Adaptive, hands-off stream mining. In *29th International Conference on Very Large Data Bases VLDB,* 2003.
14. M. A. Sharaf, J. Beaver, A. Labrinidis, and P. K. Chrysanthis. Tina: a scheme for temporal coherency-aware in-network aggregation. In *3rd ACM international workshop on Data engineering for wireless and mobile access,* pages 69–76, 2003.
15. Tinyos operating system. *http : //webs.cs.berkeley.edu/tos/.*
16. B.-K. Yi, N. D. Sidiropoulos, T. Johnson, H. V. Jagadish, C. Faloutsos, and A. Biliris. Online data mining for co-evolving time sequences. In *16th International Conference on Data Engineering,* page 13. IEEE Computer Society, 2000.

Progress Based Localized Power and Cost Aware Routing Algorithms for Ad Hoc and Sensor Wireless Networks

Johnson Kuruvila, Amiya Nayak, and Ivan Stojmenovic

SITE, University of Ottawa, Ontario K1N 6N5, Canada.
{kuruvila, anayak, ivan}@site.uottawa.ca

Abstract. In this article we propose several new progress based, localized, power and cost aware algorithms for routing in ad hoc wireless networks. Node currently holding the packet will forward it to a neighbor, closer to destination than itself, which minimizes the ratio of power and/or cost to reach that neighbor, and the progress made, measured as the reduction in distance to destination, or projection along the line to destination. In localized algorithms, each node makes routing decisions solely on the basis of location of itself, its neighbors and destination. The new algorithms are based on the notion of proportional progress. The new power and cost localized schemes are conceptually simpler than existing schemes, and have similar or somewhat better performance in our experiments. Our localized schemes are also shown to be competitive with globalized shortest weighted path based schemes.

1 Introduction

Due to its potential applications in various situations such as battlefield, emergency relief, environment monitoring, etc., wireless ad hoc networks [1,2] have recently emerged as a premier research topic.

Stojmenovic and Lin [5] described the first localized power and cost aware routing protocols for ad hoc networks. In this article, we revisit the problem and propose several new such protocols. We compare the performance of the new power and/or cost aware algorithms with these described in [5].

The rest of this article is organized as follows. In Section 2, we briefly present related work in this area. Section 3 presents the *Power Progress, Cost Progress, Power-cost Progress* algorithms and iterative variants. It also presents the *Projection Power Progress, Projection Cost Progress, Projection Power-cost Progress* algorithms and the iterative versions. Results of our simulation and performance study of these algorithms are given in Section 4. In Section 5, we provide concluding remarks and list some open problems in this area.

I. Nikolaidis et al. (Eds.): ADHOC-NOW 2004, LNCS 3158, pp. 294–299, 2004.

2   Related Work

There exist a vast amount of literature devoted to localized routing in ad hoc networks. A survey of power optimization techniques for routing protocols in wireless networks can be found in [9].

Existing Power aware Routing algorithms: Rodoplu and Meng [6] proposed a general model in which the power consumption between two nodes at distance d is $u(d) = d^\alpha + c$ for some constants α ($2 < \alpha < 5$) and c. In their experiments, they adopted the model with $u(d) = d^4 + 2 * 10^8$, which we refer as the *RM Model*.

Heizelman et al. [8] used signal attenuation to design an energy efficient routing protocol for wireless micro-sensor networks, where destination is fixed and known to all nodes. We refer to their model as the *HCB Model*, where power needed for transmission and reception at distance d is $u(d) = d^2 + 2 * 1000$.

If nodes have information about position and activity of all other nodes, then optimal power saving algorithm, that will minimize the total energy per packet, can be obtained by applying Dijkstra's single source shortest weighted path algorithm, where each edge has weight $u(d) = ad^\alpha + c$, where d is the length of the edge. We refer to it as *SP Power* algorithm, which is used to compare the performance of our new localized algorithms.

Localized power aware algorithms have been proposed by Stojmenovic and Lin [5]. In Section 4 of this article, we refer to their algorithm as the *Power* algorithm. The authors in [5] also proposed the *NC* (Nearest Closer) algorithm, where node, currently holding the message, will forward it to the nearest neighbor. Only nodes closer to destination than the current node are considered.

Finn [4] proposed localized *greedy* scheme, where node, currently holding the message, will forward it to the neighbor that is closest to destination. Only nodes closer to destination than the current node are considered.

Existing Cost and Power-cost Routing Algorithms: Singh et al. [7] proposed to use a function $f(A)$ to denote node A's reluctance to forward packets, and to choose a path that minimizes the sum of $f(A)$ for nodes on the path. As a particular choice for f, [7] proposes $f(A) = 1/g(A)$ where $g(A)$ denotes the remaining lifetime ($g(A)$ is normalized to be in the interval $[0, 1]$). Thus reluctance grows significantly when lifetime approaches 0. The reluctance $f(A)$ is used as a weight in a shortest weighted path algorithm. In our experiments, we refer to this optimal algorithm as the *SP Cost* algorithm in Section 4.

A cost aware localized algorithm, assuming constant power for each transmission, was proposed in [5]. The algorithm *Cost-ii* [5], is used to compare the performance of our new cost based algorithms. The authors also proposed several ways to combine power and cost metrics into a single power-cost metric, based on the product and sum of two metrics, respectively. We use *PowerCost2* [5] algorithm in our comparison study.

An efficient globalized power-cost route can be computed by applying Dijkstra's shortest path algorithm where the node's *powercost* value is transferred to the edge leading to the node. In Section 4, we refer to this optimal algorithm as the *SP Power*Cost* algorithm for our comparisons.

3   New Power and Cost Aware Algorithms

Let the current node holding the packet be S, destination be D, and A be a neighbor of S. Let $|SA| = r$, $|SD| = d$, and $|AD| = x$, with $x < d$. The new *Power Progress* algorithm is based on the notion of proportional progress. Let us measure the proportional progress as the power used to make a portion of the progress. The power needed to send from S to A is $r^\alpha + c$. The portion of progress made with it is $(d-x)$. With similar advance continuing, there would be $d/(d-x)$ such steps, and the total cost would be $(r^\alpha + c)d/(d-x)$. Therefore the neighbor A that would minimize $(r^\alpha + c)/(d - x)$ will be selected for forwarding the message. This means that the selected neighbor minimizes the power spent per unit of progress made, in terms of getting closer to the destination. Power metrics can be similarly replaced by a cost or power-cost metric to define cost or power-cost per unit of progress made. This leads to the *Cost Progress* and *Power-Cost Progress* algorithms that select forwarding neighbors that minimize $f(A)/(d - x)$, and $f(A)(r^\alpha + c)/(d - x)$, respectively.

The *Projection Power Progress, Projection Cost Progress* and *Projection Power-Cost Progress* algorithms differ from the the power progress algorithms in terms of the measure of proportional progress made. Node S will forward to a neighbor A that minimizes $(r^\alpha + c)/(SD \cdot SA)$, where $SD \cdot SA$ is the dot product of two vectors.

The *Iterative Power Progress* algorithm is an improvement of the *Power Progress* algorithm. It can be described as follows. As in *Power Progress*, a node S currently holding message will first find a neighbor A that minimizes $(r^\alpha + c)/(d - x)$. Then, an intermediate node B (closer to destination than S, if exists) is found (that is neighbor to both S and A) which satisfies $(power(SB) + power(BA)) < power(|SA|)$ and has the minimum $(power(SB) + power(BA))$ measure. If found, such node B replaces A as selected neighbor, and the search for even better intermediate node B repeats. This process is iteratively repeated until no improvement is possible. Node S will forward the message to the selected neighbor A which than applies again the same scheme for its own forwarding.

Iterative Power-Cost Progress algorithm can be similarly defined. There is no iterative version defined for the *Cost Progress* algorithm, since the overall cost increases by adding intermediate nodes on a path.

The iterative projection progress scheme is very similar to the *Iterative Power Progress* scheme, except that the first candidate node A is found using the projection progress method, (minimizes $(r^\alpha + c)/(SD \cdot SA)$), instead of power progress scheme.

4   Experimental Results

Evaluation of Power Aware Algorithms. First, we studied the effect of different squares (placement areas) on various algorithms. We used squares of size $50, 300, 500, 5000$ and 50000 (for the *HCB Model*, $\alpha = 2, c = 2000$). However, for the *RM Model* ($\alpha = 4, c = 2 * 10^8$), we used square sizes $500, 1000, 5000, 10000$

and 50000 to derive various relative performances with respect to selected α and c. Randomly generated connected graphs of density $d = 20$ were used with number of nodes fixed at $n = 250$. The network density d is defined as the average number of neighbors per each node. Dijkstra's shortest path scheme was used to test network connectivity, and only connected graph were used in measurements. We used dense graphs because the success rate of all the algorithms then remained close to 100%, therefore the success rate was not a variable in these measurements. For sparser graphs, success rates of new and existing power aware localized algorithms was similar.

Algorithm	n =: 250 C = 2000, Alpha = 2					n =: 250 C = 2*10^8, Alpha = 4					n =: 250 C = 0	
	Square Sizes					Square Sizes					Alpha Values	
	50	300	500	5000	50000	500	1000	5000	10000	50000	2	4
SP Power	1	1	1	1	1	1	1	1	1	1	1	1
Power Progres	1.02	1.03	1.03	1.16	1.18	1.03	1.05	1.79	1.85	1.82	1.17	1.84
I Power Progress	1.02	1.03	1.03	1.12	1.13	1.03	1.04	1.65	1.68	1.69	1.13	1.69
Proj. Progress	1.05	1.03	1.03	1.15	1.16	1.06	1.05	1.71	1.75	1.76	1.16	1.76
I Proj. Progress	1.05	1.03	1.03	1.12	1.13	1.06	1.05	1.63	1.67	1.69	1.13	1.68
Power	1.02	1.03	1.03	1.21	1.28	1.03	1.05	1.71	1.74	1.72	1.27	1.75
NC	3.96	2.54	1.80	1.26	1.28	3.68	2.07	1.72	1.75	1.72	1.27	1.75
SP	1.11	1.03	1.04	1.80	1.85	1.10	1.22	7.67	7.98	7.81	1.84	7.92
Greedy	1.02	1.05	1.13	2.10	2.16	1.03	1.43	10.26	10.76	10.65	2.17	10.74
	A: HCB Model					B: RM Model					C: RM/HCB Models	

Fig. 1. Power dilation **A:** HCB Model **B:** RM Model **C:** HCB/RM Models with c=0

The performance metric we have used to compare the performance of various algorithms is the total amount of power required to successfully route a message from the source to the destination (averaged over a huge number of randomly selected connected unit graphs and source destination pairs in each graph). We define *power dilation* as the ratio of the power requirement of the specific algorithm to that of *SP Power*, which is optimal performing algorithm. We simulated, and compared the performance of *power progress, iterative power progress, projection power progress* and *iterative projection power progress* with that of the *power* and *NC* algorithms from [5], greedy scheme [4] and global shortest path algorithms (with and without power weights).

Figure 1:A&B present power dilations for various algorithms, with different square sizes, for the *HCB Model* and *RM Model* respectively. For $\alpha = 2, c = 0$ and $\alpha = 4, c = 0$ values, the power dilations remained stable for all square sizes. Figure 1:C shows the average power dilations for $\alpha = 2\&4$, $c = 0$.

We can observe that the performance of the newly proposed localized routing algorithms are better than the performance of existing power aware scheme [5] for $\alpha = 2, c = 2000$, and comparable for $\alpha = 4, c = 2 * 10^8$. The iterative power progress method performed somewhat better for $\alpha = 4, c = 2 * 10^8$ than power scheme [5]. Overall, all proposed schemes have very encouraging performance with respect to the optimal global *SP Power* algorithm, which is a significant achievement. The newly proposed methods reduced the excess power from 28% to 13% for the largest square size considered for $\alpha = 2, c = 2000$. The excess power for $\alpha = 4, c = 2 * 10^8$ is reduced to 69% for the largest square size considered. In all cases, the excess power rate has apparently slow increase with further square size increase. Similar comparison is obtained for $c = 0$. Our algorithms outperform the greedy and *SP* methods in all cases. For $c > 0$, the *NC* method is very inefficient at low square values, but comparable with other localized algorithms at high square values. For $c = 0$, the *Power* [5] and *NC* methods are identical.

Evaluation of Cost/Power-cost Algorithms. The performance metric used to compare the algorithms is the number of routing tasks that can be successfully performed before the power of any one of the nodes in the network reaches zero. We used squares of size $50, 300, 500$, and 5000 for the *HCB Model*. However, for the *RM Model*, we used square sizes $500, 1000, 5000, 10000$ and 50000 to derive various relative performances with respect to selected α and c. Graphs were generated as in the previous case. Each node is assigned with an initial energy value, constant to all nodes in the network. A random source destination pair is selected and the route is calculated from the source to the destination using each algorithm. Once the path is computed, actual power required is charged to the nodes that are part of the calculated path. This process is repeated for the same graph, until a node has insufficient power to perform assigned task. The reported results are averaged over 10 graphs.

Algorithm	n = 250 C = 2000, Alpha = 2 Square Sizes				n = 250 C = 2*10^8, Alpha = 4 Square Sizes					n = 250 C = 0 Alpha Values	
	50	300	500	5000	500	1000	5000	10000	50000	2	4
SP Power	0.348	0.278	0.2687	0.399	0.305	0.321	0.300	0.298	0.250	0.262	0.251
SP Cost	0.999	0.890	0.728	0.587	0.979	0.202	0.047	0.0391	0.041	0.454	0.056
Cost Progress	0.629	0.619	0.509	0.443	0.710	0.131	0.035	0.038	0.041	0.362	0.050
Proj. Cost Progress	0.612	0.593	0.520	0.469	0.674	0.179	0.032	0.029	0.040	0.341	0.049
SP Power*Cost	1	1	1	1	1	1	1	1	1	1	1
Power*Cost Progress	0.652	0.638	0.568	0.596	0.726	0.414	0.142	0.119	0.166	0.364	0.160
I Power*Cost Progress	0.652	0.638	0.568	0.596	0.726	0.414	0.142	0.119	0.166	0.364	0.160
Proj Power*Cost Progress	0.610	0.649	0.624	0.573	0.708	0.447	0.142	0.129	0.165	0.372	0.180
I Proj. Power*Cost Progress	0.685	0.663	0.629	0.573	0.818	0.462	0.142	0.129	0.165	0.376	0.180
Cost-ii	0.868	0.858	0.721	0.672	0.944	0.139	0.0437	0.046	0.047	0.473	0.081
Power*Cost2	0.889	0.862	0.789	0.783	0.969	0.558	0.310	0.156	0.197	0.322	0.180
	A: HCB Model				B: RM Model					C: HCB/RM Models	

Fig. 2. Iteration Dilation A: HCB Model B: RM Model C: HCB/RM Model with c=0

We compared the performances of global algorithms *SP-Power*, *SP-Cost*, *SP-Power*Cost*, newly proposed local algorithms *Cost Progress*, *Power-Cost Progress*, *Iterative Power-Cost Progress*, *Projection Cost Progress*, *Projection Power-cost Progress*, *Iterative Projection Power-cost Progress*, *Cost-ii* [5] and *PowerCost2* [5]. We define the *iteration* dilation as the ratio of the number of iterations of the algorithm to that of *SP Power*Cost*. Figure 2:A&B shows the results (expressed as iteration dilation ratios) for the *HCB Model* and the *RM*

Model respectively. For $c = 0$, the iteration dilation ratios remained relatively same for different square sizes, for different α values. Fig 2:C shows the average iteration dilation ratios for the different square sizes. The newly proposed cost and power-cost aware localized algorithms are competitive with several schemes already proposed in [5].

5 Conclusion

This paper described several localized routing algorithms that attempt to minimize the total energy per packet and/or life time of each node. The localized nature of the protocols avoids the energy expenditure and communication overhead needed to build and maintain the global topological information. Our simulation results show that the performance of our localized algorithms is close to the performance of the shortest path algorithms, which require global knowledge.

It appears that we have achieved an optimal design for power aware localized algorithms. However, our experiments do not give an ultimate answer on the selection of approach that would give the most prolonged life to each node in the network. There is a need to consider further localized cost and power-cost aware routing schemes. In our future work, we also plan to address power and cost aware localized routing with a realistic physical layer.

References

1. S. Basagni, M. Conti, S. Giordano, I. Stojmenovic (eds.), "Mobile Ad Hoc Networking," IEEE Press/Wiley, 2004, to appear.
2. G. Giordano, "Mobile Ad Hoc Networks," in *Handbook of Wireless Networks and Mobile Computing*, I. Stojmenovic (ed.), Wiley, 2002, pp. 325-346.
3. I. Stojmenovic, "Location updates for efficient routing in ad hoc networks," in *Handbook of Wireless Networks and Mobile Computing*, I. Stojmenovic (ed.), Wiley, 2002, pp. 451-472.
4. G.G. Finn, "Routing and addressing in large metropolitan-scale internetworks," *ISI Research Report*, ISU/RR-87-180, March 1987.
5. I. Stojmenovic, Xu Lin, "Power aware localized routing in ad hoc networks," *IEEE Transactions on Parallel and Distributed Systems*, Vol. 12, No. 10, October 2001, pp. 1023-1032.
6. V. Rodoplu, T. H Meng, "Minimum energy mobile wireless networks," *IEEE Journal on Selected Areas in Communications 1999* 17(8); 1333-1344.
7. S. Singh, M. Woo, C.S Raghavendra, "Power-aware routing in mobile ad hoc networks," *Proceedings MOBICOM, 1988*; 181-190
8. W.R Heinzelman, A. Chandrassekaran, H. Balakrishnan "Energy efficient routing protocols for wireless microsensor networks," *Proceedings Hawaii International Conference on System Sciences*, January 2000.
9. S. Lindsey, K. Sivalingam, C.S Rahgavendra "Power optimization in routing protocols for wireless and mobile networks," *Wireless Networks and Mobile Computing Handbook*, Stojmenovic I (ed.), John Wiley & Sons: 2002; 407-424.

Application of Eigenspace Analysis Techniques to Ad-Hoc Networks

Sumeeth Nagaraj, Stephen Bates, and Christian Schlegel

High Capacity Digital Communications Lab.
University of Alberta, Edmonton, Canada.
{snagaraj, sbates, schlegel}@ece.ualberta.ca

Abstract. In this paper we apply eigenspace analysis techniques to Ad-Hoc networks. We develop the concept of the dominant eigenvector and show that the values in this vector are related to a node's connectivity. We present some illustrative examples and show how information in the dominant eigenvector can be used to improve the fairness of a Media Access Control (MAC) protocol commonly used in wireless systems.

1 Introduction

An Ad-Hoc network is a collection of nodes that communicate with each other over multiple hops with no backbone or infrastructure. Each node acts as a router for other nodes by forwarding packets to their destination.

One issue for Ad-Hoc networks is that the total network-capacity grows only as $O(\sqrt{n})$, where n is the number of nodes in the network [1]. This low network-capacity scaling is due to the fact that nodes must limit their transmission range since each transmission blocks all other nodes within the transmit radius from transmitting themselves. However a converse criteria is that the transmission range must be large enough to ensure the network is strongly connected. The resultant is that on average each communicated packet must be routed by several intermediate nodes before it reaches its destination. Therefore routing and routing algorithms are of critical importance in Ad-Hoc networks. In addition, nodes with high connectivity can be required to route many more packets than they self-generate. This is a fairness issue and needs to be addressed otherwise nodes in this position may decide not to participate in the Ad-Hoc network.

Chapter 2 of [2] deals with counting walks and graph spectra. Theorem 2.2.4 of [2] gives relation between number of walks of length k and eigenvectors. In this paper we develop the concept of the dominant eigenvalue and its associated eigenvector, which contains information about the connectedness of each node in the network. In doing so, we take the nodes perspective of the network over k hops. We can use this information in a number of ways related to media access, routing and scheduling. As an example we show how the dominant eigenvector values can be used to increase the fairness of a random Medium Access Control (MAC) scheme.

I. Nikolaidis et al. (Eds.): ADHOC-NOW 2004, LNCS 3158, pp. 300–305, 2004.

2   Eigenspace Analysis of Ad-Hoc Networks

This section of the paper is divided into two parts. We begin in Section 2.1 by summarizing the main results of the paper and discussing the implications for Ad-Hoc networks. In Section 2.2 we present the formal theorems and proofs associated with our results.

2.1   Discussion and Results Summary

The application of eigenspace analysis techniques to the adjacency matrix $\mathbf{A}$ of the graph (V, E) yields the following interesting results.

1. If the graph (V, E) is not strongly connected then $\mathbf{A}$ is reducible. In this case $\mathbf{A}$ can be split into m irreducible adjacency matrices which comply with Results 2 and 3. The individual irreducible matrices can be thought of as m strongly connected subnets within the graph.
2. If the graph (V, E) is strongly connected there exists only one dominant, positive, eigenvalue of $\mathbf{A}$ with a corresponding positive eigenvector.
3. If the graph (V, E) is strongly connected the ith value of the dominant eigenvector gives the relative connectedness of the ith node in the network.

The first result prove that any non-strongly connected graph can be decomposed into several sub-graphs, each of which is strongly connected. This is equivalent to stating that the adjacency matrix corresponding to a non-strongly connected graph is reducible and can be permuted into a number of smaller irreducible matrices. The number of smaller matrices is equal to the number of isolated subnets in the original network. We can use the these results to test for a strongly connected network by applying eigenspace decomposition and noting that if $\mathbf{A}$ is irreducible then $\mathbf{A}^{n-1}$ is positive. However the complexity of this test does not compare favorable with other techniques [3] and is merely a by product of our analysis.

The second and third results show that we can determine if a node is well connected to the rest of the network. Those nodes with a lot of connectivity will have large dominant eigenvector values. Those nodes which are isolated have a value of zero with respect to the eigenvector element for that node.

2.2   Theoretical Analysis

Definition 1. *A path $\rho_{ij}^{(k)}(\mathbf{A})$ is a set of k elements from an adjacency matrix $\mathbf{A}$ such that*

$$\prod_{p,q:a_{pq}\in\rho_{ij}^{(k)}(\mathbf{A})} a_{pq} = 1$$

and the set of $2k$ indices of $\rho_{ij}^{(k)}(\mathbf{A})$ can be labeled μ_0 to μ_{2k-1} such that

$$\begin{aligned}
\mu_0 &= i, \\
\mu_{2k-1} &= j, \\
\mu_t &\neq i,j \ \forall t \notin \{0, 2k-1\}, \\
\mu_{t+1} &= \mu_t \ \forall t \ odd.
\end{aligned}$$

The set $\Omega_{ij}^{(k)}(\mathbf{A})$ is the set of all unique $\rho_{ij}^{(k)}(\mathbf{A})$. That is the set of all unique paths from i to j of length k. Note that by definition a path $\rho_{ij}^{(k)}(\mathbf{A})$ cannot contain rings on either i or j but can contains rings on any other node.

Theorem 1. *Any irreducible non-negative matrix $\mathbf{A}$ has a single positive eigenvalue which is also the dominant eigenvalue. The corresponding eigenvector is positive.*

Proof. This proof is part of the Perron-Frobenius Theorem (see for example [4] and also [2]).

Theorem 2. *Any reducible non-negative matrix $\mathbf{A}$ has m $(2 \leq m \leq n)$ positive eigenvalues and associated non-negative eigenvectors where m is the number of isolated strongly connected subnets in the network. If the eigenvector $\mathbf{x}_i$ is associated with subnet i then it will have positive elements at all positions corresponding to nodes in that subnet and zeros elsewhere.*

Proof. Since $\mathbf{A}$ is reducible, there exists a permutation matrix $\mathbf{P}$ such that

$$\mathbf{P}^{\mathbf{T}}\mathbf{AP} = \begin{bmatrix} \mathbf{A}_1 & \cdots\cdots & 0 \\ & \mathbf{A}_2 & \cdots\cdots \\ \vdots & \ddots & \vdots \\ & & \cdots\ddots \\ 0 & \cdots & 0 \ \ \mathbf{A}_m \end{bmatrix}$$

where $\mathbf{A}_1, \mathbf{A}_2, \ldots, \mathbf{A}_m$ are irreducible matrices obtained simply by relabeling the nodes. Since $\mathbf{A}$ is reducible then $m \geq 2$. We can then apply Theorem 1 to each block matrix which proves the existence of m positive eigenvalues and positive eigenvectors corresponding to each of the blocks $\mathbf{A}_1, \mathbf{A}_2, \ldots, \mathbf{A}_m$. However since $\mathbf{P}^{\mathbf{T}}\mathbf{AP}$ is a block diagonal its eigenvalues equal the eigenvalues of the individual blocks and the eigenvalues are the same with zeros inserted into the rows outwith that block's domain. Since these eigenvectors were positive and we added zeros they must now be non-negative. □

Theorem 3. *Element i of the dominant eigenvector $\mathbf{x}_1$ of an irreducible non-negative square matrix $\mathbf{A}$ has a value that is proportional to the connectivity of node i in the network.*

Proof. The elements of $\mathbf{A}^k$ can be written as

$$\begin{aligned} a_{ij}^{(k)} &= \sum_t a_{it}^{(k-1)} a_{tj}, \\ &= \sum_t \sum_s a_{is}^{(k-2)} a_{st} a_{tj}, \\ &= \sum_z \cdots \sum_t \sum_s a_{iz} \cdots a_{st} a_{tj}. \end{aligned}$$

Where $t, s, z \in \{1, 2, \cdots, n\}$. Since $a_{ij} \in \{0, 1\}$ the products inside the summations can only equal zero or one. If they equals one then the terms in that product form an instance of a path $\rho_{ij}^{(k)}(\mathbf{A})$ or an instance of $m \geq 2$ paths

$$\{\rho_{ii}^{(\nu_0)}, \rho_{ii}^{(\nu_1)}, \cdots, \rho_{ii}^{(\nu_{\varsigma-1})}, \rho_{ij}^{(\nu_\varsigma)}, \rho_{jj}^{(\nu_{\varsigma+1})}, \rho_{jj}^{(\nu_{m-1})}, \cdots, \rho_{jj}^{(\nu_{m-1})}\} \tag{1}$$

such that $\sum_i \nu_i = k$. Since the summations are exhaustive and unique over k, $a_{ij}^{(k)}$ must contain $|\Omega_{ij}^{(k)}(\mathbf{A})|$. However $a_{ij}^{(k)}$ is not exhaustive over $\{1, 2, \cdots, k-1\}$ because by definition $|\Omega_{jj}^{(1)}(\mathbf{A})| = 0$, therefore $a_{ij}^{(k)}$ cannot contain $|\Omega_{ij}^{(k-1)}(\mathbf{A})|$. In addition it is obvious from (1) that the summation is not unique over $\{1, 2, \cdots, k-1\}$. Therefore we can only express the relationship as a bound

$$a_{ij}^{(k)} \geq \sum_{k' \in \sigma_{ij}^{(k)}(\mathbf{A})} |\Omega_{ij}^{(k')}(\mathbf{A})|. \tag{2}$$

Where $\sigma_{ij}^{(k)}(\mathbf{A})$ is the set $\{\mu : |\Omega_{tt}^{(k-\mu)}(\mathbf{A})| > 0$ where $t \in \{i, j\}\}$. Note from above that that the set $\sigma_{ij}^{(k)}(\mathbf{A})$ can never contain $k - 1$.

From the eigenvalue decomposition of $\mathbf{A}$

$$a_{ij}^{(k)} = \lambda_1^k \mathbf{x}_{1,i} \mathbf{x}_{1,j} + \lambda_1^k \sum_{t \neq 1} \left(\frac{\lambda_t}{\lambda_1}\right)^k \mathbf{x}_{t,i} \mathbf{x}_{t,j}$$
$$\to \lambda_1^k \mathbf{x}_{1,i} \mathbf{x}_{1,j} \text{ as } k \to \infty.$$

Summing over $j \neq i$ gives an indication of the connectivity of node i to the entire network, we denote this $\Gamma_i^{(k)}(\mathbf{A})$.

$$\Gamma_i^{(k)}(\mathbf{A}) = \sum_{j \neq i} a_{ij}^{(k)} = \lambda_1^k \mathbf{x}_{1,i} \sum_{j \neq i} \mathbf{x}_{1,j} + \lambda_1^k \sum_{t \neq 1} \left(\frac{\lambda_t}{\lambda_1}\right)^k \mathbf{x}_{t,i} \sum_{j \neq i} \mathbf{x}_{t,j}. \tag{3}$$

From (3) it is obvious that $\Gamma_i^{(k)}(\mathbf{A})$ varies linearly with $\mathbf{x}_{1,i}$. We can also see that for any given positive dominant eigenvector with values $\mathbf{x}_{1,1} \geq \mathbf{x}_{1,2} \cdots \geq \mathbf{x}_{1,n}$ that $\Gamma_1^{(k)}(\mathbf{A}) \geq \Gamma_2^{(k)}(\mathbf{A}) \cdots \geq \Gamma_n^{(k)}(\mathbf{A})$. Also note the asymptotic result

$$\Gamma_i^{(k)}(\mathbf{A}) \to \lambda_1^k \mathbf{x}_{1,i} \sum_{j \neq i} \mathbf{x}_{1,j} \text{ as } k \to \infty.$$

We conclude the proof by noting from (2) and (3) that

$$\Gamma_i^{(k)}(\mathbf{A}) \geq \sum_{j \neq i} \sum_{k' \in \sigma_{ij}^{(k)}(\mathbf{A})} |\Omega_{ij}^{(k')}(\mathbf{A})|. \tag{4}$$

$\square$

Theorem 3 is not as powerful as we might wish since (2) gives a lower bound and we have not yet developed any theorems determining the tightness of this bound. It is clear from (1) that $a_{ij}^{(k)}$ includes paths with cycles. These cycles can only be on i and j if they occur at the beginning or end of the path respectively.

The authors are currently attempting to improve upon this theorem by noting the behavior of the cycles in (1) and removing them to increase the tightness of (2). The simulations in Section 3 suggest that this bound is already tight enough to make the dominant eigenvector values useful.

3   Application: Fairness in IEEE 802.11

Several suggestions have been made to improve the fairness of the IEEE 802.11 standards. A distributed example was given by Vaidya et al in [5]. Their improved technique was termed the distributed fair scheduling (DFS) scheme. DFS assigns bandwidths for packet transmissions among various nodes based on some weights. However, the fundamental question of choosing these weights was not addressed. We attempt to correct this by using the same algorithm but assigning the weights determined from the dominant eigenvector.

Simulation Model. The network simulation tool NS-2 [6] was used to perform the network simulations. When the eigenspace techniques were applied to the network in Figure 1 the dominant eigenvector was found to be

$$\mathbf{x}_1^T = \begin{pmatrix} 0.576 \ 0.198 \ 0.377 \ 0.198 \ 0.377 \ 0.520 \ 0.179 \end{pmatrix}.$$

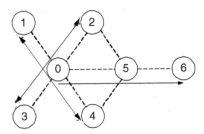

Fig. 1. The contrived network used in the NS-2 simulations to investigate increasing the fairness of the 802.11 MAC protocol. Connectivity is given by dashed lines, traffic flow is given by solid lines with arrows.

We define the fairness index to be the standard deviation of the throughput distribution. The standard deviation indicates how much, on average, each of the per node throughput deviates from the mean throughput and can be used as a fairness index.

Figure 2a shows that the throughput of the nodes with biased backoff and without. The results for biased backoff have a much smaller standard deviation and this indicates that the biased case a fairer scheme. This result holds for both low loads and high loads. The high loads case is the one in which unfairness tends to manifest itself since a large number of packets are competing for finite transmission resources.

Figure 2b shows the ratio of the self-generated throughput obtained by node 0 over the number of packets forwarded by it. It can be seen that without the backoff biased, node 0 spends most of the time forwarding packets and hence gets a fairly low throughput itself. However with biasing based on the eigenvector values node 0 has nearly a constant ratio of routed traffic to self-generated traffic over different traffic loads.

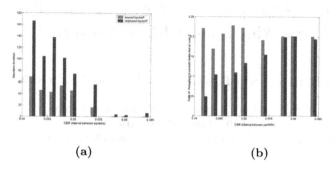

(a)   (b)

Fig. 2. (a) The fairness of the throughput for all the flows versus load. The fairness metric used is the standard deviation of the flows.(b) Ratio of the routed frames versus self-generated frames successfully passed through (or from) node 0.

4 Conclusion and Future Work

In this paper we have introduced the concept of eigenspace Analysis of Ad-Hoc networks. We have extended the spectral analysis of non-negative matrices to include the concept of the dominant eigenvector and shown how the values in this vector related to the connectedness of the nodes. We have also shown how the dominant eigenvector values can be used to increase the fairness in a Medium Access Control scheme commonly used in wireless communication systems.

This work presents many opportunities for future work. A main focus of ongoing work is considering the impact of both capacities and asymmetric links in the analysis. Since the capacity of the link may depend on some underlying physical model we can incorporate this into our adjacency matrix. Now terms are not longer only zeros and ones and in the case of asymmetry the adjacency matrix is no longer symmetric. Once capacities are included this work becomes more applicable to routing since decisions can be made between two or more paths between any source and destination nodes.

References

1. P. Gupta and P. R. Kumar, "The capacity of wireless networks," *IEEE Trans Information Theory*, vol. 46, pp. 388–404, Mar. 2000.
2. D. Cvetkovic, P. Rowlinson, and S. Simic, *Eigenspaces of Graphs*, 1st ed., G. Rota, Ed. Cambridge, UK: Cambridge University Press, 1997.
3. R. E. Tarhan, "Depth-first search and linear graph algorithms," *SIAM J. Comput.*, vol. 1, pp. 146–160, 1972.
4. P. Lancaster, *Theory of Matrices*. New York: Academic Press, 1985.
5. N. H. Vaidya, P. Bahl, and S. Gupta, "Distributed fair scheduling in a wireless LAN," in *Mobile Computing and Networking*, 2000, pp. 167–178.
6. S. McCanne and S. Floyd. Ns (network simulator).

A Rate-Adaptive MAC Protocol for Low-Power Ultra-Wide Band Ad-Hoc Networks*

Ruben Merz, Jean-Yves Le Boudec, Jörg Widmer, and Božidar Radunović

EPFL, School of Computer and Communication Sciences
CH-1015 Lausanne, Switzerland
{ruben.merz,jean-yves.leboudec,joerg.widmer,bozidar.radunovic}@epfl.ch

Abstract. Recent theoretical results show that it is optimal to allow interfering sources to transmit simultaneously, as long as they are outside a well-defined exclusion region around a destination, and to adapt the rate to interference. In contrast, interference from inside the exclusion region needs to be controlled. Based on these theoretical findings, we design a fully distributed rate-adaptive MAC protocol for ultra-wide band (UWB) where sources constantly adapt their channel code (and thus their rate) to the level of interference experienced at the destination. To mitigate the interference of sources inside the exclusion region, we propose a specific demodulation scheme that cancels most of the interfering energy. Through simulation we show that we achieve a significant increase in network throughput compared to traditional MAC proposals.

1 Introduction

Emerging pervasive networks assume the deployment of large numbers of wireless nodes embedded in everyday life objects. For such networks to become accepted, it is important that the level of radiated energy per node be kept very small; otherwise environmental and health concerns will surface. Ultra-wide band (UWB) is a radio technology for wireless networks, which has the potential to satisfy this requirement. The radiated power per node depends on technological choices; it is of the order of 0.1 mW to less than 1 μW per sender. We are interested in *very low power* UWB, by which we mean that the radiated energy per node does not exceed 1μW ($= -30$dBm). With currently planned technology, it is possible with such very low power to achieve rates of 1 to 18 Mb/s per source at distances on the order of tens of meters. These rate values are reduced when several nearby UWB sources transmit concurrently. Our protocol avoids much of the rate reduction through a joint design of MAC and physical layer.

It was shown in [1] that the optimal wide-band signaling consists of sending short infrequent pulses. Consequently, our physical layer is based on the widely used proposal of [2]. It is a multiple access physical layer using Time-Hopping Sequences (THS) with a chip time $T_c = 0.2$ ns (which corresponds roughly to 5 GHz) and Pulse Repetition Period $PRP = 280$. Hence, the average radiated power $P_{rad} = P_{pulse}/(PRP \cdot T_c) = 1 \mu$W

* The work presented in this paper was supported (in part) by the National Competence Center in Research on Mobile Information and Communication Systems (NCCR-MICS), a center supported by the Swiss National Science Foundation under grant number 5005-67322

where $P_{pulse} = 0.28$ mW [3].[1] Additionally, we use the so-called rate-compatible punctured convolutional (RCPC) channel codes [4,5] which provide a variable encoding rate. In particular we use the RCPC codes from [5].

Existing MAC protocols [6,7,8,9,10] are either based on mutual exclusion (no other communication is possible within the same collision domain) or on a combination of power control and mutual exclusion. All of these proposals either have a fixed rate or allow the users to choose between a very small number of fixed rates. A largely unexploited dimension is to let the rate vary with the level of interference. A mathematical analysis of an optimal MAC design including exclusion, power control, and rate adaptation is given in [11]. It is proven that the optimal MAC layer should not use power control but should send at full power whenever it sends. Furthermore, it is optimal to allow interfering sources to transmit simultaneously, as long as they are outside a well-defined *exclusion region* [11] around a destination. In contrast, interference from inside the exclusion region should be combatted. We base our MAC layer design on these findings.

Instead of enforcing exclusion within the exclusion region (a difficult problem), we propose a different form of interference management called *interference mitigation*. At a receiver, interference is most harmful when pulses from a close-by interferer collide with those of the sender. Even though the probability of collisions is fairly low (below 1% with PRP = 280 and one interferer) collisions with strong interferers do cause a significant rate decrease. Inspired by the work of [12], we use a *threshold* demodulator at the receivers that detects when the received energy is larger than some threshold (i.e. when high energy pulses from one or more strong interferers collide with pulses from a source). In such a case, the chip is skipped and an erasure is declared. The loss incurred by those erasures can mostly be recovered by our channel codes and therefore translates into a small reduction of the rate. In addition, this technique reduces the size of the exclusion region to a negligible value.

Hence, what remains is (1) adapt the rate to the varying channel conditions and (2) enforce exclusion between sources that simultaneously send to the same destination. This is solved by means of *dynamic channel coding* and a *private MAC* protocol (DCC-MAC) respectively. Our design moves the complexity of the MAC protocol away from global exclusion between competing sources (a difficult problem) to the combination of channel coding (a private affair between a source and a destination) and a collection of independent private MAC protocol instances (one instance per destination). Problems like hidden or exposed nodes naturally disappear. Simulation results (Section 3) show a significant increase in throughput compared to traditional MAC protocol designs.

2   Joint PHY/MAC Protocol for UWB

2.1   Dynamic Channel Coding and Incremental Redundancy

To make the best use of the channel, the rate needs to be constantly adapted to the highest rate that still allows successful reception of the data packet at the receiver.

---

[1] Note that the maximum achievable rate is still equal to $\frac{1}{T_c \cdot PRP} = 18$ Mb/s.

In our case, a variable encoding rate is achieved by puncturing [4], where a high-rate code is created from a low-rate code by removing coded bits from the lowest rate block of coded bits. Let $R_0 = 1 > R_1 > R_2 > \ldots > R_N$ be the set of rates offered by the channel code. The rate compatibility feature [4] implies that a block of coded bits with rate R_n is a subset of the block of coded bits with rate R_{n+1}. Hence these codes permit the use of incremental redundancy (IR) using a typical hybrid-ARQ protocol as explained below.

ARQ procedure. To communicate with a destination D, a source S has to perform the following steps:

- S adds a CRC to the packet and encodes it with the lowest rate R_N.

- S punctures the encoded data to obtain the desired code rate R_i and sends the packet. The punctured bits are stored in case the decoding at D fails.

- Upon packet reception, D decodes the data and checks the CRC. If the decoding is successful, an acknowledgement (ACK) is sent back to S. Otherwise, a negative acknowledgement (NACK) is sent.

- As long as S receives NACKs, further packets of punctured bits (IR Data) are sent, until transmission succeeds or no more punctured bits are available. In the latter case, S may attempt another transmission at a later time (see below).

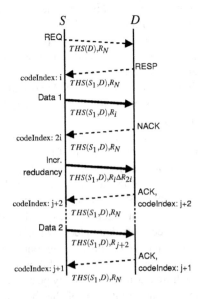

Fig. 1. Dynamic Channel Coding

Note that if the receiver cannot even detect reception of data it cannot send a NACK. In this case the sender will time out and retry communication with a more powerful code.

Rate selection. When nodes communicate for the first time, it is necessary to bootstrap the code adaptation mechanism. The first data packet is encoded with the most powerful (lowest rate) code R_N. If the destination can decode the data packet successfully, it will estimate which higher rate $R_j, j \leq N$ would still allow to decode the data. Decoding of the data packet with rate R_N is performed by step-wise traversal of the trellis of the Viterbi decoder [13]. The packet is then reproduced from the bits corresponding to the sequence of selected branches. Hence, as soon as the outcome of a decoding step for a higher rate code $R_i > R_N$ differs, code R_i can be eliminated. Because of the rate compatibility feature of RCPC codes, this allows to also eliminate all codes R_i', with $R_i' > R_i$. Senders maintain a cache of channel codes for a number of destinations. If a sender does not communicate with a receiver for a certain amount of time, the corresponding cache entry times out and the sender bootstraps the code selection procedure with code R_N as described previously.

In summary (Figure 1), the algorithm for the selection of code is as follows. Remember that a large code index means a small rate.

- S keeps in a variable codeIndex the value of the next code index to use. Initially or after an idle period, codeIndex$= N$.
- When D sees that a packet is sent but cannot decode it, it sends a NACK to S.
- When S receives a NACK, it sets codeIndex to min($2*$codeIndex, N).
- When D can decode, it computes the smallest code index, say j, that could have been used, and returns a codeIndex attribute in the ACK equal to $j + 2$.
- When S with codeIndex$= i$ in the cache receives an ACK with codeIndex$= i'$, if $i' < i$ then S sets codeIndex to $i + 1$, else it sets codeIndex to i'.
- When S times out on a packet it sent, it sets the corresponding codeIndex to min($2*$codeIndex, N) and resumes a transmission request/reply exchange.

The heuristic used by a destination of increasing the optimal rate index by 2 is used to cope with the fact that the channel may vary until the next data transmission.

2.2 Private MAC

The goal of the private MAC protocol is to enforce that several senders cannot communicate simultaneously with one destination.

We cannot use traditional carrier sensing scheme since carrier sensing is not possible with our UWB physical layer. We solve the problem by a combination of receiver-based and invitation-based selection of THSs.

Contention for a destination uses the *public* THS of the destination, but an established communication uses the *private* THS to a source-destination pair. The public THS of the user with MAC address S is the THS produced by a pseudo-random number generator (PRNG) with seed S. The private THS of users S and D is the THS produced by the PRNG with a seed equal to the binary representation of the concatenation of S and D.

As shown in Figure 2, a successful data transmission consists of a transmission request (REQ) by the sender S_1, a response (RESP) by the receiver D, the actual data packet, and an acknowledgement (ACK). Assume S_1 has data

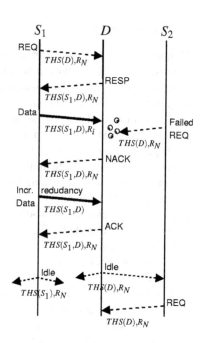

Fig. 2. Private MAC

to transmit to D, and no other node is sending data to D. When D is idle, it listens on its own public THS. As soon as S_1 wants to communicate with D, it sends a REQ on D's public THS using the lowest possible rate R_N. D answers with a RESP using the private THS of the pair S_1-D coded with rate R_N. This response contains the channel code $R_i \geq R_N$ to be used for subsequent data packets dictated by the channel code assignment procedure. When S_1 receives the reply, it starts transmitting the data packet on the THS private to S_1 and D. After the transmission, S_1 listens for an ACK sent by D on the private THS with rate R_N. If a negative ACK is received, S_1 sends incremental redundancy until a positive ACK is received (which marks the end of the packet transmission). Together with the previous data, this results in a code of rate R_j with $R_i > R_j \geq R_N$.

If no feedback is received, S_1 retries transmission after a random backoff, up to a certain retry limit. After a transmission (either successful or unsuccessful), both sender and receiver issue a (short) idle signal each on their own public THS to inform other nodes that they are idle.

Assume now that S_2 wishes to communicate with D while D is receiving a packet from S_1. It sends out a REQ on D's public THS; this may create some interference but will usually not disrupt the private communication between S_1 and D since it is on a different THS. S_2 then switches to D's public THS and listens for the idle signal. When it hears the idle signal, it waits for a random small backoff time. If the timer expires without the node overhearing a REQ from another node, a REQ is sent. Otherwise, the node defers transmission and pauses the backoff timer until it hears the idle signal again.

3   Simulations

We implemented DCC-MAC in the network simulator ns-2.[2] We compared our protocol to a CSMA/CA-like exclusion-based protocol as well as a power control protocol. The exclusion based protocol is similar to the 802.11 MAC layer. The transmit power is fixed but dynamic channel coding is used. The power-control protocol is based on the CA/CDMA protocol [10] and uses a fixed channel coding.

Note that *all* MAC layers have the same UWB physical layer in common as well as the same maximum power limit. We consider a network with 32 parallel sender-receiver pairs. The distance between sender and receiver varies from 1m to 40m. Simulation results are depicted in Figure 3 along with the result for a single sender-receiver connection without any interference. We also performed similar simulations for random topologies. Due to space constraints we omit the simulation results but again, DCC-MAC performance is far superior to the performance of power control and the exclusion-based protocols.

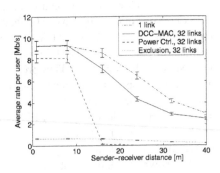

Fig. 3. Throughput vs. sender-receiver distance in the parallel link scenario

4   Conclusion

In this paper, we presented a MAC protocol for low-power UWB ad-hoc networks. To the best of our knowledge, this the first MAC protocol based on dynamic channel coding

---

[2] We redesigned the physical layer support in ns-2 to account for varying interference over the course of a packet transmission. Bit error rates and transmission rates are obtained by interpolation from lookup tables created by offline Matlab experiments. The channel model is the one from [14]

where the rate is constantly adapted to the level of interference. Its design is closely coupled with the physical layer. It is based on the assumptions that all nodes have simple receivers and transmitters (single user decoding, one transmitter per node) and all have the same PRP. No common channel is necessary. Also, due to the constant adaptation to varying channel condition, mobility is well supported

We investigated the performance of our design through analysis in Matlab and simulation in ns-2. The results show that it outperforms exclusion-based protocols as well as protocols based on power control. For future research, we plan to investigate optimal policies for channel code adaptation and under which conditions it is better to directly send the data packets instead of having a request/response packet exchange beforehand (i.e., when the additional overhead of the REQ).

References

1. Verdu, S.: Spectral efficiency in the wideband regime. IEEE Transactions on Information Theory **48** (2002) 1319–1343
2. Win, M.Z., Scholtz, R.A.: Ultra-wide bandwidth time-hopping spread-spectrum impulse radio for wireless multiple-access communications. IEEE Transactions on Communications **48** (2000) 679–691
3. Hélal, D., Rouzet, P.: ST Microelectronics Proposal for IEEE 802.15.3a Alternate PHY. IEEE 802.15.3a / document 139r5 (2003)
4. Hagenauer, J.: Rate-compatible punctured convolutional codes (RCPC codes) and their applications. IEEE Transactions on Communications **36** (1988) 389–400
5. Frenger, P., Orten, P., Ottosson, T., Svensson, A.: Rate-compatible convolutional codes for multirate DS-CDMA systems. IEEE Transactions on Communications **47** (1999) 828–836
6. IEEE: 802.15 WPAN high rate alternative PHY task group 3a (TG3a). (http://www.ieee802.org/15/pub/TG3a.html)
7. Cuomo, F., Martello, C., Baiocchi, A., Fabrizio, C.: Radio resource sharing for ad hoc networking with UWB. IEEE Journal on Selected Areas in Communications **20** (2002) 1722–1732
8. Hicham, A., Souilmi, Y., Bonnet, C.: Self-balanced receiver-oriented MAC for ultra-wide band mobile ad hoc networks. In: IWUWBS'03. (2003)
9. Kolenchery, S., Townsend, J., Freebersyser, J.: A novel impulse radio network for tactical military wireless communications. In: IEEE MILCOM'98. (1998) 59–65
10. Muqattash, A., Marwan, K.: CDMA–based MAC protocol for wireless ad hoc networks. In: Proceedings of MOBIHOC'03. (2003) 153–164
11. Radunovic, B., Le Boudec, J.Y.: Optimal power control, scheduling and routing in UWB networks. (Accepted for publication in IEEE Journal on Selected Areas in Communications)
12. Knopp, R., Souilmi, Y.: Achievable rates for UWB peer-to-peer networks. In: International Zurich Seminar on Communications. (2004)
13. Proakis, J.G.: Digital Communications. 4th edn. McGraw–Hill, New York,NY (2001)
14. Ghassemzadeh, S., Tarokh, V.: UWB path loss characterization in residential environments. In: IEEE Radio Frequency Integrated Circuits (RFIC) Symposium. (2003) 501–504

Lifetime-Extended, Power-Aware, Non-position Base Localized Routing in Hybrid Ad-Hoc/Infrastructure Network

Xueqing Li and Roshdy H.M. Hafez

Dept of System and Computer Engineering, Carleton University, Ottawa, ON Canada
{xueqing,hafez}@sce.carleton.ca

Abstract. This paper presents a localized routing protocol for both hybrid ad-hoc/infrastructure networks and conventional ad-hoc networks. The goal is to extend the network lifetime and to reduce power consumption. Several performance metrics are proposed considering node's transmission power and/or remaining battery capacity. The proposed protocol avoids entire network flooding to reduce interference, delay and excessive overhead. The protocol does not need position (GPS) assistance. Simulation results show that the proposed protocol improves efficiency in both hybrid and conventional ad-hoc networks. The simulation also shows that there is an optimal search radius (power) that the protocol achieves its best performance in terms of network lifetime.

1 Introduction

Wireless ad-hoc networks are typically isolated, autonomous and without infrastructure. Such a network has limited applications. A more general model would include infrastructure access points that have connections to the external networks/services. This paper refers to this general model as the hybrid ad-hoc/infrastructure network. In such a model, wireless terminals communicate with each other in an ad-hoc fashion. However, some terminals can have access to fixed gateways (access points). The density of gateways is assumed to be small. If the density is reduced to zero, then the hybrid network becomes a conventional (pure) ad-hoc network.

In this paper, we analyze the general model and show that the presence of access points influences the network performance and choice of routing algorithm. Our study is concerned with power-aware routing protocols. Our measure of power efficiency is the lifetime of the network. Lifetime is a statistical parameter indicating the period elapsed until certain percentages of wireless terminals are dead.

A routing protocol could be non-localized or it could be localized. Generally, all existing non-localized protocols are variations of shortest weighted path algorithm using different flooding methods. Some notable non-power/cost-aware examples include DSDV, AODV, DSR [11]. More power-aware, battery cost aware non-localized protocols are proposed in [1] ~ [5]. In general, non-localized protocols are

I. Nikolaidis et al. (Eds.): ADHOC-NOW 2004, LNCS 3158, pp. 312–317, 2004.

not scalable. They bring excessive overhead and high interference during route discovery and maintenance procedure.

This paper focuses on non-position based localized routing where the sender needs only find the most appropriate next hop terminal within a limited search radius. Many existing localized routing protocols are position-based (e.g., GPS). Paper [6] provides a survey on several well-known position-based, non-power/cost-aware localized routing protocols, such as MFR, NFP, NC, GEDIR, and compass routing DIR[8]. Paper [7] proposes power, cost, power-cost GPS based localized routing protocols.

However, position-based localized protocols have the following drawbacks: 1) GPS positioning may not be accurate enough, especially in environment with poor coverage or reception. 2) GPS-free positioning technique [9] brings a lot of overhead. 3) Accuracy of destination node location is a serious problem [10].

The objective of this paper is to develop a localized, non-position based routing protocol that maximizes the network lifetime and reduces power consumption in both hybrid and conventional ad-hoc networks. New normalized cost metrics are proposed to solve the problem of unevenly weight the impact of the power and battery cost on the route decision. Simulation results show that there exists an optimal search radius (power) that the network lifetime can be extended the most.

2 Description of Proposed Localized Routing Protocol

In hybrid network model, traffic is categorized into internal and external traffic.
1) Internal traffic is a data flow from a wireless node to another wireless node within the network service area.
2) External traffic is a data flow from a wireless node to a gateway (uplink traffic) or from a gateway to a wireless node (downlink traffic). The downlink case will not be further discussed in this paper, since it is simply the forwarding of traffic directly from a gateway to the destination.

We introduce two level neighborhoods: small ring and large ring. A small ring is a route request packet transmission range of a sender using small transmission power which we call *reference power* P_{ref}. A large ring is a sender's transmission range using its maximum transmission power. A sender starts to search for the destination within the small ring two-hop neighborhood first. If it is found, an optimal complete route is selected. If not, a new search within the sender's large ring two-hop neighborhood will be initiated and an optimal next hop node that has high potential to reach the destination is chosen. Then data packet will be forwarded to this node, and the procedure will be repeated again from using a small ring search until the destination is reached, if possible. The route decision is made after applying cost metric (described in Section 3).

The major protocol steps are:
- The sender issues a request to its immediate neighbors (small ring). Any terminal within this ring that either has access to a gateway, or has the destination in its list will respond. If more than one response is received, the sender applies the cost function and chooses the most power efficient hop among the responders. The packet is delivered to that node.

- If the first step fails (i.e. no response), the terminal sends the same request with maximum power (large ring). Any terminal within the large ring that either has access to a gateway or has the destination in its list would respond. The senders again calculate the energy cost metrics for all responders and chooses the most efficient.
- If no response is received. Then the sender sends a packet to a terminal in its large ring that has the smallest cost. This new terminal can repeat the process from a small ring search.

3 Cost Metrics

Eight cost metrics are designed for the proposed routing protocol. They can be classified into three groups: a) consider Tx power cost only. b) Consider node's battery cost only. c) Consider both, which includes two subcategories: Add & Product. Normalized cost functions belong to group c). Each cost metric has a set of two cost functions, which respectively apply to the two cases in the routing algorithm: small ring search and large ring search. Proposed routing protocol uses the following cost metrics to pick the optimal (lowest cost) next relay node r or R respectively for these two cases.

Below is the detail of the eight cost metrics. Note that r denotes a small ring one hop neighbour of sender s and R denotes a large ring one hop neighbour of sender s. d is the destination, P is transmission power and B is battery residual time.

1) LPE: consider Tx power only

$\quad$ *small ring* : $C_{lpe}^{s} = P_{s \to r} + P_{r \to d}$ $\qquad\qquad\qquad$ (1)

$\quad$ *large ring* : $C_{lpe}^{l} = P_{s \to R}$

2) LBLN: consider node's battery cost only

$\quad$ *small ring* : $C_{lbln}^{s} = B_{o}(r) / B_{r}(r)$ $\qquad\qquad\qquad$ (2)

$\quad$ *large ring* : $C_{lbln}^{l} = B_{o}(R) / B_{r}(R)$

Normalized cost metrics considering Tx power and Br:

3) COSA

$\quad$ *smallring:* $C_{cosa}^{s} = (C_{s}^{B})_{N} - \cos*(C_{sr}^{P})_{N} + (C_{r}^{B})_{N} - \cos*(C_{rd}^{P})_{N}$ $\quad$ (3)

$\quad$ *large ring:* $C_{cosa}^{l} = (C_{R}^{B})_{N} - \cos+(C_{sR}^{P})_{N}$

4) COSP

$\quad$ *smallring:* $C_{cosp}^{s} = (C_{s}^{B})_{N} - \cos*(C_{sr}^{P})_{N} + (C_{r}^{B})_{N} - \cos*(C_{rd}^{P})_{N}$ $\quad$ (4)

$\quad$ *large ring:* $C_{cosp}^{l} = (C_{R}^{B})_{N} - \cos*(C_{sR}^{P})_{N}$

5) SIGA

$\quad$ *smallring:* $C_{siga}^{s} = (C_{s}^{B})_{N} - sig*(C_{sr}^{P})_{N} + (C_{r}^{B})_{N} - sig*(C_{rd}^{P})_{N}$ $\quad$ (5)

$\quad$ *large ring:* $C_{siga}^{l} = (C_{R}^{B})_{N} - sig+(C_{sR}^{P})_{N}$

6) SIGP

$\quad$ *smallring:* $C_{sigp}^{s} = (C_{s}^{B})_{N} - sig*(C_{sr}^{P})_{N} + (C_{r}^{B})_{N} - sig*(C_{rd}^{P})_{N}$ $\quad$ (6)

$\quad$ *large ring:* $C_{sigp}^{l} = (C_{R}^{B})_{N} - sig*(C_{sR}^{P})_{N}$

non-normalized cost metrics considering Tx power and Br:

7) INVA

smallring: $C_{inva}^{s} = (B_o(s)/B_r(s)) * P_{s \to r} + (B_o(r)/B_r(r)) * P_{r \to d}$ (7)

large ring: $C_{inva}^{l} = B_o(R)/B_r(R) + P_{s \to R}$

8) INVP

smallring: $C_{invp}^{s} = (B_o(s)/B_r(s)) * P_{s \to r} + (B_o(r)/B_r(r)) * P_{r \to d}$ (8)

large ring: $C_{invp}^{l} = B_o(R)/B_r(R) * P_{s \to R}$

4 Simulation Result Analysis

4.1 Simulation Model and Assumption

In both the hybrid and conventional network model, wireless nodes are randomly allocated according to uniform distribution in a network service area. Gateways in a hybrid network are evenly distributed in a service area. Static model is assumed for the simulation since mobility is believed to have less impact on localized non-position based protocols. The network service area size and path loss model are described as below. Two-ray shadowing model is the path loss model using two-ray model to predict the mean path loss and considering shadowing effect.

- *Hybrid ad-hoc/infrastructure:* service area: 2km * 2km. (2-ray model)
- *Conventional ad-hoc network:* service area: 100m * 100m (free space model)

The following assumptions are made:
1) All wireless nodes have the same initial battery and maximum Tx Power of 50mw.
2) We don't not consider concurrent traffic.
3) Power consumed by overheads (route discovery etc) is not considered for all the protocols simulated.

4.2 Results and Performance Evaluation

Each of the eight proposed cost metrics is applied to the proposed routing protocol individually. Their performances are evaluated to provide comparison.

Figure1 shows the network lifetime of the network with 70 wireless nodes, 2 gateways without external traffic. Figure2 shows the case with 4 gateways. Performances of Min-Hop and Min-Eng are not related to Pref, therefore they are the fixed values as shown in the box in each of the figures. The eight curves represent the performance of eight cost metrics. They show the same pattern that there exists an optimal reference power Po at which the network lifetime can be extended the most. Too small Pref causes low node degree, which leads to frequent large ring searches. Degree of a node is the number of neighbors of a node. Detoured, less cost-effective routes may be discovered with small Pref, which will shorten the network lifetime. Routes discovered by a too large Pref tend to a direct transmission, which apparently results in short network lifetime. Thus the performances of eight curves are close when Pref is large. As shown in Figure 1 and 2, Po is smaller in a gateway-dense network. Lifetime of a gateway-dense network is longer than that of a gateway-sparse network. If we observe the lifetime at Po, cost metrics that consider both Tx power and Br: (SIGA, SIGP, COSA, COSP, INVA, INVP) present competitive performance in comparison with the ones who consider power or Br only (LPE, LBLN). Simulation results also show that SIGP provides the longest lifetime when the network lifetime is defined as 20% of nodes dead time. Compared with Min-Hop, proposed protocol shows dramatic performance improvement in terms of network lifetime. Also, the network lifetime at optimal reference power Po of each curve reaches approximately 80% of that of Min-Eng.

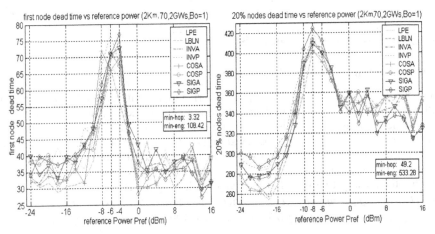

Fig. 1. Network lifetime: 70 wireless nodes, 2 gateways, 0% external traffic

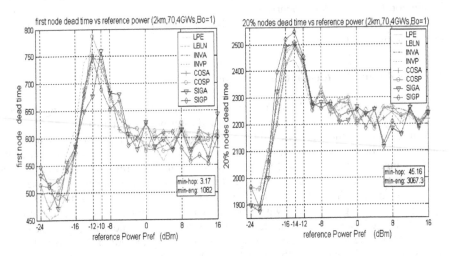

Fig. 2. Network lifetime: 70 wireless nodes, 4 gateways, 0% external traffic

5 Conclusion and Future Work

A non-position based, localized routing protocol for both hybrid and conventional ad-hoc networks is proposed in this paper. It reduces the routing overhead greatly in comparison with non-localized and position-based localized protocols, and it can effectively prolong the network lifetime. Limited simulation is done without considering the overhead. Network lifetime, total transmission power and other performances are investigated. Proposed protocol demonstrates excellent or comparable performance in comparison with some existing protocols. Cost metrics, especially normalized ones that consider both Tx power and battery reserve are proven to present better performance than the ones that only rely on one of those two factors. An optimal route request reference power level Po exists for each network model. To adjust a node's small ring transmission power to the optimal level is the key to achieve optimal performance.

References

1. S. Singh, et al. "Power-aware routing in mobile ad hoc networks". Mobicom 98. Dallas Texas, USA
2. C.-K. Toh. "Maximum Battery Life Routing to support ubiquitous mobile computing in wireless ad hoc networks". IEEE communication magazine, June 2001.
3. A. Misra, S. Banerjee. "MRPC: maximizing network lifetime for reliable routing in wireless environments". WCNC, Mar. 2002.
4. J. Chang, L. Tassiulas. "Energy conserving routing in wireless ad-hoc networks". IEEE InfoCom 2000.
5. A. Michail, A. Ephremides. "Energy efficient routing for connection-oriented traffic in ad-hoc wireless networks". Proc. 11th IEEE International Symposium on Personal Indoor and Mobile Radio Communications (PIMRC), London, UK (September 2000).
6. I. Stojmenovic. "Position-based routing in ad hoc networks". IEEE communications magazine, July 2002.
7. I. Stojmenovic, X. Lin. "Power-aware localized routing in wireless networks". IEEE transaction on parallel and distributed system, 2000.
8. E. Kranakis, et al. "Compass routing on Geometric networks". Proc. 11th Canadian Conference on Computational Geometry, 1999.
9. S. Capkun, et al. "GPS-free positioning in mobile ad hoc networks". Proc. Hawaii Int. Conf. on System Sciences, January 2001.
10. I. Stojmenovic. "Location updates for efficient routing in ad hoc networks". Handbook of wireless networks and mobile computing, 2002.
11. C-K. Toh "Ad Hoc Mobile Wireless Networks – Protocols and Systems". 2002 by Prentice Hall PTR.
12. X. Li, R. Hafez. "Lifetime-extended, Power-aware, Non-position Based Localized Routing in Hybrid Ad-Hoc/Infrastructure Network", M.A.Sc thesis, Carleton University, 2004

Routing Table Dynamics in Mobile Ad-Hoc Networks: A Continuum Model

Ernst W. Grundke

Faculty of Computer Science, Dalhousie University
6050 University Avenue, Halifax, Nova Scotia, Canada B3H 1W5
Ernst.Grundke@dal.ca

Abstract. This paper gives a simple continuum model for generic pro-active routing protocols in a mobile ad-hoc network. Rate constants characterize the routing protocol and the network mobility. We obtain analytic expressions for the rate at which routing tables approach equi-librium. The slowest rate depends only on the routing protocol and not on the mobility. The model also gives an analytic expression for the fraction of valid entries in a typical routing table, which is shown to degrade gracefully with increasing mobility. The model suggests a strategy for tuning a routing protocol for changes in mobility.

Keywords: Mobile ad-hoc networks, routing table, dynamics, continuum.

1   Introduction

The problem of routing in mobile ad-hoc networks has been the subject of numerous recent papers [1], [3], [4], [5]. In an earlier paper [2] we found that a simple continuum model of a mobile ad-hoc network is sufficient to (a) model the scaling of traffic with network size, and (b) characterize the network's routing traffic and data traffic in terms of simple dimensionless parameters. This paper extends the same approach to model a proactive routing protocol in terms of the behavior of the routing table at a typical node. We develop a general theoretical model that is not tied to any particular protocol, yet is capable of addressing questions such as: What is the effect of mobility on the routing table entries? To what extent are the entries in a routing table likely to be valid? How quickly can we expect a routing table to fill up with valid entries once a routing protocol starts running? How should we tune a routing protocol as mobility changes?

The remainder of the paper is organized as follows. In Sections 2 and 3 we model a generic proactive routing protocol and the effect of node mobility, respectively. Section 4 describes a node's routing table and the events that change its entries. In Section 5 we obtain differential equations for the rates of change of the number of various types of entries in a typical routing table. The equations are solved for the equilibrium case in Section 6 and for the dynamic case in Section 7. Conclusions and acknowledgments are given in the final sections.

I. Nikolaidis et al. (Eds.): ADHOC-NOW 2004, LNCS 3158, pp. 318–323, 2004.
© Springer-Verlag Berlin Heidelberg 2004

2   The Routing Protocol

We consider a mobile ad-hoc network using a proactive routing protocol [3], that is, each node attempts to maintain a complete routing table for every possible destination, whether or not it currently has traffic for that destination. For the sake of simplicity we assume that N, the number of nodes in the network, is constant.

   As the routing protocol reports (changing) topology data, nodes receive routing packets that let them update entries in their routing tables. To make our model as general as possible, we ignore routing protocol details, and simply consider the rate at which the protocol delivers routing data to a node. Let us assume that a typical node receives routing data about r destinations per unit time; the data need not be new to the node. We assume that the routing protocol is such that r is proportional to N.

   In order to make the mathematical model tractable, we assume that the destinations described by the routing data are randomly distributed. This simplistic assumption is compatible with the level of sophistication of the model in [2].

3   Node Mobility

When nodes move, they may enter or leave the radio ranges of other nodes. We call the making or breaking of a radio link a *link event*. We assume that such events occur at a rate p_m per unit time per node.

   When a link is broken, paths containing that link become invalid, and some entries in various routing tables may become invalid. We can determine the extent of this damage as follows. There are $N(N-1)$ possible one-way paths in the network. From [2], we conclude that any link is contained, on average, in c of these paths, where

$$c = \frac{\beta}{2} \frac{N-1}{g-1},\qquad(1)$$

$\beta = r_1(\sqrt{N}-1)/R$, $g = (R+r_1)^2/r_1^2$, R is the radio transmission range of a node, and $2r_1$ is average distance between "neighboring" nodes (see [2]). The factor β is the number of hops in a typical path. With the approximation that $N \gg 1$ and the assumption that R is as small as possible ($R \approx 2r_1$), (1) becomes $c \approx N^{1.5}/32$.

   We ignore the possibility that a link event can accidentally correct an invalid routing table entry.

4   The Routing Table

Let us identify the nodes using the integers 1 to N. For each destination d in $[1,N]$ the routing table stores either a null or a node identifier f, meaning that packets destined for d are forwarded to node f, which must be in radio contact with the current node.

   An entry in the routing table may be of three kinds:
   (a) Missing (f null). The routing protocol has not yet delivered relevant data to the node, so the node is not able to route packets to destination d.
   (b) Valid. The entry has recently been updated with (presumably valid) data

provided by the routing protocol.

(c) Invalid. Mobility has made the entry obsolete, and data packets routed according to this entry are unlikely to reach their destination. Short of sending out trial packets, the node cannot tell which entries are invalid.

We let $n_{0i}(t)$ and $n_{1i}(t)$ denote the number of invalid and valid entries, respectively, in the routing table at node i at time t. There are $n_i(t) = n_{0i}(t) + n_{1i}(t)$ entries in total, and there are $N - n_i(t)$ missing entries. For a randomly chosen destination the probability of a missing entry is $1 - n_i/N$, the probability of a valid entry is n_{1i}/N, and the probability of an invalid entry is n_{0i}/N.

Further we define $n_0(t)$ and $n_1(t)$ as the averages of $n_{0i}(t)$ and $n_{1i}(t)$ over i, that is, the average number of invalid and valid routing table entries. Similarly we define $n(t)=n_1(t)+n_0(t)$, and then $N-n(t)$ is the average number of missing entries. These averaged functions are not restricted to integer values.

Two kinds of events can change the values of $n_0(t)$ and $n_1(t)$: the arrival of routing data about a destination node, and a link event elsewhere in the network. (Of course a node receives no indication of the latter.) We assume that the events are equally likely for all destinations.

Consider what happens when node i receives routing data about a destination d. (We assume that the routing data can arrive quickly enough to be valid.) If the corresponding entry is missing, n_{1i} rises by 1. If the entry is already valid, the arriving routing packet has provided nothing new, so that n_{0i} and n_{1i} remain unchanged. Finally, if the entry is invalid, n_{1i} rises by 1 and n_{0i} drops 1.

Now consider the case of a link event at that destroys the path from node i to various destinations. If the relevant routing table entries at node i are missing or invalid, n_{0i} and n_{1i} remain unchanged. Entries that are valid will be rendered invalid, so that n_{0i} rises by the number of such entries and n_{1i} drops by the same amount. In terms of the averaged functions, link events will increase n_0 by c/N and decrease n_1 by c/N; see Equation (1).

5   Rate Equations

Table 1 summarizes the effects described in Section 4. Assuming that $n_0(t)$ and $n_1(t)$ are continuous, differentiable functions of t, the discrete changes in Table 1 become rate equations for $n_0(t)$ and $n_1(t)$:

$$\frac{dn_0}{dt} = \frac{n_1}{N} cp_m - \frac{n_0}{N} r \; , \tag{2}$$

$$\frac{dn_1}{dt} = (1 - \frac{n_1}{N}) r - \frac{n_1}{N} cp_m \; . \tag{3}$$

With the definitions $k_a = r/N$ for the data arrival rate per node and $k_m = cp_m/N$ for the effective link event (mobility) rate, these equations become

$$\frac{dn_0}{dt} + k_a n_0 = k_m n_1 \; , \tag{4}$$

$$\frac{dn_1}{dt} + (k_a + k_m) n_1 = k_a N \; . \tag{5}$$

Equations (4) and (5) are a coupled pair of first order differential equations in two unknown functions, $n_0(t)$ and $n_1(t)$. The coupling is trivial, since the second equation does not contain $n_0(t)$.

Table 1. Effect of Events on Routing Table Entries at a Typical Node

Routing table entry		Event	
Type	Probability	Routing Data (Rate r)	Link Event (Rate Np_m)
Missing	$1-(n_0+n_1)/N$	n_1++	No effect
Valid	n_1/N	No effect	n_0+= c/N, n_1-= c/N
Invalid	n_0/N	n_0- -, n_1++	No effect

6   The Routing Table in Equilibrium

The values of n_0 and n_1 are the result of a competition between the routing protocol, which tries to increase n_1, and the link events, which tend to reduce n_1. In equilibrium, although the individual entries keep changing, the *number* of entries (n_0 and n_1) will be constant. Therefore the derivative terms in (4) and (5) become zero, and in equilibrium we obtain

$$n_0 = \frac{k_m}{k_a+k_m}N, \quad n_1 = \frac{k_a}{k_a+k_m}N, \quad n = N.$$

(6)

In other words, there will be no missing entries, the invalid entries will make up a fraction $k_m/(k_a+k_m)$ of the table, the valid entries will make up the remaining $k_a/(k_a+k_m)$ of the table, and the ratio of valid entries to invalid entries is simply

$$\frac{n_1}{n_0} = \frac{k_a}{k_m}$$

(7)

We see clearly the competition between the rate of link events and the rate of routing data arrival events. The routing protocol will dominate this "tug-of-war" if k_a is significantly larger than k_m. Equations (6) and (7) gives a strategy for tuning the routing protocol: estimate k_m from the expected mobility, and adjust any protocol parameters to deliver routing data at a rate k_a that gives a tolerable percentage of invalid routing table entries in accordance with Equation (6).

Even in a highly mobile network, where $k_a \approx k_m$ or $k_a < k_m$, the situation is not catastrophic. The routing system degrades relatively gracefully, that is, linearly with the k values. This is unlike queuing systems, for example, where delays become extremely large as a limiting capacity is approached.

7   Routing Table Dynamics

In the dynamic (non-equilibrium) case, the derivative terms in (4) and (5) are non-zero, and the general solutions are

$$n_0(t) = C_0 \exp[-k_a t] - C_1 \exp[-(k_a+k_m)t] + \frac{k_m}{k_a+k_m} N , \tag{8}$$

$$n_1(t) = C_1 \exp[-(k_a+k_m)t] + \frac{k_a}{k_a+k_m} N , \tag{9}$$

and

$$n(t) = n_0(t) + n_1(t) = C_0 \exp[-k_a t] + N , \tag{10}$$

where C_0 and C_1 are arbitrary constants to be determined by the initial conditions. In the limit $t \rightarrow \infty$ the exponential functions approach 0, and we recover the equilibrium solution (6) as a special case.

In the case of initially empty tables, we would require $n_0(0) = n_1(0) = 0$, giving

$$C_0 = -N , \quad C_1 = -\frac{k_a}{k_a+k_m} N , \tag{11}$$

and therefore

$$n_0(t) = -N \exp(-k_a t) + \frac{k_a}{k_a+k_m} N \exp[-(k_a+k_m)t] + \frac{k_m}{k_a+k_m} N , \tag{12}$$

$$n_1(t) = \frac{k_a}{k_a+k_m} N (1 - \exp[-(k_a+k_m)t]) , \tag{13}$$

and

$$n(t) = n_0(t) + n_1(t) = N (1 - \exp[-k_a t]) . \tag{14}$$

Figure 1 shows these results graphically for $k_a/k_m=3$.

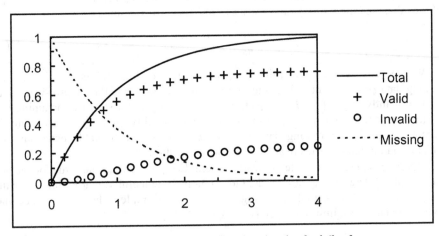

Fig. 1. Fraction of entries plotted against $k_a t$, for $k_a/k_m=3$.

Both n_0 and n_1 contain transient (decaying exponential) terms. The slowest term present is $\exp(-k_a t)$, which settles to $1/e$ of its initial value in a time $1/k_a$, where e is the base of natural logarithms. This settling time is determined by the protocol alone, and not by the mobility. The effect of mobility is to render a greater fraction of the entries invalid, not to affect the rate of approach to equilibrium.

In the special case of a static network, $k_m = 0$, and so that

$$n_0(t) = 0 , \quad n(t) = n_1(t) = N\,(1 - \exp[-k_a t]) . \tag{15}$$

8 Conclusion

The simple continuum model presented here uses two rate constants, k_m and k_a, to characterize the network mobility and a proactive routing protocol, respectively. The model yields analytic expressions for the rate at which routing tables can be expected to approach equilibrium values. The slowest of these depends only the routing protocol. The model also yields analytic expressions for the fraction of valid entries in a typical routing table. The number of valid routing table entries is shown to degrade gracefully with increasing mobility. The model has also suggested how to tune a routing protocol to changes in mobility.

Acknowledgments. The author acknowledges helpful discussions with Nur Zincir-Heywood, Nicholas Pilon and Donald Morrison.

References

1. Braginsky D., Estrin D.: Rumor Routing Algorithm for Sensor Networks, WSNA '02 September 28 (2002) Atlanta, Georgia, USA
2. Grundke E., Zincir-Heywood, N: A Uniform Continuum Model for Scaling of Ad-Hoc Networks. Ad-Hoc, Mobile and Wireless Networks, Proceedings of the Second International Conference, ADHOC-NOW 2003, Canada. Sringer-Verlag, Berlin, Heidelberg, New York (2003) pp. 96-103
3. Hong X., Xu K., Gerla M.: Scalable Routing Protocols for Mobile Ad Hoc Networks, IEEE Network, July/August (2002) 11-21
4. Johansson P., Larsson T., Hedman N., Mielczarek B., Degermark M.: Scenario-based Performance Analysis of Routing Protocols for Mobile Adhoc Networks. ACM Mobicom '99, Seattle Washington USA (1999)
5. Liang, S., Zincir-Heywood, N., Heywood, M.: The Effect of Routing under Local Information Using a Social Insect Metaphor. IEEE 2002 World Congress on Computational Intelligence, Congress on Computation (2002) 1438-1443

Realistic Mobility for Mobile Ad Hoc Network Simulation

Michael Feeley, Norman Hutchinson, and Suprio Ray

Computer Science Department, University of British Columbia, Vancouver, Canada

Abstract. In order to conduct meaningful performance analysis of routing algorithms for Mobile Ad Hoc Networks (MANETs), it is essential that the mobility model on which the simulation is based reflects realistic mobility behavior. However, current mobility models for MANET simulation are either unrealistic or are tailor-made for particular scenarios. We introduce GEMM, a tool for generating mobility models that are both realistic and heterogeneous. These models are capable of simulating complex and dynamic mobility patterns representative of real-world situations. We present simulation results using AODV, OLSR and ZRP, three MANET routing algorithms and show that mobility-model changes have a significant impact on their performance.

Keywords: Mobility Model, MANET, GEMM

1 Introduction

Mobile ad-hoc network (MANET) routing protocols have received considerable recent attention. The standard way to evaluate these protocols is simulation, key to which is a model of node mobility. Typically, a simple abstract mobility model such as random waypoint [6] or random walk [1] is used [5,7]. By virtue of their simplicity, these random models do not attempt to reflect real human mobility. The hope, however, is that a simple model captures enough of the key characteristics of human mobility to make protocol evaluations meaningful.

Humans, of course, rarely move randomly. Consider, for example, a typical public park. Park users will be unevenly distributed over this landscape. Some of them will be stationary and others will move at different characteristic speeds: walkers, joggers and bikers, for example. The course that mobile users take will not be random. Some will move to attraction points such as snack bars, restrooms, play areas, etc.

While some previous works have observed that routing algorithm performance may be influenced by choice of mobility model [9], most research continues to use random waypoint. In order to provide a framework for producing more realistic mobility models, we have developed a tool called GEMM, that generates mobility scenarios capturing some key features of typical human mobility. We use GEMM to re-examine three MANET routing algorithms: AODV [3], OLSR [2] and ZRP [4], representative of the three popular routing approaches.

I. Nikolaidis et al. (Eds.): ADHOC-NOW 2004, LNCS 3158, pp. 324–329, 2004.

We compare algorithm performance for random waypoint and several more realistic models generated by GEMM. Our evaluation shows significant performance differences between random waypoint and the alternatives, thus confirming that this random approach is insufficient for accurate simulation.

2 Related Work

Numerous studies have evaluated MANET routing protocols using the *random waypoint* mobility model. Maltz et. al. compared DSDV, TORA, DSR and AODV [5,6]. Das et. al. evaluated DSDV, TORA, DSR and AODV with exponential distance distribution with mean 5m and no pause time [7]. Variations of the *random waypoint* model have been proposed such as *random walk* [1], *random direction* and *boundless simulation area* mobility model. *Gauss-Markov* is a model in which a node's next position is determined incrementally based on its current position and velocity [8].

Several researchers have explored the use of customized models designed to be more realistic than random waypoint. Johanson et. al. proposed three mobility scenarios: *conference*, *event coverage* and *disaster area* [9]. Sanchez explored several custom-made group mobility models such as *column* , *pursue* and *nomadic community* mobility model [10]. Although tailor-made models are useful, significant development effort is required for each simulation scenario and each scenario has limited applicability.

3 GEMM

When setting out to model human mobility realistically, one must consider how humans move and what features of this behavior are important enough to capture.

3.1 Developing a Realistic Mobility Model

Motivated by studies of human walking [11], we observe that mobility can be fairly expressed using four characteristics: attraction points, activities, roles and group behavior.

Attraction points. An attraction point is a destination of interest to multiple people. On a university campus, for example, students may tend to move among locations such as classrooms, cafeterias, pubs, etc. Attraction points are described by a 5-tuple consisting of *x-y coordinates, popularity, radius* and *type*.

Activities. An activity is the process of moving to an attraction point and remaining there for some period of time. Activities can be parameterized by by a 6-tuple consisting of *minimum and maximum trigger time and duration, destination* and *type*.

Roles. A role characterizes the mobility tendencies intrinsic to different classes of people. For example, university students spend much more time moving among classrooms than professors; some people bike others walk, etc. A real environment consists multiple user classes each taking on different roles. Roles are specified by a 5-tuple consisting of *weight, minimum and maximum speed* and *activity set*.

Group Behavior. Group behavior captures the way that people influence each others mobility. People may tend to cluster in groups, match each others velocity or avoid colliding with each other [12]. This is specified by a four-tuple consisting of *collision-avoidance, inertia, velocity-matching* and *group-centering probabilities*. Each probability is expressed as the fraction of nodes that exhibit the specified group behavior.

3.2   Design and Implementation of GEMM

The basic framework is that each node is statically assigned a distinct role. This role specifies a set of activities that nodes perform, chosen randomly during the simulation. Each activity consists of moving to a new location and specifies zero or more attraction points as possible destinations.

GEMM is implemented in Java conforming to the BonnMotion API and uses its output-file format [13]. The input to GEMM is a set of parameter settings and the output is a detailed mobility scenario that can be used directly by either the NS2 [14] or Glomosim [15].

4   Evaluation

This section compares the performance of three MANET routing algorithms using random waypoint and a set of more realistic scenarios generated by GEMM.

4.1   Experimental Setup

In order to conduct comparison studies with GEMM, we select OLSR, AODV and ZRP, which are representative of *proactive, reactive* and *hybrid* routing algorithms respectively. The goal of our simulation experiments is to assess the impact of different aspects of realistic mobility scenarios on the ability of the MANET routing protocols to successfully deliver data packets. Simulations were conducted using Glomosim [15].

We adopt a simulation environment similar to [5], where 50 mobile nodes move about in an area of 1500m x 300m. Each node in the simulation has a radio transmission range of 250m. We report packet delivery ratios for pause times of 0, 30, 60, 120 and 300 seconds. The data traffic characteristics are based on constant bit rate (CBR).

4.2   Attraction Points

The goal of this first set of simulations is to illustrate the ways in which attraction points change routing algorithm behavior. In these simulations node speed is chosen uniformly between 0 and 20m/sec as in [5].

Mobility scenarios. We used GEMM to generate the following mobility scenarios.

RWP. The standard random waypoint model.

OAP-RTO. Nodes move to a single attraction point, pause and return to their original position.

TAP-RTO. Nodes randomly choose among three attraction points, pause and return to their original position.

TAP. Nodes randomly choose among three attraction points, pause and move to another randomly chosen attraction point.

Figure 1 shows the packet delivery ratios for OLSR, AODV and ZRP respectively, for all four mobility scenarios. The first and most important thing to notice is that there are substantial differences among the mobility scenarios. Furthermore, each algorithm reacts differently to mobility-model changes. These differences indicate that the choice of mobility has a big impact on comparisons among competing algorithms.

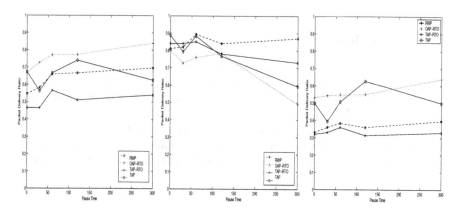

Fig. 1. Attraction-point packet delivery ratio for OLSR and AODV and ZRP

4.3   Impact of Speed Variation

We now turn to simulations using realistic human walking and bicycling speeds of 1-1.6 m/s and 4-11 m/s respectively. We add the suffix **HS** to model names to signify walking speed and **MIXED** to signify a mix of 20% bikers and 80% walkers.

Figure 2 shows the normalized packet delivery ratio attained by three routing algorithms, OLSR, AODV and ZRP, for these six scenarios (OAP-RTO-HS, OAP-RTO-MIXED, TAP-RTO-HS, TAP-RTO-MIXED, TAP-HS and TAP-RTO-MIXED). The normalization is done with respect the earlier scenarios with nodes having maximum speed 20m/s. For instance, the packet delivery ratio for

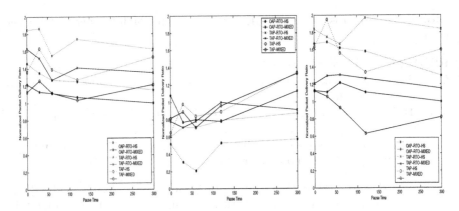

Fig. 2. Speed-variation normalized delivery ratio for OLSR and AODV and ZRP

OAP-RTO-HS is normalized with respect to OAP-RTO. Again the key thing to observe from these graphs is that performance differs substantially among the mobility models.

4.4   Group Mobility Behavior

Finally we turn to group mobility. For this simulation we generated group mobility variants of the mixed walking/biking scenarios. For these models, we set all four group parameters to 100%, indicating that every node acts with all four group behaviors. We add a **-GM** suffix to these models names.

The results are shown in Figure 3, which gives packet delivery ratio normalized to the non-group variant of each scenario. Again we see that group settings seem to matter, though less so than for speed variation, particularly for AODV.

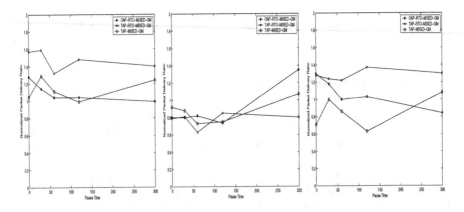

Fig. 3. Group mobility normalized packet delivery ratio for OLSR and AODV and ZRP.

5   Conclusion

This paper describes a tool we have built, called GEMM, that generates mobility scenarios suitable for MANET routing algorithm simulation using either NS2 or Glomosim.

We simulated three algorithms, AODV, OLSR and ZRP, using a variety of mobility scenarios designed to be more realistic than random waypoint. Our results show that the algorithms we studied behaved significantly differently under the models generated by GEMM than under random waypoint. We have also shown that GEMM can be used to generate models that are arguably more realistic.

We believe that generic approaches such as GEMM provide substantial benefit with the evaluation of MANET protocols. Our work also points out the danger of choosing a routing algorithm based on unrealistically simple mobility models.

References

1. A. Bar-Noy, I. Kessler, M. Sidi.: Mobile users: To update or not to update? IEEE INFOCOM, 1994.
2. T.H. Clausen, G. Hansen, L. Christensen and G. Behrmann : The Optimized Link State Routing Protocol, Evaluation through Experiments and Simulation. IEEE Symposium on Wireless Personal Mobile Communications, 2001.
3. Charles E. Perkins and Elizabeth M. Royer : Ad hoc On-Demand Distance Vector Routing. IEEE Workshop MCSA, 1999.
4. Z.J. Haas and M.R. Pearlman : The Zone Routing Protocol (ZRP) for Ad Hoc Networks. Internet Draft, 1999.
5. Josh Broch, David A. Maltz, David B.Johnson, Y.-C. Hu, and J. Jetcheva : A Performance Comparison of Multi-Hop Wireless Ad Hoc Network Routing Protocols. ACM/IEEE Mobicom, 1998
6. David B. Johnson and David A. Maltz : Dynamic Source Routing in Ad Hoc Wireless Networks. Mobile Computing, Kluwer Academic Publishers, 1996.
7. S. R. Das, R. Castañeda and J. Yan : Simulation-based performance evaluation of routing protocols for mobile ad hoc networks. Mobile Net. and Appl. 2000.
8. B. Liang and Z. Haas : Predictive distance-based mobility management for PCS networks. IEEE INFOCOM, 1999.
9. P. Johansson, T. Larsson, N. Hedman, B. Mielczarek and M. Degermark : Scenario based Performance Analysis of Routing Protocols for Mobile Ad Hoc Networks. Mobicom 1999.
10. Sanchez M. Mobility Models : http://www/disca.upv.es/ misan/mobmodel.htm
11. David C. Brogan and Nicholas L. Johnson : Realistic Human Walking Paths. CASA-2003, 2003.
12. D. S. Tan, S. Zhou, J. Ho, J.S. Mehta and H. Tanabe : Design and Eval. of an Indiv. Simulated Mobility Model in Wireless Ad Hoc Networks. CNDSMS, 2002.
13. BonnMotion: http://www.cs.uni-bonn.de/IV/BonnMotion/
14. The Network Simulator - ns2 : http://www.isi.edu/nsnam/ns/
15. GloMoSim : http://pcl.cs.ucla.edu/projects/glomosim/

On Timers of Routing Protocols in MANETs

Kiran K. Vadde and Violet R. Syrotiuk*

Computer Science & Engineering, Arizona State University, Tempe, AZ 85287-8809

Abstract. Routing protocols enable multi-hop communication in mobile ad hoc networks. Most routing protocols make use of timers to respond to link failure caused by network dynamics. Despite their importance, the duration of these timers is typically set in a trial-and-error manner. We apply techniques from design and analysis of experiments to show the effect of timers on response variables and to identify factor interactions involving timers in AODV. Such techniques are a first step toward a formal methodology for timer value optimization.

1   Introduction

Routing is a fundamental problem in *mobile ad hoc networks* (MANETs). Since a MANET has no fixed infrastructure or centralized control, each node functions as a router. As each node has a limited transmission range, each packet may be forwarded over multiple hops to reach its destination. Node mobility and the varying wireless channel characteristics make route set-up and maintenance a challenging problem.

Soft-state, i.e., state valid for a fixed time, is commonly used by routing protocols. Timers associated with route entries may be refreshed by packet arrivals or periodic updates. If a path breaks from node mobility, the timer associated with the route at the node preceding the break eventually expires. It may also trigger a route repair. The advantage of soft-state is that route deletion does not require explicit control messages, reducing overall control overhead.

Perkins et al. [9] present guidelines for setting timer values for the AODV routing protocol in MANETs. These guidelines take into account parameters of the network such as diameter, per-hop time, and the number of nodes. Chin et al. [3] implemented AODV on a test-bed. Timer values are set and adjusted using trial-and-error according to the network configuration; they conclude that a formal methodology for setting timer values is needed.

We present a methodology to address this need, applying techniques from *design and analysis of experiments* (DOE). In complex systems where many factors interact, DOE provides methods to identify and optimize significant and interacting factors. These techniques were first applied to MANETs by Barrett et al. [1,2] to study the effect of routing and MAC protocols on throughput and delay. Vadde et al. [10] examine whether individually QoS-aware protocols impact the overall ability of a MANET to support QoS.

* This work was supported in part by NSF grant ANI-0240524.

I. Nikolaidis et al. (Eds.): ADHOC-NOW 2004, LNCS 3158, pp. 330–335, 2004.

While this work identified interactions between protocols at different layers in the protocol stack, we look to identify factors *within* a protocol that interact and affect performance in the context of soft-state in AODV. Understanding how design decisions of an algorithm interact with each other and with other factors impact individual as well as cross-layer protocol design, ultimately yielding improved network performance.

2   Timers in the AODV Routing Protocol

AODV [8] is a reactive routing protocol that makes use of four primary timers for route set-up and maintenance. AODV discovers routes from a source to a destination using route request (RREQ) packets, incrementally increasing the time-to-live (TTL) field until it receives a route reply (RREP). The timer used to bound the time a node backs off in sending RREQ packets is the MAX_ROUTEREQ_WAIT_TIMEOUT (W).

A value for MY_ACTIVE_ROUTE_TIMEOUT (M), an initial timeout for the route, is specified by the destination to the source in the RREP. Each node, on receipt of a RREQ, maintains a REV_ROUTE_LIFE (R) timer. This allows it to propagate a RREP to the source.

Once a route is established, an ACTIVE_ROUTE_TIMEOUT (A) timer is associated with the route at each intermediate node. As long as packets of the flow continue to arrive and the next hop is reachable, the timer is refreshed.

3   Factors Affecting Timers

We are interested to identify the main effects and interactions of AODV timers, packet arrival rate, and node speed on a number of performance metrics. The packet arrival rate specifies the rate at which a source node generates data packets to be routed. Node speed is the mean speed of the nodes in the network.

We consider the following response variables: the average number of RREQs generated (this includes requests for both local route repair and end-to-end repair), the average end-to-end delay of data packets, and the throughput of the flow defined as the average number of bytes transmitted per second.

Table 1. The factors, and levels of each factor, in our study.

Factor	Levels
MY_ACTIVE_ROUTE_TIMEOUT (M) ACTIVE_ROUTE_TIMEOUT (A) REV_ROUTE_LIFE (R) MAX_ROUTEREQ_WAIT_TIMEOUT (W)	2 and 4 seconds 1 and 3 seconds 2 and 4 seconds 6 and 10 seconds
Packet arrival rate (P) Node speed (S)	1 and 0.33 pkts/s 10 and 90 m/s

Table 2. Effects table for each response variable.

Number of RREQs		
Source	Sum of Squares	Percentage Contribution
A	0.001782	21.061199
W	0.000195	2.304300
P	0.002513	29.704201
S	0.000120	1.417930
AW	0.000093	1.099850
AP	0.000131	1.549770
AS	0.002798	33.080002

Delay		
Source	Sum of Squares	Percentage Contribution
W	1.731000	15.888000
S	8.652000	79.400002

Throughput		
Source	Sum of Squares	Percentage Contribution
P	184.453995	70.559097
S	68.273399	26.116600

We use `ns-2` [7] to run our experiments. We generate scenarios with 30 nodes, connected to their peers via a shared 11 Mbps interface, distributed over a 300×1500 m area. Nodes are distributed according to the *stationary distribution* of the random waypoint mobility model [6]. IEEE 802.11 is the MAC protocol and AODV is the routing protocol. We consider a single flow from a source to a destination and use the idea of *mobility patterns*, introduced in [5], to limit the variance in the data due to scenario variability. A two factorial model is used for our experiments. For each point, we average the simulations over 10 runs, each run for 300s using a different seed. Table 1 shows the factors and their levels we used in our experiments.

4   Statistical Analysis

We use `Design-Expert` [4] for the statistical analysis of the simulation results. As we have six factors, the number of k-way, $k = 1, \ldots, 6$, interaction terms are 6, 15, 20, 15, 6, and 1, respectively. Due to the large number of terms, we show only those that contribute at least 1% to each response variable.

4.1   Effects Tables

An *effects table* shows the contribution of each factor and interaction term on the response variable as a percentage (see Table 2). This allows us to identify the main factors and/or factor interactions to optimize a given response. The packet arrival rate contributes almost 30% to the number of RREQs generated. One reason for this is that timers at the intermediate nodes are refreshed based on the packet arrival rate. If packets of the flow do not arrive within the timer interval, the route is deleted even if the physical route exists. If another packet of the flow arrives after the route entry is deleted, the intermediate node either sends a route error (RERR) to the source or tries to repair the route locally. If a route is deleted prematurely this generates unnecessary control packet overhead contributing to congestion in the network.

The other main contributor to the number of RREQs generated is the A timer. If this timer value is too short, RREQs are generated even when the route is in place. Having too long a value causes the routing protocol to conclude the route exists even if the physical link is broken; this will not change until the MAC protocol informs the routing protocol of the breakage. Long values for the A timer may cause intermediate nodes to generate intermediate RREPs to some other node requesting a route to the same destination when the link is already broken. This leads to delay in establishing the route as when the source node learns of the false route it restarts route discovery.

The AS factor interaction contributes 33% to the number of RREQs generated. The A timer interacts with node speed; that is, the mobility level determines the extent of the effect of the A timer on the number of RREQs generated. The A timer also interacts with the W timer and packet arrival rate, although the contribution of these factor interactions is quite low.

Table 2 also shows the effects table for delay. For this response, node speed is a major factor (79%). As the node speed increases, links break frequently causing routes to be re-established. This increases the delay for data packets. The W timer contributes almost 16% to the delay. Recall that this timer controls the time a node backs off after failing to get a response from the destination. This affects the delay, as setting a large timeout value causes a node to miss the opportunity to form a route when there is actually a feasible route between the source and destination. On the other hand, having a short back-off timer triggers more RREQs; this might lead to congestion in the network. Factor interactions are insignificant for the delay response variable.

The effects table for throughput is also given in Table 2. The main effect for throughput is packet arrival rate, with 70% contribution; node speed contributes 26%. The number of bytes transmitted clearly depends on the packet arrival rate; this is no surprise. The node speed also affects throughput. As the nodes move, routes require maintenance. This causes packet loss at intermediate nodes which affects the throughput. Similar to delay, factor interactions do not play a role for throughput.

Table 3. ANOVA table for response variables.

	Number of RREQs		
Source	Sum of Squares	Mean Square	F-Value
Model	0.008092	0.001011	151.32
A	0.001782	0.001782	266.549988
W	0.000195	0.000195	29.160000
P	0.002513	0.002513	375.929993
S	0.000120	0.000120	17.950001
AW	0.000093	0.000093	13.920000
AP	0.000131	0.000131	19.610001
AS	0.002798	0.002798	418.660004
R^2: 0.9565			

	Delay		
Source	Sum of Squares	Mean Square	F-Value
Model	10.84	1.35	1234.03
W	1.73	1.730000	1577.25
S	8.65	8.650000	7882.20
R^2: 0.9945			

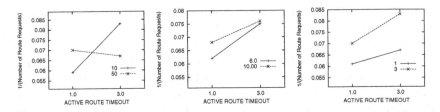

Fig. 1. Interaction graph for A and (a) S; (b) W; (c) P; on RREQs generated.

4.2 ANOVA Analysis

The *ANOVA table* shows the analysis of variance between means of different groups, and tests if the means of groups formed by values of the independent variable or combination of multiple independent variables are different enough not to have occurred by chance. These values indicate the effectiveness of the model. Of importance is the R^2 value as it gives the percentage variability explained by the major factors. For average delay (Table 3), an F-value of 1234.03 implies the model is significant; there is only 0.01% chance that a model value this large could occur due to noise. 99.45% of the variability in the data is explained by just two factors. An experimenter can focus attention on these factors to achieve optimum performance for delay.

For number of RREQs generated, 95.65% of the variability in the data is explained by seven factors. As the effects table for throughput shows, timers did not have major impact on throughput, hence we do not show the ANOVA table or interaction graphs for throughput.

4.3 Interaction Graphs

An *interaction graph* presents the interaction between two factors. It shows how changing the level of one factor affects the response variable at different levels of other factors. Figure 1 shows the interaction of the A timer with the node speed, W timer, and packet arrival rate, respectively, for the response variable of number of RREQs. The x-axis shows the levels of the factor and the y-axis has the scaled value of the response variable. Scaling of response variable is done to stabilize the response variance and to improve the fit of the model to the data. Note that the y-axis label is the inverse of number of RREQs; the larger this fraction the fewer the number of RREQs.

The A timer interacts with node speed. The A timer value has to be set based on average node speed to minimize the number of RREQs generated. Looking at the interaction of the A timer value with W timer, we see that timers not only interact with network and flow parameters but also with other timers. For the levels 6 and 10 seconds of the W timer, the A timer needs to be 3 seconds to minimize the average number of RREQs generated. While individually the A timer and packet arrival rate affect the response variable significantly, their interaction is not too significant. Since there are no strong interactions involving timers we do not show any interaction graphs for delay.

4.4 Discussion

It should be made clear that results of DOE and the statistical analysis are dependent on the factors selected for the study (among which there are many) and the simulation set-up. Nevertheless, the techniques are general and allow for identification of the main effects and factor interactions on response variables of interest. Joint optimization of factors for a given response is also possible; this forms part of our ongoing work.

5 Conclusions

Routing protocols in MANETs commonly use soft-state to reduce control packet overhead. AODV uses timers extensively for route set-up and maintenance. We apply techniques from design and analysis of experiments to show the effect of timers on response variables and to identify factor interactions involving timers. Understanding the contribution of factors and their interactions are a first step toward a formal methodology for timer value optimization.

References

1. C.L. Barrett, et al., "Characterizing the Interaction Between Routing and MAC Protocols in Ad-Hoc Networks," *Proc. of MobiHoc,* pp. 92–103, 2002.
2. C.L. Barrett, et al., "Analyzing Interaction Between Network Protocols, Topology and Traffic in Wireless Radio Networks," *Proc. of WCNC,* pp. 1760–1766, 2003.
3. K.W. Chin, et al., "Implementation Experience with MANET Routing Protocols," *ACM Computer Communications Review,* Vol. 32, No. 5, pp. 49–59, Nov. 2002.
4. Design-Expert Software, StatEase Inc. http://www.statease.com
5. G. Holland et al., "Analysis of TCP Performance over Mobile Ad Hoc Networks," *Proc. of MobiCom,* pp. 219–230, 1999.
6. W. Navidi et al., "Improving the Accuracy of Random Waypoint Simulations Through Steady-State Initialization," CSM TR *MCS-03-08,* June 2003.
7. The Network Simulator — ns-2. UC, Berkeley. http://www.isi.edu/nsname/ns/
8. C.E. Perkins et al., "Ad hoc On-Demand Distance Vector Routing," *Proc. of MCSA,* pp. 90–100, 1999.
9. C.E. Perkins, et al., "Ad Hoc On Demand Distance Vector (AODV) Routing," *IETF Internet draft, Work in progress,* Oct. 2003.
10. K.K. Vadde et al., "Factor Interaction on Service Delivery in Mobile Ad Hoc Networks," To appear in *IEEE JSAC,* accepted March 2004.

UAMAC: Unidirectional-Link Aware MAC Protocol for Heterogeneous Ad Hoc Networks*

Sung-Hee Lee, Jong-Mu Choi, and Young-Bae Ko

College of Information and Communication, Ajou University, South Korea
shlee@dmc.ajou.ac.kr, {jmc, youngko}@ajou.ac.kr

Abstract. In heterogeneous ad hoc networks, high-power nodes could potentially interfere with any on-going transmissions of low-power nodes. In this paper, we propose a unidirectional-link aware MAC protocol (UA-MAC) which prevents such an interference problem by reserving the wireless channel of unidirectional high-power node selectively over one-hop. Our preliminary performance results show that the UAMAC works well in these heterogeneous environments with unidirectional links.

1 Introduction

Recent research in ad hoc networks has focused on medium access control (MAC) and routing problems, resulting in many different protocols proposed [1]. Most of such protocols typically assume that the transmit power capability of nodes in networks is homogeneous. In reality however, it is reasonable to assume that nodes in ad hoc networks have some degree of heterogeneity with various power capabilities and transmission ranges. For instance, there are many handheld devices that have low-power capabilities running with battery. On the other side, there are more powerful devices such as large file servers. Usually these devices are equipped with more powerful network components than handheld devices.

The IEEE 802.11 [2] MAC protocol has been popular for use in mobile ad hoc networks. According to this protocol, a hidden node problem [3] is prevented by exchanging small RTS and CTS control frames. It works pretty well in the homogeneous network. However, its performance may degrade significantly, when used in the heterogeneous network environment [4, 5]. The main reason for this degradation is that in certain scenarios high-power nodes cannot overhear the exchange of the low-power RTS or CTS frames. In such a case, they can simply initiate their own transmissions to another node, causing an increased number of collisions. Furthermore, due to this effect, low-power nodes may suffer from lack of sufficient channel reservation time to communicate with others successfully. Hence, more efficient MAC protocols are needed to handle unidirectional links properly in ad hoc networks.

---

* This research was supported by University IT Research Center Project and grant No. R05-2003-000-10607-0(2004) and M07-2003-000-20095-0 from Korea Science & Engineering Foundation.

I. Nikolaidis et al. (Eds.): ADHOC-NOW 2004, LNCS 3158, pp. 336–341, 2004.

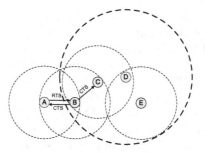

Fig. 1. An example of heterogeneous ad hoc networks, a high-power node D cannot hear RTS/CTS exchange sent by low-power nodes A and B

In this paper, we focus on the efficient MAC protocol design in heterogeneous ad hoc networks. We propose the UAMAC (Unidirectional-link Aware MAC) protocol that detects unidirectional links and prevents any possible interference caused by 'blind' high-power nodes. Our UAMAC can reduce collisions occurred by high-power nodes and improve throughput at the MAC layer.

2   Related Works

Before addressing related works, we will examine the problem of the IEEE 802.11 MAC protocol in the heterogeneous networks more closely. The problem may occur in ongoing transfer of low-power nodes when some blind high-power nodes (that unable to hear the RTS/CTS exchange between the low-power nodes) attempt to transmit to another node. In the Fig.1, RTS/CTS exchange of low-power nodes A and B is not reached to high-power node D, because node D currently exists outside the propagation range of nodes A and B. Let us assume that, while the data exchange between A and B is in progress, the high-power node D tries to initiate its own transmission to another node, for example node E. In this case any receiving packets at node B can be lost due to interference from high-power node D. The simulation results in previous works [4] show that the performance of low-power nodes degrades significantly in the heterogeneous network as compared to that in a homogeneous network.

In [4], the authors have attempted to solve this problem by enlarging CTS propagation range over one-hop. Thus, any nodes that hear CTS frames are required to broadcast further once or twice more in forms of Bandwidth Reservation (BW_RES) control frames. The objective of this broadcast is to let any high-power nodes in the neighborhood know about ongoing RTS/CTS exchange between the sender and the receiver so that they would in turn inhibit their own transmissions for the duration in the BW_RES frame. Although a fairness of low-power nodes can be improved, throughput at the MAC layer will be more degraded since the increased overhead incurred in propagating the BW_RES frames.

[5] proposed another scheme to solve this problem of performance degradation in [4]. The authors point out that the reason of this degradation is due to a collision of multiple BW_RES frames. More than two nodes simultaneously forwarding BW_RES may cause a collision. To prevent this situation, when any node needs to forward BW_RES frame (note that it is renamed as FCTS(Forward-CTS) in [5]), it first senses the wireless channel. Only if the channel is not busy, the node is allowed to transmit the FCTS. In order to further reduce excessive BW_RES frame overhead, only high-power nodes forward the BW_RES frames. In addition, [5] also proposed that high-power nodes receiving RTS framed transmit a Forward-RTS (FRTS) frame. It is to prevent any high-power node interfering other nodes that are supposed to receive ACK packets. In this approach however, the amount of increasing control frames holds practically.

All of these previous works use a flood-based broadcast algorithm for enlarging propagation range of RTS/CTS frames. However these approaches could increase control message overheads. We propose more efficient protocol that detects unidirectional link nodes and selectively broadcasts additional control messages to unidirectional high-power nodes. Eventually our scheme prevents interference of high-power node with the minimized control message overhead.

3   UAMAC (Unidirectional-Link Aware Mac)

In UAMAC (Unidirectional-link Aware MAC), a node detects the presence of unidirectional links to its neighbors by utilizing some distance estimation scheme. The UAMAC assumes that there are specific frames carrying additional information to estimate a link status between a transmitter and a receiver. This information would be carried in periodic beacon frames or in RTS/CTS frames only exchanged when there are data packets to transmit. Due to the excessive overhead of periodic beacon exchange, we choose the later approach in our simulation.

The basic idea behind the UAMAC is represented by example. In Fig. 2(a), when low-power node A exchanges RTS/CTS frames with another low-power node B, node A compares its maximum transmission range to an estimated distance between transmitter (node A) and receiver (node B). If the value of estimated distance from node A to B is larger than the transmission range of node A, node A considers its wireless link to B as a unidirectional link and marks it in the neighbor node list. Otherwise, a link towards B, node A marks as bidirectional in its neighbor node list. Every nodes receive RTS/CTS frames, they update a link status for their neighbors.

Using these neighbor lists, nodes can perform following operations. In Fig. 2(a), node B receives RTS frame from node A and sends CTS frame with its unidirectional link node D's ID, for example MAC address. When node C receives CTS frame, it checks the neighbor node list to see if node D is in the list (i.e. determines whether D is its neighbor or not.) If D is found, then node C will forward an additional control frame called BTS (Block-To-Send) to D. Consequently, node D will block itself for this time equal to duration in BTS

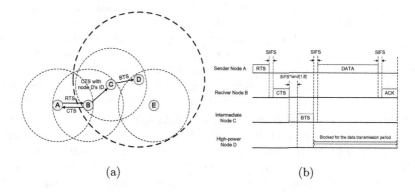

(a)

(b)

Fig. 2. BTS frame propagation and the extended reservation scheme

frame. Therefore, nodes A and B can communicate without any interference from high-power node D.

We have extended the RTS/CTS reservation scheme for adding BTS frame. Each node hearing the CTS frame determines whether it sends a BTS or not. If a node decides to send the BTS frame, it waits until random value (between 1 and 6) of the SIFS units before transmitting the BTS frame. This minimizes collisions caused by multiple simultaneous BTS transmissions from neighbors that hear same CTS frame. In addition, if the transmitter of the CTS frame has no bidirectional link, it sends the CTS frame with unidirectional link node field as blank. When sender node A receives this frame, it transmits data frame immediately without waiting for BTS scheduling period. It helps to minimize unnecessary waiting time in the extended reservation scheme. The whole timing diagram of this extended reservation scheme is shown in Fig. 2(b).

4   Unidirectionality Detection

The question is how to calculate an estimated distance between two communicating nodes, so that it can be compared with the radio transmission range to determine whether the links between two nodes are bidirectional or unidirectional. This can be done as described below.

We can utilize a wireless channel propagation model, i.e., the two-ray ground path loss model that is designed to predict the mean signal strength for arbitrary transmitter-receiver separation distance. In wireless networks, if we know the transmitted signal power (P_t) at the transmitter and a separation distance (d) of the receiver, the receiver power (P_r) of each frame is given by the following equation:

$$P_r = \frac{P_t \times G_t \times G_r \times (h_t^2 \times h_r^2)}{d^4 \times L} \qquad (1)$$

G_t and G_r are transmitter and receiver antenna gain, and h_t and h_r are transmitter and receiver antenna height. The above Eq.(1) states that the distance between two communicating nodes can be estimated at a receiver side, if the transmitted power level P_t of the transmitter and the power received at the receiver P_r are known. To implement this method, the transmitter should make the transmitted power information available to the receiver, by putting the power information either on a specified message, for example RTS/CTS frames.

With this estimated distance, we detect unidirectionality between two neighbor nodes. If the estimated distance is shorter than receiver node's communication range with its maximun power, two nodes can communicate each other with bi-directional link. If not, however, a unidirectional link is made between two communication nodes. We can achieve more realistic direcionality assumption by comparing received power level, which is calculated by same channel propagation equation, with a receving threshold.

5   Performance Evaluation

In this section, we evaluate the performance of UAMAC scheme. We performed simulations using ns-2 for a two-level heterogeneous network at nodes' transmission range of $250m$ and $125m$. We use grid topology with total 49 nodes – 4 high-power nodes and 45 low-power nodes. Distance of each node is uniform $100m$ and the whole network size is 700×700 unit square grid. In our simulations, high-power nodes are intentionally located at the four corners separately, because we want they have independent radio propagation area. Traffic pattern we used consists of 8 CBR connections running on UDP, half from high-power nodes and half from low-power nodes. We use static and relatively simple topology as our preliminary performance evaluation. This is because in random topology and mobility, the situations that we want to show in the simulation are easily collapsed by topology change.

We evaluate our schemes in terms of successful data transmission rate and throughput. We define the successful data transmission ratio as percentage of successful data frame transmission after a successful RTS/CTS handshaking. Fig. 3(a) shows the performance of successful data transmission rate with varying data traffic. In the IEEE 802.11 MAC, low-power nodes degrade its successful data transmission rate up to 40% versus high-power nodes. But UAMAC reduces this degradation within 5%. This improvement is achieved through reducing data frame collision caused by interference of high-power nodes.

And then we examine the throughput. We define throughput as total size of received data frame. In Fig. 3(b), it is found that throughput performance of UAMAC is higher than the IEEE 802.11 MAC as much as maximum 19%. This is because UAMAC reduces unnecessary data frame retransmissions through the reducing collisions (as we can see above analysis).

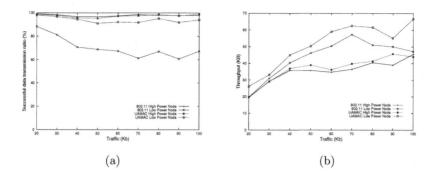

(a)   (b)

Fig. 3. Successful data transmission ratio and Throughput

6 Conclusions

In this paper we consider the performance degradation of the IEEE 802.11 MAC protocol in the heterogeneous ad hoc networks. To overcome this problem, we propose Unidirectional-link Aware MAC (UAMAC) scheme. It detects the unidirectional link by the distance estimations, and having this information, it prevents interference of high-power nodes during communication period of low-power nodes. We show that the use of UAMAC can improve throughput up to 19% and alleviate the unfairness cause by the legacy IEEE 802.11 MAC protocol. Future work would include simulation results with mobility scenarios and more realistic environment.

References

1. C. E. Perkins, "Ad Hoc Networking," *Addison Wesley*, 2001.
2. "Draft International Standard ISO/IEC 8802-11, IEEE P802.11/D10," *LAN/MAN Standards Committee of the IEEE Computer Society*, Jan. 1999.
3. F. A. Tobagi and L. Kleinrock, "Packet Switching in Radio Channels: Part II - The Hidden Terminal Problem in Carrier Sense Multiple-Access and Busy-Tone Solution," *IEEE Tran. on Communications*, pp. 1417-1433. Dec. 1975.
4. N. Poojary, S. V. Krishnamurthy, and S. Dao, "Medium Access Control in a Network of Ad Hoc Mobile Nodes with Heterogeneous Power Capabilities," *Proc. of the IEEE International Conference on Communications*, 2001.
5. T. Fujii, M. Takahashi, M. Bandai, T. Udagawa and I. Sasase, "An Efficient MAC Protocol in Wireless AdHoc Networks with Heterogeneous Power Nodes," *Proc. of the 5th International Symposium on Wireless Personal Multimedia Communications*, Oct. 2002.
6. J. P. Monks, V. Bharghavan and W.-M. Hwu, "Transmission Power Control for Multiple Access Wireless Packet Networks," *Proc. of the 25th Annual IEEE Conference on Local Computer Networks (LCN)*, Nov. 2000.

Author Index

Lecture Notes in Computer Science

For information about Vols. 1–3035

please contact your bookseller or Springer-Verlag